ECOLOGICAL MODELLING OF RIVER-WETLAND SYSTEMS

A Case Study for the Abras de Mantequilla Wetland in Ecuador

Maria Gabriela ALVAREZ MIELES

The front cover belongs to the collection 'Naturancestral '(ancestral nature) by **Fernando Alvarez Mieles**. The collection combines different elements of nature with the Pre-Columbian art style using the stencil technique, and is in preparation to be exhibited at the Anthropological Museum of Contemporary Art (MAAC) in Guayaquil, Ecuador.

"Nature is the source of all true knowledge. She has her own logic, her own laws, she has no effect without cause nor invention without necessity"

Leonardo da Vinci

ECOLOGICAL MODELLING OF RIVER-WETLAND SYSTEMS

A Case Study for the Abras de Mantequilla Wetland in Ecuador

DISSERTATION

Submitted in fulfilment of the requirements of

the Board for Doctorates of Delft University of Technology

and of

the Academic Board of the IHE Delft Institute for Water Education

for the Degree of DOCTOR

to be defended in public

on Tuesday, 28 May 2019 at 12:30 hours

in Delft, the Netherlands

by

Maria Gabriela ALVAREZ MIELES

Master of Science in Environmental Sciences,

UNESCO-IHE Institute for Water Education, Delft, the Netherlands

born in Guayaquil, Ecuador

This dissertation has been approved by the promotors
Prof.dr.ir. A.E. Mynett and Prof.dr. K.A. Irvine

Composition of the Doctoral Committee:

Rector Magnificus TU Delft	Chairman
Rector IHE Delft	Vice-Chairman
Prof.dr.ir. A.E. Mynett	IHE Delft / Delft University of Technology, promotor
Prof.dr. K.A. Irvine	IHE Delft / Wageningen University, promotor

Independent members:

Prof.dr.ir. W.S.J. Uijttewaal	Delft University of Technology
Prof.dr. M.E. McClain	IHE Delft / Delft University of Technology
Prof.dr. J.T.A. Verhoeven	Utrecht University
Prof.dr. F. Martinez-Capel	Universidad Politécnica de Valencia, Spain
Prof.dr.ir. J.A. Roelvink	IHE Delft / Delft University of Technology (reserve member)

CRC Press/Balkema is an imprint of the Taylor & Francis Group, an informa business

Published by:
CRC Press/Balkema
Schipholweg 107C, 2316 XC, Leiden, the Netherlands
Pub.NL@taylorandfrancis.com
www.crcpress.com – www.taylorandfrancis.com
ISBN: 978-0-367-34450-4

"The highest education is that which does not merely give us information
but makes our life in harmony with all existence"

Rabindranath Tagore

A mi familia
por su apoyo infinito

ACKNOWLEDGMENTS

'Levantate y anda' (Stand up and walk, *Facundo Cabral*), is probably the best phrase to describe this PhD adventure. It has been a long trajectory, with wide and narrow, straight and curved roads, achievements and challenges. That is what happens when a 'natural sciences human being' enters the world of physics and numerical calculations and when unexpected health issues occur.

First and foremost, I would like to express my deepest gratitude to my promotors, Professor Arthur Mynett and Professor Kenneth Irvine. Prof Mynett, your support, guidance and specially your perseverance have been essential to accomplish this research. Thanks for sharing with me your technical knowledge but also for continuously asking me to think in the 'big picture'. Many thanks for coming to our home at some distance away from Delft, to work together during difficult periods; for our numerous Skype meetings and last but not least for providing me a certificate that proved pivoting in a court case with the Belanstingdienst. Professor Irvine, my sincere appreciation for your valuable guidance in the biological component, fieldwork design, and for checking my manuscripts in so much detail with specific and to the point suggestions, always guiding me to a deeper thinking and reflecting on every analysis.

I would also like to express my gratitude to Prof. Ann van Griensven who allowed me to join the WETwin project with a case study from my own country, initially as a special programme student which lead me to apply for a PhD fellowship. Thanks Ann for your support and guidance during the initial period of my PhD, and for providing me with funds for conferences and fieldwork campaigns as part of the WETwin project.

To my sponsors: NUFFIC with the NFP fellowships programme; the EU FP7 WETwin project for the fieldwork funding; the WETwin Ecuadorian partner (ESPOL) for facilitating the use of the laboratory for macroinvertebrates classification during my first monitoring campaign.

To MSc. Vilma Salazar, my dear aquaculture teacher from Universidad de Guayaquil: thank you for your supporting letter to apply for the NFP fellowship. To Antonio Torres: many thanks for your important collaboration with the fish sampling and for sharing your knowledge about Ecuadorian fish communities; your enthusiasm about science made the sampling campaigns very pleasant. To the Biology students: Andreina Morán, Veronica Araujo, Wilson la Fuente for your valuable assistance with the macroinvertebrates identification. To Enrique Galecio for doing his master research under the framework of this PhD thesis: many thanks Enrique for your valuable work in developing the hydrodynamic model.

I am very grateful to the Instituto Nacional de Pesca (INP) in Guayaquil for providing the staff and logistics for assisting in the monitoring campaigns. The support from the whole team made it possible to sample so many variables at once. Thanks for the identification of the plankton samples and for allowing me to use the laboratories for macroinvertebrates identification.

To Efficacitas consulting and CEO Juan Carlos Blum in Guayaquil for facilitating using the equipment for the sampling campaigns, and for the hospitality to work in the office during the periods when I was in Ecuador, making me feel like I never left. Many thanks Juan Carlos for all the environmental experience I gained while working with the Efficacitas team and for your support when applying for the first adventure to study abroad for the MSc degree, and for your continuous encouraging words throughout this PhD journey.

My gratitude for the local inhabitants of the wetland at 'El Recuerdo': Telmo España, Jimmy, Angel. Simon Coello, for transporting us with your canoes, and to Don Abdón Moran for facilitating the boat motors for the sampling. Many thanks Don Telmo, the "cacique" of the AdM wetland: your accurate coordinate system allowed us to reach all our GIS sampling points. Thanks for sharing your ancient knowledge about Abras de Mantequilla wetland and telling all these stories about birds and monkeys when we were coming back tired after a whole day of sampling. You offered me to celebrate at El Recuerdo village once I would finish, so I will come soon to fulfil that promise.

My sincere gratitude to Ir. Leo Postma for coming to IHE to discuss Water Quality Modelling, for your time checking my Eco-model grid and set up, and for your motivating and calm way of sharing your infinite knowledge. To DELTARES staff: Dr. Hans Los and Dr. Tineke Troost for your feedback on the plankton components of the model, Dr. Claudette Spiteri for your time discussing the outputs of the model time series, to Ir. Jos van Gils for the inspection of the mass balances and to Cristophe Thiange for your willingness to answer all my FAQ. To all, many thanks for your feedback in this complex world of ecohydraulics modelling.

To Dr. Gerald Corzo of IHE, for our interesting and extensive talks about hydrology, time series analysis, habitat modelling, even outside working times; for your patience introducing this 'biological species' into the world of MATLAB, and especially for your continuous support and friendship.

To all the administrative staff from IHE, that in one way of the other helped me during my PhD period. Thanks Jolanda Boots, Silvia Stylen for renovating my visas and all the administrative issues in time, also when I was on maternity leave; to Martine and Tonneke for assigning me a place to sit and for practicing with me my first attempts in de 'Nederlandse taal'; to Maria Laura for your happy spirit and our talks about our Latin-American roots. To Anique Karsten for solving all the administrative issues towards the defence date.

To the IHE tribe already graduated: Veronica Minaya, Ma.Fernanda Reyes, thanks for your support, company and nice moments we shared throughout this IHE experience. To Heyddy Calderon 'my valid interlocutor', many thanks for our intellectual, straightforward, no buffered conversations, about science, life and philosophy after an IHE working day. To Mijail Arias: thanks my fellow country man for sharing the initial paths of this PhD adventure that started with your WETwin project, including the sampling challenges in the wetland, and specially for our solid friendship. To all the Latin PhD fellows from IHE: for the good times we shared outside working times. The IHE experience has been very fulfilling thanks to the intercontinental PhD fellows that over the years have shared with me their friendship, support and culture.

To my dear friends Saira and Claudette: thanks for all the moments we shared together, your friendship and company supported me enormously through the time I was living on my own in Delft.

In Ecuador: to my friends from life 'amigos queridos', university 'Biologuitos', my cousins 'Mieles y asociados', to my sister Rocio for being present at the distance, sending me messages that make me feel connected with my roots, during those times that living abroad can be difficult, and for receiving us with open arms when we came to Ecuador.

To our family friends in the Netherlands: los Schuurman-Duque, los Carpay, los Briere-Spiteri, los Verheijen-Kulqui, Marlene, with all of you we shared so many experiences and nice moments also with our kids. You are like family for me, making my life abroad more enjoyable.

This achievement is not only mine, it also belongs to my 'lieve' partner Remco Rozendaal and the treasure of our life, our son Rafael; thanks boys for your patience in those weekends that I needed to stay home working. Thanks for taking care of me during the unexpected difficult health periods. For all your support, you deserve half of this degree. To opa en oma: Tanja en Wim Rozendaal, for taking care of Rafael during Mondays and receiving me in your family.

To my parents, who invested all what they have in our education. To my mother, for supporting me through the bachelor studies, for sending me to the English school for 2 years: without this key investment, I would probably not have been able to cross the ocean of graduate education, establish a family, and get to know friends from all continents! And for educating us with a magic mix of humanity, generosity, culture, passion, curiosity, and especially perseverance in whatever we start.

To my brother Ricardo who always had encouraging words: you kept on repeating 'you can do it Gabichi', you will be the first PhD in the family'. To my brother Fernando for designing the cover page of this thesis, related to bird fauna and water (main reason why the AdM wetland is a Ramsar site); and to both, for taking care of my mom in Ecuador: infinite thanks for that.

"When you want something, all the universe conspires in helping you to achieve it"

Paulo Coelho

SUMMARY

Wetlands are among the most productive environments in the world. Around 6% of the Earth's land surface is covered by wetlands, which are key to preserving biodiversity. Wetlands provide multiple services like a source for water supply and a shelter for numerous species of fauna and flora. Wetlands are therefore of immense socio-economic as well as ecological importance. In this research the focus was on the Abras de Mantequilla (AdM) wetland, a tropical wetland system that belongs to the most important coastal river basin of Ecuador. It was declared a Ramsar site in 2000 and was the South American case of the EU-FP7 WETwin project, which provided the starting point of this thesis. A range of tools and approaches was used to develop a knowledge base for the AdM wetland. The research involved a combination of primary data collection (two fieldwork campaigns), secondary data acquisition (from literature), multivariate analyses, and numerical modelling approaches to explore the characteristics of the wetland system in terms of hydrological conditions, hydrodynamic patterns, biotic communities, chemical and ecological processes and fish-habitat suitability.

The AdM wetland is subject to hydrological conditions that exhibit a clear seasonal variability. Annual precipitation may vary from relatively dry conditions to extremely wet events during El Niño years. Moreover, there are clear connections between the AdM wetland and the contributing river system with its tributaries. As a consequence, water depth and inundated area in the wetland exhibit extreme changes during the year: from low depths and almost stagnant conditions during the dry season (May-December) to a very dynamic system during the wet season (January-April). The main source of inflow into the wetland was found to be the Nuevo River (86%). Also, the timing of peak discharges was seen to vary from year to year, but occurred usually during the months of February and March. The inundation volumes and areas were seen to vary by more than a factor of three between dry and wet years. As a result, the wetland is experiencing large variations in inundation area (from 5 to 27 km^2), water depth (from 0.4 to 9 m) and flow velocities (up to 0.9 m/s). Overall, it can be concluded that the wetland is a highly dynamic system in terms of its hydrological forcing and hydrodynamic response.

Main physico-chemical and ecological wetland processes were identified by performing Principal Component Analyses (PCAs). Key variables for the water column were found to be temperature, total suspended solids, DO, turbidity, alkalinity, nitrogen and phosphorus (organic and inorganic), as well as the Redfield ratio (N/P). During high inundation conditions, silicates and flow velocity were also found to be of relevance. For sediments, sand and silt, nitrogen and phosphorus content (inorganic and organic), organic matter and organic carbon were the most influencing, due to their higher correlations with the PC components. The system shows a clear environmental gradient, divided into river sites with higher concentrations of DO, TSS, organic phosphorus, higher N/P ratios and flow velocities and wetland sites with higher concentrations of organic nitrogen, alkalinity, chlorophyll-a, turbidity.

Dominant key species in the AdM wetland system were obtained from field measurement campaigns by evaluating the densities and distribution of the taxa collected for the different biotic assemblages. Clear differences in densities were observed between sites located in the wetland area itself (lentic sites) and in the inflow areas (lotic sites). Higher densities of zooplankton, macro-invertebrates and fish were observed in the middle area where higher retention times occurred. Higher nutrient concentrations were observed at the inflow areas. Phyto- and zooplankton communities showed an inverse pattern: at the inflows, phytoplankton had high densities, while zooplankton had low densities, while in the middle area, zooplankton densities as well as macro-invertebrates and fish were found to be higher.

Fish was found to be dominated by the family *Characidae* during both campaigns. Species of this family are largely widespread in the neotropics and are mainly omnivorous and of small size. The dominance of these omnivorous fish species is important because they are a source of food for carnivorous fish and important for migratory birds. In general, the wetland is dominated by few macro-invertebrate species, a pattern that was observed as well for both phytoplankton and zooplankton assemblages. A range of 4 to 8 species usually contributed more than 70% of the total community density, while a high number of species are present in percentages lower than 3%. This dominance pattern has been observed in other tropical areas as well. The fact that small organisms dominate the zooplankton community reflects that the community is experiencing a high predation rate by fish. This was confirmed when secondary temporal data of zooplankton was analysed, showing that small specimens also dominated the community during other months of the year. This shows the importance of zooplankton supporting the next trophic level of fish, which in turn is an important food source for endemic and migratory birds. The importance of the AdM wetland as a bird sanctuary supporting the bird fauna was a central motivation to declare this area a Ramsar site in 2000.

From a clustering and ordination analysis, the distribution patterns of the biotic communities were found to show a clear separation between river and wetland sites. However, the similarity levels varied according to the biotic community. Similarity levels that produce these two main clusters (river/wetland) were generally around 20% for all communities during both conditions. Nevertheless, a more detailed inspection revealed that the similarities at which initial splits occurred for planktonic pelagic communities were always lower than the ones of littoral communities. This zonation indicates that littoral communities are more similar than planktonic communities that are driven by the flow and therefore experience more mixing.

From a SIMPER analysis, different species from different biotic communities were found to be key discriminators between wetland areas. These species are related to the particular environmental conditions (physico-chemical and hydrodynamic) in the respective wetland zones. As a key outcome it was found that average dissimilarities between wetland areas were lower during high inundation conditions than during low inundation conditions for all biotic groups but fish. This reflects a more homogeneous system in terms of species distribution when the wetland is at its maximum inundation capacity.

A multivariate analysis of biotic and abiotic variables resulted in achieving a better understanding of the most important environmental factors influencing the biotic communities distribution and the overall functioning of the river and wetland ecosystems. Flow velocity and sediment type (river or wetland) are influencing the taxa distribution, their abundance, richness and diversity. The riverine sites with sandy substrates and high velocities had lower species richness and abundance than the wetland sites with fine particle substrate (silt, clay) and low velocities. Even though both ecosystems share some species mostly because of river and wetland connectivity, the highest densities and number of taxa were found in the wetland sites.

The AdM wetland exhibits concentrations of nutrients and primary production in the range of other tropical systems and can be classified as a mesotrophic system. Temporal analyses indicate that generally the wet season is characterized by higher concentrations of nutrients, primary producers and consumers. Spatial analyses indicate that nutrient concentrations in the wetland areas are influenced by the nearest inflows. Thus, upper and middle wetland areas are more affected by the discharge of the El Recuerdo River, and lower wetland areas by the Nuevo River.

A mass balance analysis implemented with the eco-model was a key tool to describe the main processes ruling the wetland functioning. Processes such as denitrification were found to be not important compared with external loads, perhaps due to the constant oxygenated conditions, gaining slightly in importance only during driest scenarios. Sedimentation processes for nutrients and primary producers were found to be low, most probably influenced by the dynamics of the system in combination with the high grazing rates. Processes associated with primary production indicate that grazing is the key processes controlling algae biomass in the water column. Algae sedimentation and mortality also play a role but to a lesser extent. Results of numerical simulations also indicate that nutrient availability does not appear as a limiting factor for algae growth. Thus, algae limitation was more linked to growth limitation due to grazing pressure, rather than nutrient availability. Therefore, results suggested that this wetland system might be governed by a top-down force (grazing) rather than by bottom-up nutrient availability.

The spatio-temporal variability of fish was explored by performing a habitat suitability analysis for the overall fish community in AdM wetland. Major environmental variables defining the presence of fish communities in water systems are the hydrodynamic variables: water depth and flow velocity. Response curves for these variables were built based on field sampling and literature survey. The suitable areas were calculated for different hydrological conditions and scenarios. Spatial zonation defined the areas close to the main inflow as the ones providing better habitat conditions, and areas related to Chojampe subbasin as the ones that will require special attention in terms of wetland management. Based on the results of the present study, it is recommended to secure the timing and magnitude of natural flows especially during periods with higher percentage of suitable areas (high flows during the wet season), since this period is crucial to foment the spawning and development of fish community. Although hydrodynamic variables were useful for an initial fish-habitat assessment, other physical, chemical and biotic variables do play an important role as well and therefore should be included in an integrated ecological habitat assessment. In this regard, the habitat tool developed for this study is quite flexible for adding more variables and their corresponding rules.

On a general note, numerical models were crucial in understanding the hydrodynamics and natural inundation variability of this wetland system. The relative importance of the different inflows can be derived and different hydrological conditions explored. From the chemical perspective, numerical models have shown that comparing concentrations of water chemistry variables was not enough to identify changes due to different inflow conditions. Assessing the system in terms of *yearly mass balances* provided a more clear perspective how different inflow conditions affect the different water variables. Numerical modelling results revealed that the AdM system is dominated by top down zooplankton grazing, rather than bottom-up nutrient availability. Sedimentation and mortality of algae are secondary processes influencing the algae standing biomass. The combination of field measurements with numerical models were extremely useful and relevant during this research and confirmed that they complement each other to obtain a better understanding of the dynamics of freshwater river–wetland systems.

The implementation of management measures for the AdM wetland as proposed by the WETwin project have not yet started. Local authorities are not involved and some local farmers have even developed unfriendly measures against 'birds spots' known as 'El Garzal', which are a type of floating islands where aquatic birds build their nests. Apparently, a couple of these spots were destroyed with the use of chemicals and were not penalized by any authority. On the other hand, there is a group of local farmers that is aware of the ecological importance of the wetland and performs fisheries activities that are sustainable with the environment, e.g. using nets with special mesh sizes in order not to capture the smaller fish. Ecotourism is still a main activity for a few farmers in the main locality named 'El Recuerdo'. Considering flows and habitat conditions for fish communities, an initial measure could be to maintain the timing and magnitude of the natural flow variability especially during the periods with higher suitable habitat areas (February and March). This period is crucial to promote the spawning and development of fish species.

The perception of local farmers about the upstream Baba dam is that it has not affected the area as expected. The management of Abras de Mantequilla wetland requires that not only local but also national authorities be involved in the management of this valuable area. Studies like the present research can be used as a way to develop more awareness about the environmental services of the wetland, but will only be of minimal help if authorities themselves are not aware of the importance of this wetland as a flora and fauna sanctuary. Awareness and cooperation from all stakeholders is mandatory to work towards the sustainable management of this valuable Ramsar site.

SAMENVATTING

Wetlands behoren tot de meest productieve gebieden van de wereld. Zo'n 6% van het aardoppervlak bestaat uit wetlands die van cruciaal belang zijn voor het in stand houden van biodiversiteit. Wetlands dienen meerdere functies waaronder het voorzien in zoet water en het bieden van onderdak aan talloze soorten van fauna en flora. Vandaar dat wetlands van enorm socio-economisch en ecologisch belang zijn. In dit onderzoek lag de nadruk op de Abras de Mantequilla (AdM) wetland, een van de belangrijkste tropische stroomgebieden nabij de kust van Ecuador. Sinds 2000 is AdM een Ramsar wetland en het diende als Zuid-Amerikaanse toepassing in het EU-FP7 project WETwin, waarop het onderzoek in dit proefschrift voortbouwt. Hierin is een scala aan technieken en benaderingen gebruikt om een kennissysteem te ontwikkelen voor AdM. Daarbij is gebruik gemaakt van een combinatie van primaire data collectie (twee meetcampagnes onder verschillende condities), secundaire data collectie (uit de literatuur), multivariate analyses en numerieke modellering teneinde de karakteristieke eigenschappen van het wetland vast te stellen in termen van hydrologische condities, hydrodynamische respons, chemische en ecologische processen, biotische structuren, en leefomgeving voor diverse vissoorten.

Het AdM wetland is onderhevig aan hydrologische condities die een duidelijke seizoensinvloed vertonen. Op jaarbasis kan de neerslag variëren van relatief droge condities tot extreem natte omstandigheden gedurende El Niño jaren. Het AdM wetland is sterk verbonden met het omringende riviersteem. Ten gevolge daarvan vertonen waterdiepte en overstromingsoppervlakte sterke variaties gedurende het jaar: van lage waterstanden en bijna stilstaand water in droge jaargetijden (Mei-December) tot een zeer dynamisch systeem gedurende het natte seizoen (Januari-April). De belangrijkste instroom van rivierwater komt van de Nuevo River (86%). Hoewel de piekafvoeren per jaar kunnen verschillen vinden deze gewoonlijk in de maanden Februari en Maart plaats. De overstromingscondities kunnen gemakkelijk een factor drie verschillen tussen droge en natte jaren: in oppervlakte van 5 tot 27 km², in waterdiepte van 0,4 tot 9 m, en in stroomsnelheid van 0 tot 0,9 m/s. Algemeen kan gesteld worden dat het wetland systeem een sterk dynamisch gedrag vertoont in termen van hydrologische condities en hydrodynamische respons.

De belangrijkste chemisch-fysische en ecologische processen werden vastgesteld op basis van Principal Component Analyses (PCAs). Als belangrijkste processen in de waterkolom werden gevonden: temperatuur, totaal opgeloste stoffen, zuurstofgehalte, troebelheid, zoutgehalte, stikstof en fosfor (organisch en anorganisch) alsmede de zogenaamde Redfield Ratio (N/P). Bij hoge overstromingscondities bleken ook silicaten en stroomsnelheden van belang. Voor het sediment bleken met name van belang: zand- en slibgehalte, stikstof en fosfor (anorganisch en organisch), gehalte aan organisch materiaal en organisch koolstof. Er is een duidelijke milieugradiënt aanwezig die het systeem verdeelt in een rivierdeel met hogere concentraties DO, TSS, organisch fosfor en hogere Redfield (N/P) verhoudingen en stroomsnelheden, en een wetland deel met hogere concentraties organisch stikstof, alkaliteit, chlorofyl-a en troebelheid.

De belangrijkste taxa konden worden bepaald aan de hand van de veldmetingen door dichtheden en verdelingen vast te stellen voor de verschillende biotische assemblages. Daarbij werden duidelijke verschillen geconstateerd tussen lentische gebieden (midden in het wetland) en lotische gebieden (nabij de instromingen van het rivier systeem). In het midden van het wetland waar het water een relatief lange verblijftijd heeft, werden de hoogste concentraties zoöplankton, macro-invertebraten, en vissoorten waargenomen. De hoogste concentraties nutriënten werden aangetroffen nabij de instromingsgebieden. Bij phyto- en zoöplankton was het beeld juist omgekeerd: nabij de instroming was het gehalte aan fytoplankton hoog en zoöplankton laag, terwijl in het midden van het wetland de dichtheden van zoöplankton, macro-invertebraten en vissoorten hoog was.

De belangrijkste vissoort bleek te behoren tot de family der *Characidae*. Deze soorten zijn relatief klein van afmeting en behoren tot de omnivoren die veel voorkomen in neotropische gebieden. De aanwezigheid van deze vissorten is een belangrijke voedselbron voor carnivore vissoorten en trekvogels. Het wetland bevat een beperkt aantal macro-invertebraten wat ook is waargenomen voor zowel fytoplankton als zoöplankton assemblages. Tussen de 4 tot 8 soorten bepaalden veelal meer dan 70% van de totale dichtheid, met een groot aantal andere soorten van minder dan 3%. Dit patroon komt overeen met andere tropische gebieden. Het feit dat zoöplankton relatief veel kleine exemplaren bevat duidt op een hoog predatiegehalte door vis. Dit werd bevestigd door de analyse van secundaire tijdreeksen van zoöplankton die aantoonden dat kleinere exemplaren het gehele jaar door voorkwamen. Dit toont het belang van zoöplankton als voedselbron voor het hogere trofische niveau (vis), dat op zijn beurt weer van belang is voor trekvogels. Juist vanwege het belang van deze trekvogels is het AdM wetland in 2000 tot Ramsar site benoemd.

Op basis van een cluster analyse uitgevoerd in het kader van dit proefschrift ontstond een duidelijk beeld van het verschil tussen delen dichtbij het omringende rivierstelsel en de delen meer binnenin het wetland, afhankelijk van de specifieke biotoop. Voor beide gold dat de gemiddelde overeenkomst rond de 20% lag, maar uit gedetailleerde analyses bleek dat er een scherper onderscheid bestond tussen plankton soorten middenin het wetland waar het water meer gemengd wordt door de stroming, dan langs de randen. Door middel van een SIMPER analyse konden specifieke soorten in verschillende zones van het wetland worden gerelateerd aan specifieke chemisch-fysische en hydrodynamische condities. Daarbij bleek dat de verschillen voor alle biotopen behalve vis het kleinst waren bij hoge waterstanden en grote overstromingsgebieden, omdat het systeem in dat geval meer homogeen wordt.

Door gebruik te maken van een multivariate analyse kon een beter inzicht worden verkregen in de belangrijkste factoren die de biotische populaties beïnvloeden en daarmee het gedrag van de ecosystemen in het wetland met omringende riviersystemen. Stroomsnelheden en type sediment (in de rivier of in het wetland) zijn bepalend voor de verdeling en diversiteit van de taxa. Nabij de rivier met zandige bodem en hogere stroomsnelheden komen minder rijke soorten voor dan in het wetland met zijn slibachtige bodem en lage stroomsnelheden. Hoewel beide ecosystemen een aantal soorten gemeen hebben vanwege hun open verbinding, werden de hoogste concentraties en aantallen taxa toch in het wetland gevonden.

Het AdM wetland bevat vergelijkbare concentraties nutriënten en primaire productie als de meeste andere tropische gebieden en kan worden geclassificeerd als een mesotrofisch systeem. Het natte seizoen wordt gekarakteriseerd door hogere concentraties nutriënten en primaire productie. Ruimtelijke analyses geven aan dat hogere concentraties aan nutriënten worden bepaald door de dichtstbijzijnde instroom. Dit betekent dat de bovenste en middelste wetland gebieden het meest beïnvloed worden door de El Recuerdo River en de onderste wetland gebieden door de Nuevo River.

Inzicht in de belangrijkste processen in het wetland werd verkregen door massa balansen op te stellen. Daaruit bleek dat sommige processen zoals denitrificatie van minder belang waren dan de externe belasting, waarschijnlijk vanwege de overwegend constante zuurstofvrije condities, behalve misschien gedurende aanhoudende droogte. Sedimentatie van nutriënten en primaire productie bleken laag, zeer waarschijnlijk vanwege de dynamiek van het systeem in combinatie met hoge begrazing die de biomassa van algen in de waterkolom begrenst. Sedimentatie en mortaliteit van algen zijn daarbij ook van belang, maar in mindere mate. Op basis van numerieke simulaties blijkt dat de beschikbaarheid van nutriënten niet een beperkende factor is voor algengroei. Algengroei wordt dus beperkt door de druk van grazers en niet door de beschikbaarheid van nutriënten. Vandaar dat kan worden geconcludeerd dat dit wetland systeem wordt gedreven door begrazing van bovenaf en niet door nutriënt beperking van onderop.

De tijd-ruimte variabiliteit van vis is onderzocht op basis van een Habitat Suitability Analyse (HSA) voor de gehele vispopulatie in het AdM wetland. De belangrijkste factoren die de visstand bepalen zijn de hydrodynamische variabelen waterdiepte en stroomsnelheid. Overdrachtsfuncties hiervoor werden bepaald op basis van veldmetingen en literatuuronderzoek. Voor verschillende hydrologische condities en scenario's werden de meest geschikte gebieden bepaald. Daarbij bleek dat met name nabij de instroming vanuit de rivieren de meest geschikte habitat condities bestaan en dat in de buurt van de Chojampe Rivier speciaal aandacht moet worden besteed aan het beheer van het wetland. Op basis van het onderzoek in dit proefschrift wordt aanbevolen om zoveel mogelijk de natuurlijk condities na te bootsen door het verzekeren van hoge(re) stroomsnelheden in het natte seizoen, aangezien dit met name van belang is voor het behoud van de visstand. Hoewel hydrodynamische grootheden zinvol zijn voor het bepalen van een eerste schatting van geschikte visgebieden, spelen andere chemisch-fysische en biotische variabelen eveneens een belangrijke rol en moeten deze dus worden meegenomen bij een uitgebreidere HSA modelvorming. Het software instrument dat in dit onderzoek is ontwikkeld is uitermate flexibel en biedt de mogelijkheid om meer variabelen en bijbehorende toepassingsregels daarin op te nemen.

In het algemeen kan worden gesteld dat numerieke modellen een belangrijke bijdrage kunnen leveren aan het begrijpen van de natuurlijke variatie in overstromingen in dit wetland systeem. Op die manier kan de relatieve bijdrage van de verschillende instromingen worden bepaald en het gedrag onder verschillende hydrologische condities (bijvoorbeeld ten gevolge van klimaatverandering) worden nagegaan. Wat betreft de chemische omstandigheden hebben numerieke modellen aangetoond dat het vergelijken van concentraties van chemische variabelen niet afdoende is om de verschillen ten gevolge van verschillende instroomcondities te bepalen. Het blijkt beter om daartoe een jaarlijkse massabalans te gebruiken. Numeriek modellen gaven duidelijk aan dat het AdM wetland wordt gedreven door begrazing van bovenaf meer dan de beschikbaarheid van nutriënten van onderop. Sedimentatie en mortaliteit blijken secundaire processen die de biomassa van algen bepalen. De combinatie van veldmetingen met numerieke modellen bleek buitengewoon nuttig en relevant bij het onderzoek in dit proefschrift en bevestigde dat beide benaderingen elkaar aanvullen teneinde een beter inzicht te verkrijgen in de dynamiek van rivier-wetland zoetwatersystemen.

De implementatie van de management maatregelen voor het AdM wetland als voorgesteld door het WETwin project is helaas nog niet begonnen. De lokale autoriteiten zijn niet betrokken en sommige lokale boeren hebben eigenhandig onvriendelijke maatregelen genomen tegen vogelgebieden die bekend staan als 'El Garzal', een soort drijvende eilanden waar watervogels hun nesten bouwen. Verschillende hiervan zijn vernietigd met behulp van chemicaliën en hiervoor zijn geen straffen uitgevaardigd door enige autoriteit. Daarentegen is er een groep lokale boeren die zich wel degelijk bewust zijn van het ecologisch belang van dit wetland en die duurzame ecologische visserij bedrijven door bijvoorbeeld netten te gebruiken met speciale afmetingen van de mazen die de kleinere vissoorten doorlaten. Ook is ecotoerisme de belangrijkste activiteit van boeren in het gebied genaamd 'El Recuerdo'.

Een maatregel waarmee zou kunnen worden begonnen betreft het nabootsen van (over)stromingen op een zo natuurlijk mogelijke manier gedurende periodes dat de habitats daarom vragen (Februari en Maart). Deze periode is van groot belang voor het instand houden van de visstand. De beleving van lokale boeren over de aanleg van de Baba dam stroomopwaarts is dat dit minder invloed heeft gehad dan verwacht. Er zijn geen grote veranderingen waargenomen in de beschikbare hoeveelheid water en vis, waarschijnlijk omdat er nog geen bijzonder droog jaar sinds de dam is aangelegd.

Het beheer van het waardevolle Abras de Mantequilla wetland vereist samenwerking tussen lokale bestuurders en de nationale overheid. Studies als dit proefschrift kunnen worden gebruikt als een aanzet om meer bewustzijn te creëren over het ecologisch belang van dit wetland. Maar dan dienen de autoriteiten zich ook zelf bewust te zijn van het belang van dit wetland als bron van flora en fauna. Bewustzijn en samenwerking tussen alle betrokkenen is een absolute vereiste om te komen tot een duurzaam beheer van deze unieke en waardevolle Ramsar site.

TABLE OF CONTENT

ACKNOWLEDGMENTS VII

SUMMARY XI

SAMENVATTING.. XVII

TABLE OF CONTENT.. XXIII

1   INTRODUCTION 1

    1.1 THE ROLE OF WETLANDS.. 2

    1.2 WETLAND CONSERVATION... 2

    1.3 ECOSYSTEM SERVICES 3

    1.4 WETLAND DYNAMICS, STRUCTURE AND FUNCTION... 5

    1.5 THE WETWIN PROJECT... 6

    1.6 THE ABRAS DE MANTEQUILLA WETLAND IN ECUADOR 9

    1.7 CAPABILITIES OF MATHEMATICAL MODELLING TOOLS 12

    1.8 OVERALL RESEARCH APPROACH 12

    1.9 SPECIFIC RESEARCH QUESTIONS 13

    1.10 THESIS OUTLINE 15

2   HYDRODYNAMICS OF THE TROPICAL ADM RIVER-WETLAND SYSTEM................................... 17

    2.1 GEOGRAPHICAL CONDITIONS.. 18

        2.1.1 Basin topography... 18

        2.1.2 Land use composition 19

        2.1.3 Soil properties 21

    2.2 HYDROLOGICAL AND METEOROLOGICAL CONDITIONS 21

        2.2.1 Annual precipitation 21

        2.2.2 Seasonal variability.. 23

        2.2.3 Discharge and water level data 25

        2.2.4 Hydrology of the Chojampe subbasin 27

        2.2.5 Regional infrastructure projects 28

    2.3 THE ADM RIVER-WETLAND SYSTEM 29

        2.3.1 Flows in main arteries and tributaries 29

        2.3.2 Inundation modelling of the AdM wetland system.. 31

        2.3.3 Model verification 37

        2.3.4 Model performance 37

    2.4 NATURAL VARIABILITY IN HYDRODYNAMIC CONDITIONS 40

        2.4.1 Boundary conditions for extremes... 40

        2.4.2 Initial conditions.. 40

        2.4.3 Variability in water depth 41

        2.4.4 Variability in inundation area 42

        2.4.5 Variability in flow velocities 44

        2.4.6 Variability in inundation area and volume 46

2.5 Water balance and relative contributions of inflows 48
    2.5.1 Water balance 48
    2.5.2 Relative contributions of inflows 52
    2.5.3 Residence times 55
2.6 Conditions during measurement campaigns 2011&2012 57
    2.6.1 Boundary conditions 57
    2.6.2 Initial conditions and water level 57
    2.6.3 Flow velocities 59
    2.6.4 Water balance 61
    2.6.5 Temporal inundation patterns 63
    2.6.6 Spatial inundation patterns 63
2.7 Discussion 64
    2.7.1 Natural variability of hydrodynamic conditions 64
    2.7.2 Inflows assessment 65
    2.7.3 Spatial analysis, inflows contribution and residence times 65
    2.7.4 Conditions during the sampling campaigns compared to historical conditions 66

3   ENVIRONMENTAL VARIABLES AND SPATIAL PATTERNS 67

3.1 Background 68
3.2 Field measurement campaigns 69
    3.2.1 Selection of environmental variables... 69
    3.2.2 Identification of sampling sites and inundation conditions......................... 70
    3.2.3 Sampling procedure for water body and sediment 72
    3.2.4 Data analysis of environmental variables... 74
3.3 Sampling results within the water body... 74
    3.3.1 Low inundation conditions... 74
    3.3.2 High inundation conditions... 76
    3.3.3 Combined analysis 79
    3.3.4 Measured concentrations and spatial distribution..................................... 81
3.4 Sampling results of bottom sediment 83
    3.4.1 Low inundation conditions... 83
    3.4.2 High inundation conditions... 84
    3.4.3 Combined analysis 86
3.5 Concentrations, gradients and key variables 89

4   COMMUNITY STRUCTURE OF BIOTIC ASSEMBLAGES 93

4.1 Background 94
4.2 Field measurement campaigns 95
    4.2.1 Sampling methods and inundation conditions 95
    4.2.2 Identification of biotic communities 96
    4.2.3 Data analysis of biotic communities.. 98
4.3 Phytoplankton 102
    4.3.1 Sampling with Niskin Bottle.. 102
    4.3.2 Sampling by horizontal tows... 107
    4.3.3 Sampling by vertical hauls 111

4.4 ZOOPLANKTON 115

    4.4.1 Sampling by horizontal tows 115

    4.4.2 Sampling by vertical hauls 118

4.5 MACROINVERTEBRATES 121

    4.5.1 Spatial patterns 123

    4.5.2 Similarities/dissimilarities 124

4.6 FISH 125

    4.6.1 Spatial patterns 127

    4.6.2 Similarities/dissimilarities 127

4.7 SUMMARY OF SIMILARITIES 129

4.8 SUMMARY OF DISSIMILARITIES 131

4.9 LINKING BIOTIC ASSEMBLAGES WITH ENVIRONMENTAL VARIABLES 132

    4.9.1 Low inundation conditions 132

    4.9.2 High inundation conditions 135

4.10 DISCUSSIONS 138

    4.10.1 Spatial patterns 138

    4.10.2 Typical species and ecological traits 139

    4.10.3 Explanatory variables 145

5 EVALUATION OF WATER QUALITY AND PRIMARY PRODUCTION DYNAMICS 147

5.1 BACKGROUND AND SCOPE 148

5.2 MODEL SET UP 150

    5.2.1 Motivation for eco-model implementation 150

    5.2.2 Model description 150

    5.2.3 Substances included in AdM eco model 151

    5.2.4 Processes included in AdM eco model 152

    5.2.5 Initial conditions, boundary conditions and observation points 153

    5.2.6 Estimation of primary producers, primary consumers, detritus and nitrogen loads 156

5.3 MODEL PERFORMANCE AND VERIFICATION 160

    5.3.1 Dissolved Oxygen and dissolved inorganic nitrogen 160

    5.3.2 Nutrients and Chlorophyll-a 161

5.4 SCENARIOS 164

    5.4.1 Hydrological conditions 164

    5.4.2 Temporal and spatial variability of key physico-chemical variables 165

    5.4.3 Temporal and spatial variations of primary producers 176

    5.4.4 Temporal and spatial variations primary consumers 182

5.5 DISCUSSION 186

    5.5.1 Temporal and spatial variability of nutrients 186

    5.5.2 Temporal and spatial variability of primary producers and consumers 188

    5.5.3 Nutrients partitioning 189

6 EVALUATION OF HABITAT SUITABILITY CONDITIONS FOR FISH **191**

6.1 BACKGROUND AND SCOPE 192

6.2 STUDY AREA 194

    6.2.1 A Ramsar site 194

    6.2.2 The hydrodynamics of AdM wetland 196

6.3 THE HABITAT SUITABILITY INDEX 198

    6.3.1 Steps for habitat index construction 198

    6.3.2 The habitat index formulation 199

6.4 RESULTS 202

    6.4.1 Natural variability of suitable areas 202

    6.4.2 Contribution of each wetland to the total wetland suitable area 203

    6.4.3 Independent analysis of the PSA per area 204

    6.4.4 Natural variability of HSI 205

    6.4.5 Independent analysis of the HSI per area 206

    6.4.6 Spatial and temporal variation of HSI 207

6.5 DISCUSSION 209

    6.5.1 The habitat index approach 209

    6.5.2 Percentage of suitable areas 209

    6.5.3 HSI scores 210

    6.5.4 Temporal availability of suitable areas 210

    6.5.5 Littoral areas and vegetation 210

    6.5.6 Fish studies in the AdM wetland and associated basin 212

    6.5.7 Overall findings 213

7 DISCUSSION AND SYNTHESIS **215**

7.1 SUSTAINABILITY OF THE AdM WETLAND HYDRODYNAMICS 216

7.2 MASS BALANCES OF NUTRIENTS 219

    7.2.1 Total Nitrogen (TN) 219

    7.2.2 Total Phosphorus (TP) 222

    7.2.3 Relative importance of internal loads 224

7.3 MASS BALANCES OF PRIMARY PRODUCERS 225

    7.3.2 Autochthonous primary production 228

    7.3.3 Primary producers and associated processes 229

7.4 NUTRIENT BALANCES AND THEIR VARIATIONS BETWEEN DIFFERENT HYDROLOGICAL CONDITIONS 231

7.5 WETLAND PRODUCTIVITY AND RELATED PROCESSES 233

8 CONCLUSIONS AND RECOMMENDATIONS **237**

8.1 RESEARCH APPROACH 238

8.2 SPATIO-TEMPORAL VARIABILITY OF THE AdM WETLAND HYDRODYNAMICS ... 238

8.3 DOMINANT AND KEY SPECIES IN THE AdM WETLAND SYSTEM 239

8.4 SPATIAL PATTERNS IN THE DISTRIBUTION OF THE ENVIRONMENTAL VARIABLES AND BIOTIC COMMUNITIES 241

8.5 MAIN PHYSICO-CHEMICAL AND ECOLOGICAL PROCESSES 242

8.6 SPATIO-TEMPORAL VARIABILITY OF FISH-HABITAT SUITABILITY 243

8.7 MANAGEMENT MEASURES FOR THE AdM WETLAND 243

8.8 NUMERICAL MODELLING AS A TOOL TO DESCRIBE WETLAND DYNAMICS 245

8.9 RECOMMENDATIONS FOR FURTHER RESEARCH 246

REFERENCES.. 247

APPENDIX A.. 271

A.1 THE 1 D DE SAINT-VENANT EQUATIONS 272

A.1.1 A 1D model application 273

A.2 THE 2D DE SAINT-VENANT EQUATION 275

A.2.1 A 2D application 276

APPENDIX B 279

B.1 THE 1D ADVECTION-DISPERSION EQUATION 280

B.1.1 The 1D equation.. 280

B.1.2 A 1D application 280

B.2 THE 2D ADVECTION-DISPERSION EQUATION 281

B.2.1 The 2D equation.. 281

B.2.2 A 2D application 282

APPENDIX C 283

APPENDIX D.. 291

APPENDIX E 311

E.1 Total Nitrogen temporal and spatial variability 312

E.2 Phosphorus - temporal and spatial variability.. 313

E.3 Total organic carbon - temporal and spatial variability 314

E.4 Chlorophyll-a - temporal and spatial variability 315

E.5 Phytoplankton biomass- temporal and spatial variability.................................... 316

E.6 Primary consumers - temporal and spatial variability.. 317

E.7 Nitrogen partitioning 318

E.8 Phosphorus partitioning.. 318

E.9 Total organic carbon partitioning 319

ABOUT THE AUTHOR... 321

"The journey of a thousand miles begins with one step"

Lao Tzu

1

INTRODUCTION

1.1 The role of wetlands

Wetlands are among the most productive of environments. Around 6% of the Earth's land surface is covered by wetlands, which are key to preserving the biodiversity of the world. Wetlands provide multiple services to mankind: they often are a source for water supply; they function as storehouses of plant genetic material; and they often have high rates of primary production upon which numerous species depend. They are of immense socio-economic as well as ecological importance. Rice, for example, a common wetland plant, is the main source of food for half of humanity.

The Ramsar Convention (Ramsar, 1971) defined wetlands as *'areas of marsh, fen, peatland or water, whether natural or artificial, permanent or temporary, with water that is static or flowing, fresh, brackish or salt, including areas of marine water the depth of which at low tide does not exceed six metres'*. Furthermore, the Ramsar Convention declared that *'for the purpose of this Convention waterfowl are birds ecologically dependent on wetlands'*.

Classification of wetlands is a complex issue, mainly because they are in an intermediate position between terrestrial and aquatic ecosystems, and hence include many kinds of habitats. Nevertheless, classification is an essential prerequisite for any wetland inventory. International agencies like the Ramsar Convention, the International Waterfowl and Wetlands Research Bureau (IWRB), and the International Union for Conservation of Nature (IUCN), are key institutions to establish a committee that develops an international classification system for the wetlands of the world (Finlayson and van der Valk, 1995).

The formal definition by the international treaty Ramsar Convention it is an agreed and political definition that has received worldwide recognition (Mitsch and Gosselink, 2007). The Ramsar definition includes both freshwater and coastal systems, including coral reefs, as well as man-made wetlands like wastewater treatment ponds. Wetland sizes can vary from a local pond to the extensive Pantanal wetlands in Brazil (135,000 hectares) as elaborated by Ramsar (2014).

1.2 Wetland conservation

Awareness about wetland conservation has increased significantly over the past decades. The historical MAR conference in 1962 (Matthews, 1993) followed by the Ramsar convention in 1971 provided a strong foundation for wetland research, management, and conservation. The Ramsar Convention increased in member states over the past four decades, from 28 Contracting Parties in 1980 to 168 in 2014, covering a total area of 2.1 million km² distributed over 2168 sites (Ramsar, 2014).

Close cooperation with other international, intergovernmental and non-governmental organizations has been fundamental to achieve the mission of the Convention (Ramsar, 2013). In Latin America, several institutions have benefited from the Wetlands for the Future (WFF) initiative (1996). This initiative was supported by the Ramsar Secretariat, the United States State Department, and the United States Fish and Wildlife Service,

Floodplain wetlands are sites of extraordinary biodiversity that are sensitive to long-term ecological effects of dams and water diversions (Kingsford, 2000). The declaration of the World Wetlands Day has been a keystone in Ramsar's public visibility. Every 2nd of February since 1997, government agencies, non-governmental organizations, and groups of citizens undertake actions aimed at raising public awareness of wetland values and benefits in general, in support of the Ramsar Convention (Ramsar, 2013).

Recent assessments by independent environmental legal experts have shown that in Africa and North America the designation of wetlands as Ramsar Sites of International Importance has contributed considerably to the conservation status of these wetlands (Ramsar, 2016a). Real benefits occurred: public awareness grew, participation by stakeholders increased, funding for research and conservation increased, ecotourism was promoted (Ramsar, 2013). Currently, 40 years after signing, the Ramsar Convention has been instrumental in worldwide action towards protection at the governmental level for conservation and wise use of wetlands, with a nine-fold increase in member states (Ramsar, 2010a).

During the first 25 years, the Convention played a crucial role in promoting awareness of wetlands and provided technical support to governments for conservation of ecosystems (Halls, 1997). The global extent of wetlands is estimated to be 12.8 million km^2 (MEA, 2005). The Ramsar convention suggested that the total area of wetlands is 7.2 million km^2, but the Convention acknowledged that some wetland types were not included (Mitsch and Gosselink, 2007). Despite the extensive areas estimated, currently, just 2.1 million km^2 have been registered with the Ramsar Convention (Ramsar, 2014) .

1.3 Ecosystem services

Ecosystem services are defined as *'the benefits people obtain from ecosystems'* (MEA, 2005). These include: providing food and water; regulating floods, droughts; reducing land degradation; providing coastal protection; supporting soil formation and nutrient cycling; providing cultural, recreational, spiritual, religious services;

conserving biodiversity (MEA, 2005; Ramsar, 2010c). Due to their multiple regulation services, wetlands can be called *'nature's shock absorbers'* (Ramsar, 2016b). Wetlands host a great array of flora and fauna, from invertebrates to fish, waterfowls and even large vertebrates and are essential areas for fish reproduction and for hosting migratory birds. Some 30% of all known fish species live in wetland areas (Gopal, 2009).

Despite all the services wetlands provide, these systems have been subject to human disturbance since historical times, as well as newer threats from climate change. In early days, wetlands were often considered wastelands that bore diseases and were obstacles for development. Their habitats were often disregarded, drained, filled and degraded. During the twentieth century, extensive wetland areas have disappeared (Davidson, 2014; Halls, 1997; Matthews, 1993).

A review of around 200 studies of change in wetland area determined that there has been a faster rate of wetland loss during the 20th and the beginning of 21st century, with a loss of wetlands between 64 and 71 % since 1900 (Davidson, 2014). A global meta analysis of more than 100 case studies on wetlands indicated that agriculture has been the main proximate cause of wetland conversion, while economic growth and increase in population were identified as the most frequent underlying causes (Van Asselen et al., 2013)

Wetland degradation is more rapid than that of other ecosystems, with the status of wetland species also declining faster than those of other ecosystems. Direct drivers for wetland loss and degradation include infrastructure development, land conversion, water withdrawal, eutrophication and pollution, overharvesting and overexploitation, and the introduction of invasive alien species. Indirect drivers comprise population growth and economic development.

Since wetlands are an integral part of river basins, wetland degradation has led to the disruption of natural hydrological cycles. Thus, wetlands face an increase in the frequency and severity of floods, droughts and also pollution. Wetland degradation caused economic and social problems to populations that were used to have access to the wetland ecosystem services (Ramsar, 2010b). Floodplain wetlands are sites of extraordinary biodiversity and their loss will continue until there is a widespread understanding of the long-term ecological effects of dams and diversions (Kingsford, 2000).

1.4 Wetland dynamics, structure and function

Wetlands, encompassing a broad range of ecosystems, are sensitive to hydrological conditions and human influence (Mitsch and Gosselink, 2007). Climate and geomorphology are key factors influencing wetland hydrology. Hydrologic conditions are key drivers for wetland dynamics, structure and function, directly affecting abiotic factors, e.g. oxygen, nutrient availability, sediment transport that determine conditions for biota development. For instance, changes in species composition have been observed, caused by only slight changes in hydrologic conditions. However, when hydrologic patterns remain similar from year to year, the structure and functional integrity of the wetland's biota can persist for longer time (Figure 1-1) (Mitsch and Gosselink, 2007).

The main driving force responsible for the productivity and interactions of the major biotic components in river-floodplain systems is the 'flood pulse' (Junk et al., 1989). Several ecological processes and interactions among a wide range of species are triggered when water arrives in a floodplain wetland. Thus, the substitution of a variable-flooding pattern with a permanent one, and loss of wet-dry cycles, has major ecological effects (Kingsford, 2000). The seasonal 'hydroperiod' characterizes each type of wetland. The hydroperiods of many wetlands are driven by surface waters coming from adjacent rivers and lakes, affecting their inflows and outflows.

Furthermore, a wetland can be seasonally or intermittently flooded (Mitsch and Gosselink, 2007; Poff et al., 2002). Floods enhance fish recruitment by providing suitable habitats, spawning areas and food. However, considering flood pulses alone as a key driver for fish recruitment is too simplistic. Other factors such as life history adaptations of the fauna and the timing of inundation are key drivers controlling the response of fish to flooding (King et al., 2003). The connection between flooding and breeding of river fish is well documented for several tropical rivers in Asia, Africa, South America and Northern Australia. Since temperature variation in tropical regions throughout the year is minimal, the hydrological regime is the dominant factor driving ecological dynamics (Humphries et al., 1999).

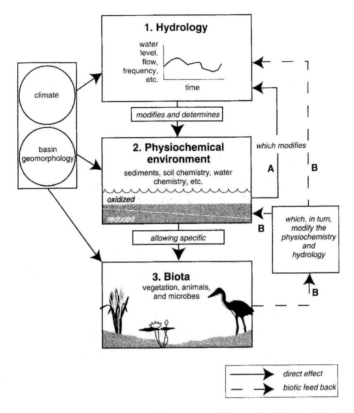

Figure 1-1 Effects of hydrology on wetland functions and biotic feedbacks (Mitsch and Gosselink, 2007)

1.5 The WETwin Project

Individual studies of wetlands have rarely considered the role of wetlands in the context of the river basin (Shamir and Verhoeven, 2013). In developing countries, although wetlands are strongly related to livelihoods, data on wetland functions, processes and values are scarce. Thus, management decisions for wetland use are frequently made without comprehensive information (Johnston et al., 2013). The need for developing approaches that involve local communities and provide reliable information on wetland services at the river basin scale in data-poor context was identified by Shamir and Verhoeven (2013).

The WETwin project, funded by the European Commission under FP7, aimed to enhance the role of wetlands in integrated water resource management. The project started in 2008 in different wetland areas located in three continents Europe, Africa and South America (Johnston et al., 2013). Seven study areas were selected in a number of river basins: the Danube in Europe, the Niger, White Nile and Olifants River in Africa, and the Guayas River Basin in South America (Figure 1-2).

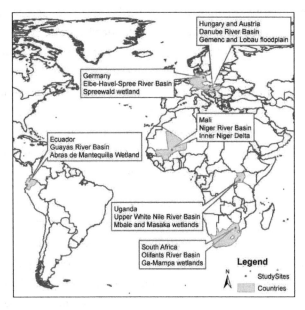

Figure 1-2 WETwin case study sites (Johnston et al., 2013)

The project developed tools for wetland assessment in a data poor context. These tools were applied and tested in the seven case studies. The main characteristic of these areas is that all of them are inland wetlands related to a river basin (Arias-Hidalgo, 2012). The term WETwin can be understood as: (a) winning the wetlands and (b) twinning the wetland studies of different parts of the world to produce common insights and studies of management options (Shamir and Verhoeven, 2013). The project aimed to improve wetland management by maximizing benefits from wetland use while maintaining ecological health (Johnston et al., 2013). More specifically, WETwin aimed to:

- improve drinking water and sanitation services of wetlands;
- improve community services while conserving or improving good ecological health;
- adapt wetland management to changing environmental conditions; and
- integrate wetlands into river basin management.

The conceptual framework of WETwin started from four basic premises of wetland management: (i) 'wise use'; (ii) adaptive management; (iii) integrated water resource management (IWRM); and (iv) participation of local communities and stakeholders.

'Wise use' acknowledges wetlands as providers of many ecosystem services important for livelihoods that should be managed properly in order to protect their ecological status. 'Adaptive management' describes the management as a continuous cyclic process, described in the Critical Path Standard approach adopted by Ramsar (Ramsar, 2010b). IWRM recognizes the fact that wetlands function within a hydrological context and are not elements separate from the catchment. This implies not only that catchment management has a direct impact on the wetland conditions, but also that the wetland management influences the functioning of the catchment. Finally, participatory planning acknowledges that involvement of the local communities and stakeholders at all stages is necessary, since they are also the beneficiaries of a sustainable management strategy (Johnston et al., 2013).

The WETwin project focus was to prepare management plans for each case study. Since wetlands provide several environmental services for multiple stakeholders, their involvement in formulating management plans for the wetlands is crucial (Shamir and Verhoeven, 2013). During the project, stakeholders participated actively in identifying and evaluating possible management options. The scope of the WETwin project was restricted to the preparatory and planning stages of the Ramsar Critical Path adaptive management cycle (Ramsar, 2010b). Implementation and monitoring of the management plans were beyond the scope of this project (Johnston et al., 2013). During the project, different tools were developed to be applied in data scarce contexts. Quantitative modelling based on technical information was combined with qualitative methods based on expert and stakeholder knowledge (Shamir and Verhoeven, 2013).

Major environmental and livelihood problems were identified in each case study of the WETwin project following the DSIR approach (Driver, Pressure, State, Impact, and Response). The core of this methodology was to assist in establishing cause-effect relationships for a particular problem, then develop measures to resolve the problem (Zsuffa and Cools, 2011). For all case studies, initial DSIR analyses identified high-level trade-offs in terms of land or water use. The approach used in WETwin had three strong points: involving stakeholders at all stages of the decision process, combining qualitative and quantitative data (allowing inclusion of poorly known and potentially important system components), and providing a relatively simple and structured approach to evaluate wetland management interventions (Johnston et al., 2013).

1.6 The Abras de Mantequilla wetland in Ecuador

The Abras de Mantequilla (AdM) wetland was one of the seven study areas selected for the WETwin project, and the only case study in South America (Figure 1-3). The wetland belongs to the Guayas River Basin and was declared a Ramsar site in March 2000, due to its important role in the conservation of bird fauna diversity (Figure 1-4). These included (i) three migratory species: *Anas discours, Chordeiles minor spp,* and *Catharus ustulatus;* (ii) three rare species and (iii) eight endemic species, *including Furnarious cinnamomeus, Veniliornis callonotus, Galucidium peruanum* and *Turdus maculirostris.* AdM also supports a significant population of indigenous fish and at the same time is a source of food, a spawning site and a development area for those species of fish that depend upon the wetland (Ramsar, 2014).

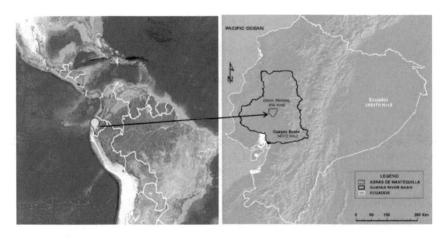

Figure 1-3 Abras de Mantequilla wetland, location in Ecuador and Guayas River Basin

Figure 1-4 Abras de Mantequilla wetland during the wet season. Source: Fieldwork Feb 2011, March 2012

The three migratory species reported in AdM wetland are originally from North America (Figure 1-5). *Anas discors* ('Blue-winged teal') is a duck species found throughout North America and a long distance migrant, some birds heading all the way to South America during the winter. Migrants typically stop in fresh water habitats (vegetated wetlands, rice fields) feeding on insects, crustaceans, snails as well as vegetation (Rohwer et al., 2002; The_CornellLab-of-Ornithology, 2018).

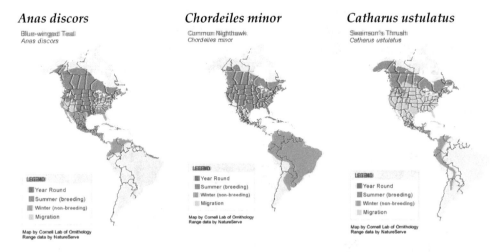

Figure 1-5 Range maps of the three migratory species from North America that migrate to South America during winter (see blue areas) (The_CornellLab-of-Ornithology, 2018). All three species have been reported in AdM wetland (Ramsar, 2014)

The DSIR methodology applied during the WETwin project evaluated that the main impacts on the habitat and ecosystem services of the AdM wetland are caused by two main activities: agriculture and water allocation projects (Zsuffa and Cools, 2011). DSIR reveals the cause-effect relationships between drivers, pressures, states, impacts, and responses within the investigated systems in a qualitative manner. For the WETwin project the DSIR framework was coupled with the Ecosystem Services approach in such a way that 'Impact' was defined as 'impact on ecosystem services'.

Some of the main 'Drivers' due to agriculture and urban development that have been identified for the AdM wetland are: food production, population growth, urbanization of riparian areas, fisheries, use of agrochemicals, inadequate waste management. These 'Drivers' cause 'Impacts' on the state (ecosystems and water quality of the wetland), which require adequate 'Responses' to be implemented (Figure 1-6)

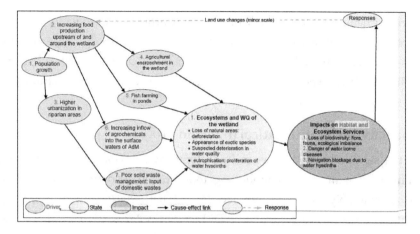

Figure 1-6 Impacts of agricultural and urban development on the ecosystems services of Abras de Mantequilla wetland in the Guayas River Basin, Ecuador. Source (Zsuffa and Cools, 2011)

Other main drivers are due to basin scale development projects related to dams and water projects located upstream of the AdM wetland. These drivers cause impacts on the states represented by the hydrology of the wetland that is directly linked with the ecosystems and water quality. To alleviate these impacts, responses such as improving water allocation schemes and introducing environmental flows, have to be taken in consideration for implementation (Figure 1-7).

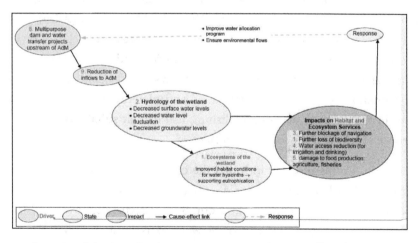

Figure 1-7   Impacts of basin-scale river management and water allocation projects on the ecosystem services of Abras de Mantequilla wetland in the Guayas River Basin, Ecuador. Source (Zsuffa and Cools, 2011).

Many tropical wetlands threatened by land use changes or modifications in hydrological regime, require effective management policies and implementation to protect them. AdM wetland is subject to two major environmental disturbances: (i) short-term agriculture (rice, maize) on the land around the wetland, and (ii) the effects of planned infrastructure works of the Baba dam in the upper catchment, which is expected to be reduced by 30% the Nuevo River flow (main AdM inflow) (Arias-Hidalgo, 2012). These activities are expected to be the main constraints for the future wetland health (Alvarez-Mieles et al., 2013).

1.7 Capabilities of mathematical modelling tools

Due to the significant link between hydrology, hydrodynamics and ecology, the development of mathematical modelling tools in the field of eco-hydraulics has evolved considerably over the past decades (Mynett, 2002; Mynett, 2003; Mynett, 2004). Integrated software systems like DELFT3D (Deltares, 2013a; Deltares, 2013b) are able to capture the behaviour of living species in the aquatic environment and determine the response of e.g. habitat suitability to river and wetland restoration, fish passage construction, or harmful algae bloom events. These are all highly complex phenomena involving many processes and interactions between physical, chemical, ecological and biological components (Mynett et al., 2007). In this thesis, the potential of using mathematical models in combination with field data analysis is explored.

1.8 Overall research approach

The overall goal of this study was to evaluate the relationship between hydrology/hydraulics and ecology of the Abras the Mantequilla (AdM) wetland in order to provide sustainable management advice. The research focuses on five key objectives: (a) assessing the variability in hydrological and hydraulic conditions; (b) assessing the physico-chemical patterns in the system; c) identifying the wetland aquatic communities and their spatial patterns; d) determining the main ecological processes in the wetland; and e) establishing fish habitat suitability conditions. For this purpose several activities were performed.

Initial activities in this research included the collection of the available information on hydrology, river hydraulics, water quality and biotic communities of the Guayas River Basin in the coastal region of Ecuador. Special focus was given to the middle catchment of the basin, where the AdM wetland is located. Following activities include the design and development of fieldwork campaigns to collect primary information on physico-chemical variables and biological communities in the river-wetland system.

Once the primary and secondary sources of data were collected, numerical model and data analysis tools were applied. The wetland (hydro)dynamic characteristics were analyzed using a numerical hydrodynamic model to assess the natural variation in inundation area, water depth, residence time of the AdM wetland (Chapter 2). Subsequently, the abiotic physico-chemical results from the sampling campaigns were analyzed and abiotic spatial patterns determined (Chapter 3).

Biotic communities of plankton, macroinvertebrates and fish collected during the fieldwork sampling were evaluated to determine main species in this wetland, spatial patterns, and their relations with abiotic variables (Chapter 4). The dynamics of the water chemistry variables and components of primary and secondary production (phytoplankton and zooplankton biomass) were evaluated using a numerical water quality and ecological simulation model (Chapter 5).

A fish-habitat analysis was performed for the fish community; one of the important functions of AdM as a Ramsar site (Chapter 6). Potential management scenarios were explored to assess the response of the wetland to different flow conditions (historical minima / maxima) with the aim to help advise authorities to develop sustainable management measures for the AdM wetland (Chapter 7). Conclusions and recommendations are summarised (Chapter 8).

1.9 Specific research questions

From the overall aim to evaluate AdM wetland using an 'integrated approach' which includes both abiotic and biotic components of the environment (i.e. hydrology, hydraulics, water quality, biology, ecology) this research focuses on whether/how such type of integrated approach can be useful to identify the interrelations between physico-chemical and biological communities in tropical regions. Can such integrated approach assess the ecological status of a river-wetland system, and how valuable can this be for the development of future management scenarios? The following specific research questions were identified with their corresponding aspects to be considered:

What is the spatio-temporal variability of the hydrodynamics in the wetland?
Using available data it is important to determine the wetland's main inflows, outflows and the natural variability in water depth and inundation area due to the (inter)annual changes in hydrological conditions. Flow patterns and typical residence times in the various wetland zones are to be determined and appropriate length and time scales to be identified to model the hydrodynamics of this wetland.

Which are the dominant species in the different aquatic communities?
Based on measurement data and literature search, the ecology of the dominant species, and their sensitivity to potential changes is to be explored.

What are the spatial patterns in the distribution of the environmental variables and biotic communities?
After identifying the typical range of water and sediments physico-chemical concentrations in the different wetland areas, the question is to be explored whether there is an environmental river-wetland gradient defining the distribution of abiotic variables, as well as biotic communities.

What are the main physico-chemical and ecological processes in the wetland?
From the temporal and spatial variability of key environmental parameters, the influence of the wetland hydrodynamics on the physico-chemical concentrations and primary producers in the wetland is to be determined.

What is the spatio-temporal variability in Fish Habitat Suitability?
From the spatial and temporal variations in hydrodynamic conditions the zonation of suitable habitat areas for the fish community are to be determined and translated into recommendations on sustainable wetland management.

What measures can be recommended for the management of the AdM wetland?
There is a particular need to explore the wetland functioning under different extreme scenarios and due to e.g. effects of upstream dam operation. Will environmental services (water availability, primary productivity, fish-habitat suitability) increase or decrease in relation to different hydrological conditions?

To what extent can numerical models help describe wetland functioning?
What lessons can be learned from the application of numerical modelling tools. How sensitive are model outcomes to variations in initial and boundary conditions. Is it possible to provide rules and guidelines for model application by non-developers (who are often non-experts in particular fields and model components).

1.10 Thesis outline

The thesis is divided into eight chapters:

Chapter 1 introduces the role of wetlands, their physical dynamics, and the importance of hydrology in wetland processes, their value as ecosystems, their main threats, and the role of the Ramsar Convention. The WETwin project is introduced, which provided the framework for the present research. The Abras de Mantequilla wetland is introduced and its role as a migration spot (Ramsar Site) is highlighted together with the main impacts encountered by this river-wetland system. The overall objectives, specific research questions, and thesis outline are presented.

Chapter 2 summarizes the long-term hydrological conditions in terms of annual precipitation, and explores the seasonal variability of the system. A 2D hydrodynamic model is used to explore the hydrodynamics of the wetland under different hydrological conditions. Changes in water depth, inundation area and residence time are investigated and inundation areas under extreme conditions (like El Niño years) are identified.

Chapter 3 summarises the results of the data collection campaigns carried out in 2011 and 2012 on environmental variables. The results of spatial pattern analyses of the various physico-chemical properties for different inundation conditions are summarised.

Chapter 4 summarises the results of the data collection campaigns carried out in 2011 and 2012 on identifying the dominant biotic communities and establishing the densities of the different biotic groups. The results of spatial pattern analyses of the various biotic communities for different inundation conditions are summarised. Relations with abiotic variables are presented.

Chapter 5 describes the temporal and spatial water quality and primary production dynamics. A water quality/ecological model is applied to evaluate the wetland dynamics.

Chapter 6 identifies the habitat suitability for the fish community in relation with particular hydrodynamic features and hydrological conditions.

Chapter 7 provides a discussion on the findings of the various chapters and provides an overall integration of relevant parts of the various components.

Chapter 8 presents the answers to the research questions and summarizes the main conclusions and recommendations.

*"Take the attitude of a student, never be too big to ask questions,
never know too much to learn something new"*

Og Mandino

2

HYDRODYNAMICS OF THE TROPICAL AdM RIVER-WETLAND SYSTEM

2.1 Geographical conditions

2.1.1 Basin topography

Further to the general introduction in Chapter 1, a detailed map including the digital elevation details at the River Basin scale is presented in (Figure 2-1). Initial topographic data for the study area was collected by Arias-Hidalgo (2012), Galecio, (2013). In the present study additional information was obtained from a Digital Elevation Map (1:50000) provided by IGM (Military Geographic Institute of Ecuador), which was published on the Internet for scientific use (http://www.rsgis.ait.ac.th/~souris/ecuador.htm). Also, maps from the Shuttle Radar Topography Mission (SRTM) were used, with an initial resolution of 90mx90m, resampled to 30mx30m, which are freely available (http//www.ambiotek.com/srtm). Moreover, a 1:10000 scale topographic map for the AdM area was refined using a local topographic survey of the area explored during the field campaign in February 2011 (including bathymetry) and merged into the existing 1:10000 DEM (Arias-Hidalgo, 2012). Finally, a new Digital Elevation Map (1:10000) that became available in February 2011 (having a 5mx5m resolution) enabled additional topographic adjustments during subsequent research (Galecio, 2013). The Digital Elevation Maps at the river basin and wetland scale are presented in Figure 2-1 and Figure 2-2.

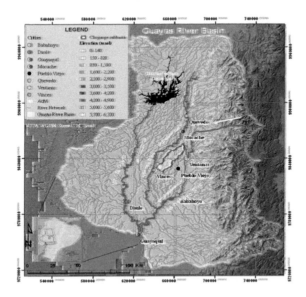

Figure 2-1 Digital Elevation Map at the Guayas River Basin scale (Arias-Hidalgo, 2012)

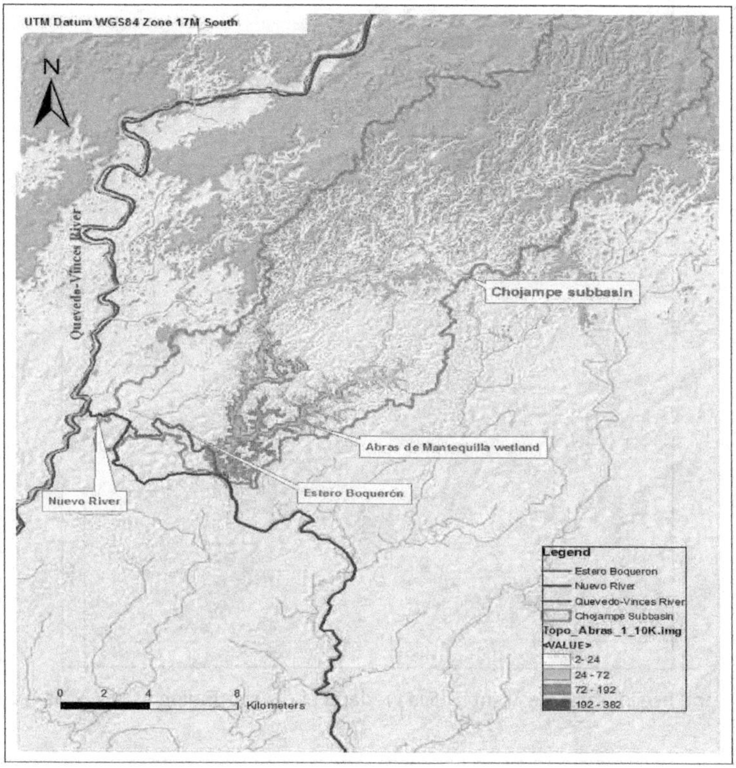

Figure 2-2 Digital Elevation Map at the AdM wetland scale.

2.1.2 Land use composition

The AdM wetland area has been subject to considerable changes in land use over the recent decades as well as the entire western region of Ecuador. These changes have been more evident since 1958 due to the increase in population, followed by an increase of roads, and gradual replacement of forest coverage (Dodson and Gentry, 1991). According to a land use map of 2008, the major part is used for short-term crops (45%) mainly maize and beans, followed by rice (26%) while the forest cover represents only around 3% of the total area of the wetland. Figure 2-3 and Figure 2-4 present the land use coverage in the wetland and surrounding areas. Due to the low coverage of natural forest, land use in the study area was scored low by the assessment carried out by the WETwin project. It was noted that intensive agriculture has led to frequent application of fertilizers and pesticides, which may have been contributing to an increase in nutrients run-off into the wetland over the last decades. On the other hand, hydrologic and geomorphologic conditions have only experienced moderate modifications due to anthropogenic activities (Arias-Hidalgo, 2012).

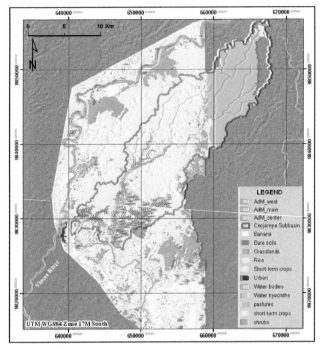

Figure 2-3 Land use in Abras de Mantequilla wetland and surrounding area (Arias-Hidalgo, 2012)

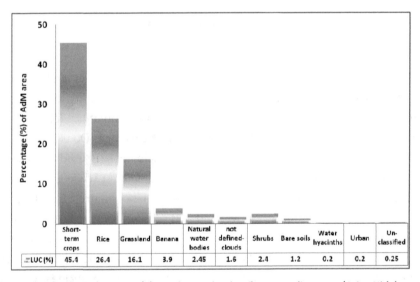

	Short-term crops	Rice	Grassland	Banana	Natural water bodies	not defined-clouds	Shrubs	Bare soils	Water hyacinths	Urban	Un-classified
LUC (%)	45.4	26.4	16.1	3.9	2.45	1.6	2.4	1.2	0.2	0.2	0.25

Figure 2-4 Land use composition (%) in AdM wetland and surrounding area (Arias-Hidalgo, 2012)

2.1.3 Soil properties

The soils of the Chojampe sub basin can be classified into four major categories: 48.2% consists of clays, loamy clays, inorganic clays; 36.4% are saturated soils and expansive clays; 13% are sandy soils; and 2.6% are sandy clays and loamy sands (Figure 2-5).

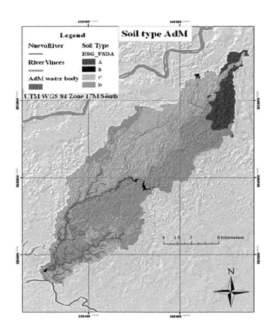

Figure 2-5   Soil type classification (Arias-Hidalgo, 2012)

2.2 Hydrological and meteorological conditions

In order to evaluate the wetland functioning and its natural variability, hydrological and meteorological data are required. Daily time series for meteorological data: precipitation, potential evapotranspiration, air temperature and relative humidity in the region were obtained from INAMHI (National Institute of Meteorology and Hydrology of Ecuador). Daily time series for discharges and water levels were also provided by INAMHI for specific locations. The data included the AdM wetland, Chojampe Subbasin and Quevedo-Vinces Basin.

2.2.1 Annual precipitation

Pluviometric data for the study area was available over the period 1963-2012 (Table 1-1) from three gauging stations as indicated in (Figure 2-6); (i) Pichilingue; (ii) Pueblo Viejo; and (iii) Vinces. The geographic position of Pichilingue station between the Upper Quevedo-Vinces basin and Chojampe basin, facilitates understanding the functioning of both basins and their interrelation.

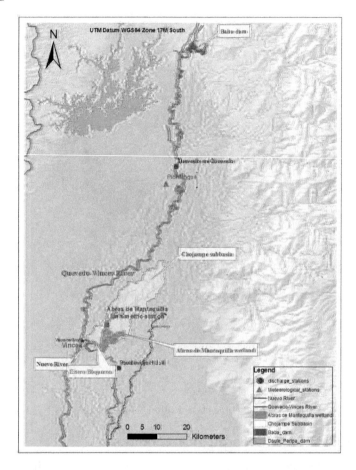

Figure 2-6 Measuring stations near the Abras de Mantequilla wetland

Table 1-1 Pluviometric data available for the study area

Station	Period	Frequency	Location
Pichilingue	1963-2012	daily	North of basin, close to Vinces River
Pueblo Viejo	1976-2012	daily	East of Abanico subbasin
Vinces	1964-2012	daily	Near the confluence of Vinces and Nuevo River

The highest annual precipitation at Pichilingue station was observed in the years 1997 / 1998 during the 'El Niño event' with values of 4736mm and 4790mm respectively, while the lowest annual precipitation was recorded in 1968, 1975 and 2005, with values between 1066mm and 1222mm (see Figure 2-7).

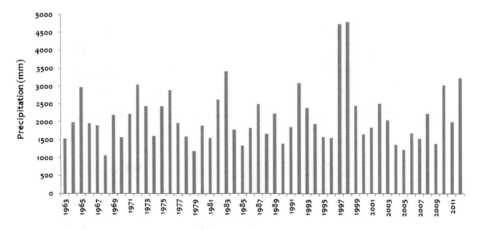

Figure 2-7 Annual precipitation at Pichilingue station (Period 1963-2012)

2.2.2 Seasonal variability

In order to assess the natural variability of the precipitation in the study area, the minimum, maximum and average monthly values over the period (1963-2012) are presented in Figure 2-8. The area is seen to typically exhibit two seasons: (i) a wet season (December-May); and (ii) a dry season (June-November). Extreme events (El Niño) cause precipitation levels to increase dramatically, extending the wet season up to 9 months (starting in November and extend until July). Maximum monthly values (up to 1100 mm) occurred during 1982-1983 and 1997-1998, which correspond to El Niño years. Minimum values during the wet season were seen to range from 47 mm to 198 mm. During the dry season, minimum values are close to zero, while maximum values can vary. Overall, the monthly average precipitation ranges from 383mm to 452mm during the wet season and between 13 mm and 53 mm during the dry season. During the wet season of a normal year, daily precipitation values were found to fluctuate from 5mm to around 23mm.

The spatial distribution of the mean monthly precipitation in Guayas River Basin (GRB) and AdM wetland are presented by the black lines and yellow lines in Figure 2-9 (Arias-Hidalgo, 2012).

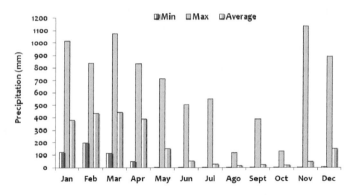

Figure 2-8 Minimum, Maximum and Average monthly precipitation in Pichilingue station (1963-2012)

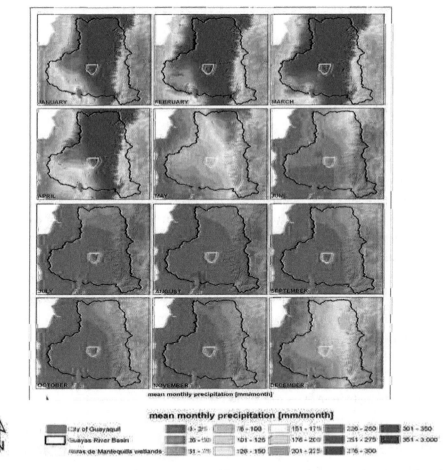

Figure 2-9 Spatial distribution of the mean monthly precipitation in Guayas River Basin (GRB) (black line) and Abras de Mantequilla (AdM) wetland (yellow line) (Arias-Hidalgo, 2012).

2.2.3 Discharge and water level data

Daily time series for discharges were available for two stations: Quevedo en Quevedo and Vinces en Vinces. Both stations have complete data (without gaps during the established period) and they are used to establish relations with other stations and to estimate the discharges of AdM wetland system. One limnometric station is located in the wetland, the AdM limnometric station (Table 2-2).

Table 2-2 Discharge and water level data

Station	Period	Frequency	Location
Quevedo en Quevedo	1962-2012	daily	North of Chojampe basin, close to the Pichilingue station.
Vinces en Vinces	1964-2012	daily	At Vinces town, downstream the diversion of Nuevo River from Vinces River
Abras de Mantequilla (limnometric)	1988-2007	daily	Near the outflow of El Recuerdo subbasin (close to measured point S1)

The main inflow into the wetland area is the Nuevo River, which is diverted from Vinces River. There are no discharge stations in the area close to the wetland. Thus, the data of the Vinces and Quevedo rivers were used to calculate the boundary conditions of the wetland inflows to describe the wetland flooding patterns. In order to illustrate the variability of the river discharges in both stations, minimum and maximum monthly discharges of both gauging stations are presented in Figure 2-10 and Figure 2-11. The minimum, maximum and average monthly values were calculated from the data available for the periods detailed in Table 2-2. The figures show a clear wet and dry season which is comparable to the precipitation pattern in Pichilingue station (Figure 2-8). During the wet season, the maximum monthly discharges can reach 910-980 m^3/s in Quevedo en Quevedo and 790-830 m^3/s in 'Vinces en Vinces'. Maximum discharges were recorded for the periods 1982-1983 and 1997-1998 during the El Niño event. Monthly average discharges in both stations can reach 500-530 m^3/s during the wet season, decreasing to 50 m^3/s during the dry season.

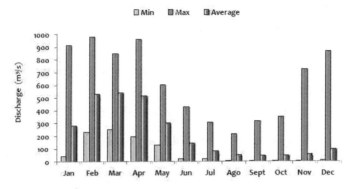

Figure 2-10 Monthly discharges in Quevedo en Quevedo station (1962-2012)

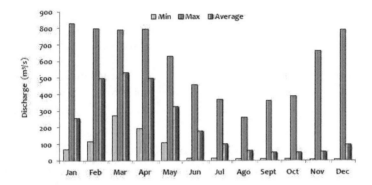

Figure 2-11 Monthly discharges in Vinces en Vinces station (1964-2012)

The water depth variation in the Abras de Mantequilla wetland was evaluated using the only available limnometric station located within the system. Monthly average water levels between 12.3 and 13.0 m were recorded for the wet season corresponding to water depths of 6.3 and 7.0 m, respectively, since the reference bed level is 6 m. For the dry season, average water levels decrease to 9.2 (2.2 m water depth). The maximum monthly water level occurs during the wet season (14.4 m level corresponding to a water depth of 8.4 m), and a minimum (8.30 m level corresponding to a water depth of 2.30 m) during the dry season (Figure 2-12a). Figure 2-12b shows the high variation in inundated area from dry to wet season.

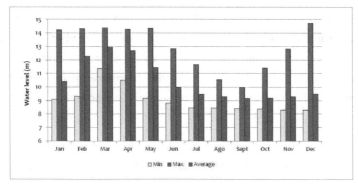

Figure 2-12a Monthly water levels in Abras de Mantequilla limnometric station (1988-2007)

Figure 2-12b Abras de Mantequilla wetland at 'El Recuerdo', same location during the dry season (left) and wet season (right).

2.2.4 *Hydrology of the Chojampe subbasin*

The AdM wetland is part of the Chojampe subbasin, which has an area of around 259 km² and is divided into seven small sub-basins (microbasins): Chojampe 1, Las Tablas, Chojampe 2, Agua Fria, El Recuerdo, Abras de Mantequilla and Abanico (Figure 2-13). The wetland is located in the southern area of this subbasin, specifically in the Abras de Mantequilla and Abanico microbasins. Each of these microbasins has a main branch, with dendritic patterns. The main tributary of the wetland is the Nuevo River, which receives water from the Vinces River (Quevedo-Vinces subbasin). El Recuerdo microbasin collects the runoff from the five upper microbasins (Upper Chojampe) representing the second hydrological tributary to the wetland. In addition, the wetland does not cover the entire area of Abras de Mantequilla and Abanico microbasins, thus three small tributaries are identified in the upper area of these two microbasins, all of them influencing locally the functioning of the wetland (Galecio, 2013). They are identified as: Abras de Mantequilla tributary 1 and 2 (AdM T1 and AdM T2), and Abanico tributary 1 (Abanico T1). All of them have a local influence in their immediate surrounding areas, but the effect of their flow on the entire wetland is minimal due to their low flow rates. Overall, the study area is strongly influenced by a wet season between January and April and a dry season between July and November, producing a clear seasonal pattern of floods and droughts. Flooding time is characterized by high precipitation, high discharges in the rivers and therefore an increase of the water levels. Dry periods show a significant decrease of the water levels due to the decrease in precipitation and discharges.

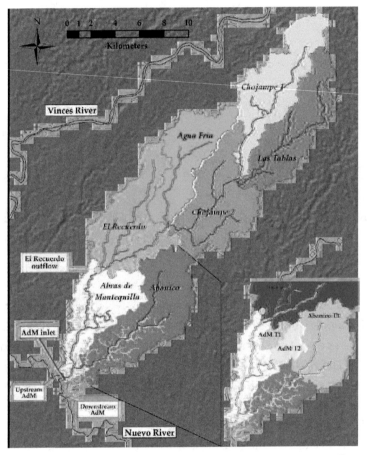

Figure 2-13 Chojampe micro-basins and tributaries to the Abras de Mantequilla wetland

2.2.5 Regional infrastructure projects

The AdM wetland is expected to face effects of infrastructure projects in the upper catchment of the Quevedo-Vinces River Basin. The multipurpose Baba Dam was constructed in the upper catchment of the Quevedo-Vinces river (Figure 2-14). The filling phase of the reservoir ended in January 2012 (Efficacitas, personal communication, 2013), and Baba dam operation started in 2013. Since the main inflow into the wetland is the Nuevo River which diverts from the Quevedo-Vinces River, Baba dam operation could divert the water supply by about 30% thereby decreasing the flow of the Nuevo River towards the AdM wetland (Arias-Hidalgo, 2012). Therefore, the dam might present a major constraint for future wetland health. These effects will be explored through potential scenario developments using numerical simulations.

2.3 The AdM river-wetland system

2.3.1 Flows in main arteries and tributaries

In order to explore the flows in the main river system, a numerical 1D river routing model for the Vinces and Nuevo Rivers was constructed by Arias-Hidalgo (2012) in order to transfer measurement information from the upstream sections of the basin (Quevedo Vinces) to the Nuevo River and the AdM wetland connection point. In this way the observed interactions between the river and wetland systems can be quantified including its seasonal variability. The 1D model was constructed using HEC-RAS, a numerical tool developed by the US Army Corps of Engineers (Brunner, 2010) for unsteady flows in the main flow direction using the (1D) De Saint-Venant equations (Appendix A).

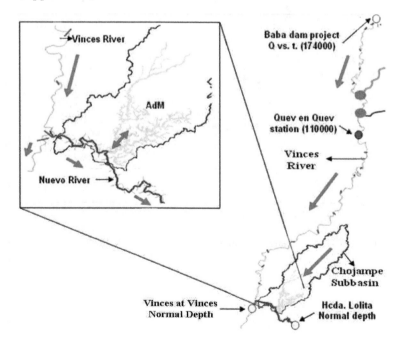

Figure 2-14 HECRAS model schematization for the main river flows including boundary conditions (yellow dots) and calibration point (blue dot). Inflows from rivers: Lulu (orange) and San Pablo (blue) (Arias-Hidalgo, 2012)

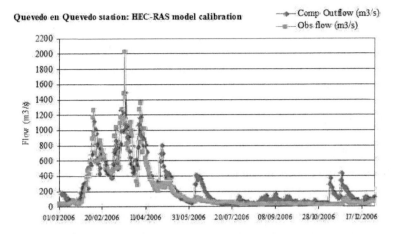

Figure 2-15 Hydrograph comparison for the HEC-RAS model
(Arias-Hidalgo, 2012)

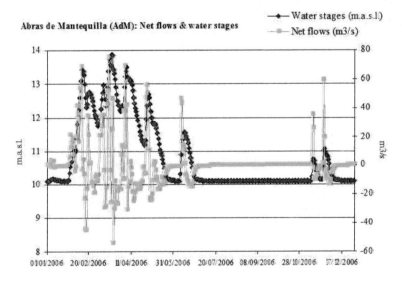

Figure 2-16 Flow exchange and stages between the Nuevo River and Main Abras
(Arias-Hidalgo, 2012)

2.3.2 Inundation modelling of the AdM wetland system

DELFT3D-FLOW

In order to explore the inundation patters in the AdM wetland system, the 1D HEC-RAS approach was useful to quantify the interactions between the river and the wetland system. However, spatio temporal variability cannot be done with a 1D channel approach, therefore, DELFT3D-FLOW, a software suite from Deltares was used to construct a 2D (horizontal x-y) model for the entire AdM wetland. This 2D model provided as output dynamic spatial inundation maps for the entire wetland area. DELFT3D-FLOW is part of the DELFT3D suite, a fully integrated computer software package for a multidisciplinary approach applicable for rivers, coasts, lakes and estuaries. The DELFT3D suite is composed of several modules grouped in a common interface that can be easily linked to each other. DELFT3D-FLOW, one of the modules of this suite, is a multi-dimensional (2D or 3D) hydrodynamic/transport simulation program that calculates non-steady flows and transport phenomena resulting from tidal and/or meteorological forcing (Deltares, 2013a).

Wetland schematization and 2D grid construction

The set up of the model started by the inspection of a 1:10000 topography. This topography is the result of an existing 1:10000 topography that was improved with a local topographic campaign developed in February 2011. The results of this local survey were merged with the available 1:10000 topography (Arias-Hidalgo, 2012). Afterwards, this merged topography was reviewed again to correct for discontinuities (Galecio, 2013). According to this topography, the wetland area recorded levels between 6 and 34 m.a.s.l. The grid for this model was built considering the wetland extension and the location of the discharges to the wetland. A 2D grid was defined for the study area, thus only one layer was considered for building the model. The geographic coordinates of the grid are presented in Table 2-3.

Table 2-3 Grid coordinates for hydrodynamic model

	WEST COORDINATES		EAST COORDINATES	
North	9836492 S	644800 W	9836492 S	652834 W
South	9825900 S	644800 W	9825900 S	652834 W

The interaction between the Nuevo River and wetland is an important issue to assess. Thus, a section of the Nuevo River located at the inlet of the wetland was considered as part of the grid to account for this interaction. The other four tributaries to the wetland (El Recuerdo, AdM T1, AdM T2, and Abanico T1) are not included in the grid but considered as discharges. The grid cell size was 75 m x 75 m, which was considered as a suitable size regarding topography characteristics and computational time. The hydrodynamic grid had a total of 7163 cells. Initial conditions, observation points and boundary conditions were considering in the set up. Boundary conditions to the wetland were represented by the two Nuevo River reaches located upstream and downstream of the wetland, and the four tributaries located in the Chojampe subbasin (El Recuerdo, AdMT1, AdMT2 and Abanico T1). The main inflow to the wetland is the Nuevo River, hence the flow conditions upstream and downstream of the inlet are the ones driving the patterns of flooding and ebbing. Six boundary conditions were defined for this model. Observation points were also defined (Figure 2-17).

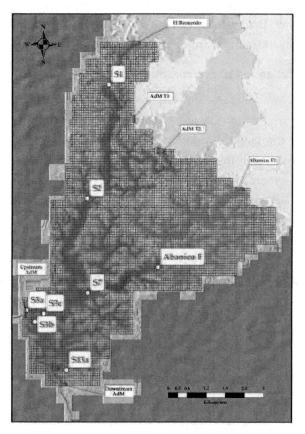

Figure 2-17 Abras de Mantequilla wetland grid; boundary conditions (red lines): Upstream AdM (Nuevo river-main inflow to the wetland); Upper Chojampe (El Recuerdo, AdmT1, AdMT2, AbanicoT1); Downstream AdM (wetland outflow). Observation points (white dots)(Galecio, 2013).

Boundary conditions

The boundary conditions for Nuevo River (inflow to the wetland) were estimated based on the HEC-RAS model for 2006 (Arias-Hidalgo, 2012) and correlations with an upstream gauging station (Quevedo en Quevedo station). The boundary conditions for the Nuevo River (outflow of the wetland) were estimated based on the total discharge flowing outside the wetland system using a rating curve. The boundary conditions for the four tributaries of Chojampe subbasin were determined using HEC-HMS, a rainfall-runoff model built for this purpose (Galecio, 2013). The procedure for both estimations is explained in the following two sections.

Nuevo River inflows

The discharges of Nuevo River (inflow and outflow) to the wetland were estimated based on the results of a 1D model developed in HEC-RAS for an average year (2006), and a correlation with Quevedo en Quevedo gauging station located in the Vinces River (Galecio, 2013). This 1D model was built for 2006 because this year was considered to represent well the general functioning of the basin (Arias-Hidalgo, 2012). Although the use of a correlation is an approximation, it was considered the most suitable procedure to overcome the absence of measure data in the area. The results of this procedure presented a good linear correlation between the HEC-RAS model results and the data measured in the Quevedo gauging station. Correlation coefficients for upstream (0.83) and downstream (0.84) of the wetland were obtained (Figure 2-18). Subsequently, discharge conditions for other years of interest were estimated using the equations (Eq 2-1 & Eq 2-2).

$$Q_{upstreamAdM} = 0.39 * Q_{Quevedo} \quad \text{(Eq 2-1)}$$

$$Q_{downstreamAdM} = 0.39 * Q_{Quevedo} \quad \text{(Eq 2-2)}$$

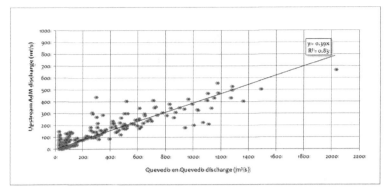

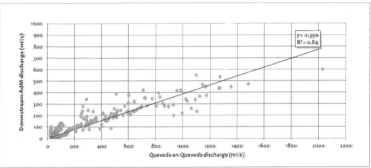

Figure 2-18 Correlation between Quevedo en Quevedo river discharge and HEC-RAS results of Upstream AdM wetland (upper panel) and Downstream AdM wetland (low panel). (Upstream=wetland inflow; Downstream=wetland outflow). Adapted from Galecio (2013)

The relation between the discharge and water level in Nuevo River (downstream AdM) was established with a rating curve (Figure 2-19). The curve was determined using the available bathymetry and the HEC-RAS model of the Vinces and Nuevo River in the cross section located downstream the wetland (Galecio, 2013). The discharges ranged from 23.2 to 603 m³/s, while the water level between 8.7 and 13.9 m. The analysis included different curves for low and high discharges to improve the relation between these two variables. Computed correlation coefficients were 0.98 and 0.99. The resulted rating curve for the cross section located downstream of the wetland was:

$$H<10.3 \rightarrow Q=3.38*(H-6.0)^{1.97} \rightarrow \qquad \text{(Eq 2-3)}$$

$$H<10.3 \rightarrow Q=2.96*(H-6.0)^{3.55} \rightarrow \qquad \text{(Eq 2-4)}$$

Where:

Q = discharge (m³/s), H = water level (m)

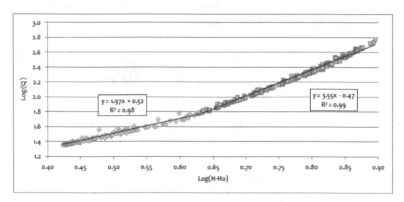

Figure 2-19 Rating curve for downstream AdM (Galecio, 2013)

Upper Chojampe subbasin

The rainfall run off model (HEC-HMS model) was developed to evaluate the contribution of this subbasin to the AdM wetland. Since there was a lack of discharge data for the tributaries of this subbasin, estimating the relation between rainfall and run off via this model was used to estimate the boundary conditions. The model was based on a previous model set-up developed by Arias-Hidalgo (2012), who distinguished five microbasins located in the Upper Chojampe subbasin: Chojampe 1, Las Tablas, Chojampe 2, Agua Fria and El Recuerdo. All these microbasins collect the rainfall from the upper part of the Abras de Mantequilla wetland generating a discharge in the outflow of 'El Recuerdo' microbasin (Figure 2-20). The objective was to calculate the discharge in this point. Precipitation data from three gauging stations were used: Pichilingue, Pueblo Viejo and Vinces. Information about soil type, topography, and size of the microbasins was evaluated. Several processes were considered in the model: water loss, transformation from rainfall to runoff, and base flow. The calibration point was the only gauging station in the wetland (Abras de Mantequilla limnometric station), located 1.5 km downstream the outflow from the El Recuerdo microbasin. In order to obtain the discharge from this gauging station, the cross sections measured in the field were combined with the assumption that the velocities in this point were comparable to the one of wetland inlet.

The year 2006 was considered as a training period for this model (Galecio, 2013), thus the present model was also calibrated with this year. The microbasins areas of these three small tributaries were: 4 km^2 for AdMT1; 7 km^2 for AdMT2; and 28km^2 for AbanicoT1 (Figure 2-20 right panel). Afterwards, boundary conditions for different hydrological years were quantified. For this purpose, the values of the parameters used for the calibrated year 2006 were maintained, but the precipitation was adjusted for each of the other years of interest selected.

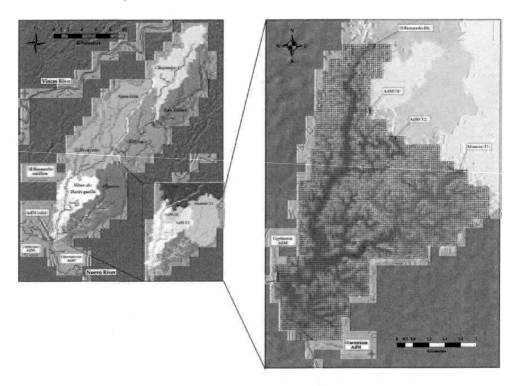

Figure 2-20 Abras de Mantequilla wetland (AdM) - main inflows and Hydrodynamic model schematization. Left panel: 'El Recuerdo' (yellow dot) collects the run off of the five contributing microbasins from Upper Chojampe Sub-basin (Chojampe 1, Las Tablas, Chojampe 2, Agua Fria and El Recuerdo). Abras de Mantequilla wetland area (in light yellow). Right panel: Hydrodynamic model schematization - Abras de Mantequilla wetland grid (from Delft3D-FLOW). Boundary conditions (in red lines). Low boundary condition (Upstream AdM) represents the main inflow to the wetland 'The Nuevo River-Estero Boquerón'. Upper boundary conditions (El Recuerdo, AdMT1, AdMT2 and Abanico T1) collect the run-off from Chojampe subbasin (Galecio, 2013).

2.3.3 Model verification

Measured and simulated values of water levels at 'El Recuerdo' (where the Abras de Mantequilla limnometric station is located) were compared to assess the performance of the model. For the verification, simulated water levels of the average year (2006) were compared with their measured levels. The magnitude of the water levels was well represented, however, simulated values showed more variation than the observed ones (Figure 2-21). Reference bed level in the wetland is 6 meters above sea level (m.a.s.l.).

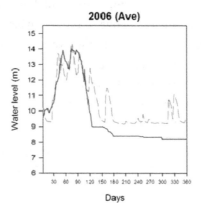

Figure 2-21 Simulated and measured water levels at Abras de Mantequilla limnometric station 'El Recuerdo' for the average year (2006). Measured (blue line), simulated (green line).

2.3.4 Model performance

The performance of the model was measured with the Nash-Sutcliffe coefficient (Nash and Sutcliffe, 1970) (Eq 2-5). This coefficient is often used for the analysis of hydrologic time-series to determine the relation between two sets of data (Eq2-5).

$$NS = 1 - \frac{\sum (Hobs - Hsim)^2}{\sum (Hobs - \overline{Hobs})^2} \qquad \text{(Eq 2-5)}$$

Where:

NS = Nash-Sutcliffe coefficient ($-\infty < NS < 1$)

$Hobs$ = observed water level (m)

$\overline{Hobs}$ = average of observed water level (m)

$Hsim$ = simulated water level (m)

Measured and simulated values of water levels at El Recuerdo (where Abras de Mantequilla limnometric station is located) were compared to assess the performance of the model for the other conditions evaluated. Results indicate a satisfactory performance of the model (NS = 0.77) and a correlation coefficient of 0.86 for the dry year and 0.83 and 0.90 for a wet year (Table 2-4). For the dry year, the model reproduces the magnitude and timing of the water levels satisfactorily, although a shift of a few days was observed. The wettest years (1992 & 1998) showed an adequate representation of the temporal pattern, although simulated values were slightly lower than the measured ones for both simulations (Figure 2-22). Reference bed level in the wetland is 6 meters above sea level (m.a.s.l.). The performance for extreme years was again measured with the Nash-Sutcliffe coefficient. Overall, results from the NS coefficient exhibit a satisfactory performance of the model, with NS coefficients ranging from 0.57 to 0.83 and correlation coefficients over 0.76. These results show that the general pattern of water level conditions for the years simulated is well represented by the model (Table 2-4).

Table 2-4 Model performance measured with the Nash- Sutcliffe coefficient

YEAR	NASH-SUTCLIFFE COEFFICIENT (NS)	R	AVERAGE HOBS
1990 (Dry)	0.77	0.86	9.59
1992 (Wet)	0.83	0.90	10.85
2006 (Average)	0.57	0.76	9.60
1998 (El Niño)	0.67	0.88	11.96

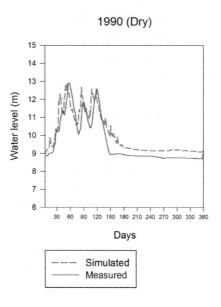

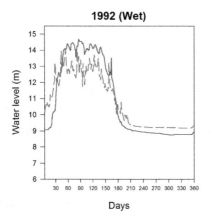

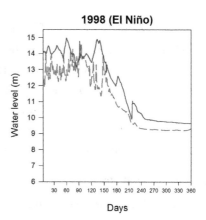

Figure 2-22   Simulated and measured water levels at Abras de Mantequilla limnometric station 'El Recuerdo'. For: dry year (1990), wet year (1992), and extreme wet year (El Niño-1998).

2.4 Natural variability in hydrodynamic conditions

The main objective of building the hydrodynamic model was to assess the wetland hydrodynamic functioning and patterns under different hydrological conditions in order to evaluate the natural variability of this wetland-river system along the last four decades. It is considered natural variability because during the period (1963-2010), there has not been an important water abstraction from the wetland system and inflows (e.g. a dam). Nevertheless, water in the surrounding inflows has been used for irrigation.

To evaluate this variability several steps were performed: first, daily time series of precipitation for the period (1963-2010) were analysed to determine the driest and the wettest year during this period. From the analysis, the driest year recorded 1397 mm of precipitation, the rainiest one 4791 mm. Secondly, the years of interest were evaluated considering their total precipitation and return period, allowing the selection of particular years that can describe the variability of this system. Three years were identified as suitable to represent extreme years: 1990 (dry); 1992 (wet); 1998 (extreme wet-El Niño) (Galecio, 2013).

2.4.1 Boundary conditions for extremes

Following to the identification of the years, the quantification of the six boundary conditions: Nuevo River inflow and outflow; El Recuerdo, AdMT1, AdMT2, AbanicoT1, for each of these years was performed following the procedure introduced earlier. For the Nuevo River, the boundaries were estimated from equations (Eq 2-1 and Eq 2-2) obtained from the correlation analysis for discharge and water level time series, respectively.

2.4.2 Initial conditions

Initial water levels set up for the model were determined for the different conditions in two areas: the wetland and the river inlet. For the wetland, the water levels were estimated from the limnometric station located in the wetland. For the river inlet, the water levels were estimated based on the water levels downstream from the wetland, which were calculated with a rating curve. Initial velocities were set to zero. Table 2-5 presents the initial water levels for the simulated years.

Table 2-5 Hydrodynamic model-Initial conditions for water level (m)
for the different simulated years

YEAR	WATER LEVEL (m)	
	WETLAND	INLET
1990 (Dry)	8.84	9.60
1992 (Wet)	9.07	9.90
1998 (El Niño)	14.17	12.83

2.4.3 Variability in water depth

Minimum and maximum simulated water depths at different wetland areas are presented in Table 2-6. One observation point at each wetland area was selected to illustrate this range. The upper wetland showed minima between 1.2–1.5 m and maxima between 5.2–7.5 m. The middle area had minima from 0.5–1.5 m and maxima from 4.2–8.6 m. At the lower area, minima ranged from 0.4–1.5 m, and maxima were up to 8.6 m. Considering the different simulated conditions, maximum values occurred during the El Niño year 1998. When analyzing the averages, it can be seen that the El Niño year was the one with higher averages in the three areas compared to the rest of simulated conditions, indicating that longer periods with higher water depths occurred during this year. Minimum water depths were observed during the dry year. Overall, the wetland experienced historical fluctuations from 0.4–8.6 m.

Table 2-6 Simulated water depths results (m) of observations points located at the upper, middle, and lower wetland areas, for the different conditions/years

	S1			S2			S7		
YEARS	MIN	MAX	AVE	MIN	MAX	AVE	MIN	MAX	AVE
1990 (Dry)	1.2	5.2	2.3	0.5	4.2	1.4	0.4	6.6	2.5
1992 (Wet)	1.4	6.4	3.1	1.2	8.0	4.0	1.2	8.0	4.1
1998 (El Niño)	1.5	7.0	3.4	1.5	8.6	4.7	1.5	8.6	4.7

2.4.4 *Variability in inundation area*

The inundation variability during the first six months of the year (January-June) for extreme and average years is presented in Figure 2-23 & Figure 2-24. The peak of the wet season mostly occurs during February-March (Figure 2-23), Results from the spatial analysis indicate that during a dry year (1990) the flooding was evident from February till May, while during the wettest year (1998) it was extended till June. The average year follows a similar pattern to the one of the dry year (Figure 2-24). Although the typical wet season is from January to May the figure also shows the month of June, which is considered a transition month in Ecuador. The representation of the dry season with a typical low inundation extent is depicted for the baseline and average year 2006 in Figure 2-25. This year was considered an average year because the precipitation during 2006 was an average precipitation from the period 1963-2010.

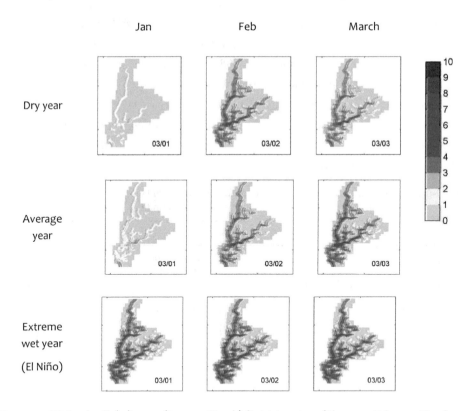

Figure 2-23 Water depth (m) maps (January-March) first trimester of the year. February-March are mostly the peak months of the wet season. For: dry year (1990), extreme wet year (El Niño-1998), and average year (2006). Scale bar indicates the water depth range (m).

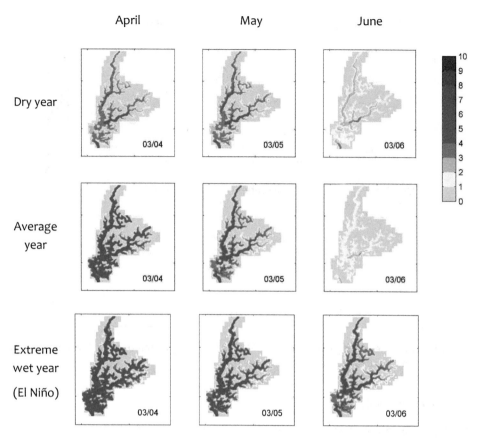

Figure 2-24 Water depth (m) maps (April-June), June is considered as transition month to the dry season. For: dry year (1990), extreme wet year (El Niño-1998), and average year (2006). Scale bar indicates the water depth range (m).

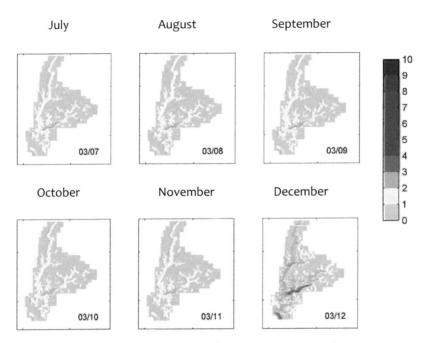

Figure 2-25 Water depth (m) maps for the DRY SEASON (July-December) for the average year (2006). Scale bar indicates the water depth range (m).

2.4.5 Variability in flow velocities

Several processes are driven by water velocity, so it is important to analyse the flow patterns occurring in both the wetland area and at the inflows, under different conditions. Measured values from campaigns 2011 and 2012 indicated that velocities in the wetland area were up to 0.4 m/s, and in the river inflow up to 0.8 m/s (Table 2-7). In the wetland area higher values were measured at observation point S7 (0.4 m/s) due to the proximity to the main inflow from Nuevo River, which showed values between 0.6–0.8 m/s. Since the campaigns took place at the peak of the wet season (February 2011 and March 2012), these measured values represent typical velocities during this period of the year at these observation points. Measured and simulated velocities were in the same range (Table 2-7 & Table 2-8). Pooled velocities from sites located in the three wetland areas showed that average simulated velocities ranged from 0.02–0.10 m/s, with maximum values up to 0.54 (1998 -El Niño year). In the Nuevo River average values were up to 0.28 m/s, and maximum ones up to 0.78 m/s (1992) (Table 2-8). Thus, measured velocities were closer to the maximum simulated velocities, confirming that the measurements were performed during the peak of the wet season.

Table 2-7 Measured velocities during campaigns 2011&2012

LOCATION	OBSERVATION POINTS	VELOCITY (m/s)
	S1	0.3
WETLAND AREA	S2	0.2
	S7	0.4
NUEVO RIVER INFLOW	S4	0.8
	S3a	0.6
NUEVO RIVER OUTFLOW	S13a	0.8

Table 2-8 Simulated velocities (m/s) for the different conditions/years (whole year simulation)

YEARS	WETLAND [(A)]			NUEVO RIVER [(B)]		
	MIN	MAX [(2)]	AVERAGE	MIN	MAX	AVERAGE
1990 (Dry)	0.00	0.39	0.02	0.00	0.42	0.18
1992 (Wet)	0.00	0.49	0.06	0.00	0.78	0.27
1998 (El Niño)	0.00	0.54	0.10	0.00	0.60	0.28

Note: (A) includes all simulated values from points: S1, S2, S7, S3c and Abanico F, (B) is represented by point S3a

More in detail, average and maximum simulated velocities at each observation point are presented in Table 2-9 and Table 2-10 to describe the differences between the wetland areas. Average velocities were between 0.01 and 0.18 m/s (Table 2-9); while maximum ones were up 0.5 m/s (Table 2-10).

Table 2-9 Simulated average velocities (m/s) of each observation points located at the Lower, Middle and Upper wetland areas for different conditions (whole year simulation)

	WETLAND AREAS				
	LOWER		MIDDLE		UPPER
Years	S7	S3c	S2	Abanico F	S1
1990 (Dry)	0.03	0.05	0.001	0.004	0.02
1992 (Wet)	0.10	0.06	0.10	0.01	0.02
1998 (El Niño)	0.18	0.07	0.18	0.01	0.04

Maximum velocities were evident in the lower area of the wetland (S7, S3c), due to the proximity to Nuevo River inflow. However, at the middle area (S2), higher values around 0.5 m/s occurred during the wettest years (1992 and 1998). At the upper wetland area, maxima were only up 0.18 m/s (Table 2-10).

Table 2-10 Maximum simulated velocities (m/s) of observation points located at the Lower, Middle and Upper wetland areas for the different conditions (whole year simulation)

| | WETLAND AREAS | | | | |
| | LOWER | | MIDDLE | | UPPER |
YEARS	S7	S3C	S2	ABANICO	S1
1990 (Dry)	0.26	0.39	0.01	0.06	0.13
1992 (Wet)	0.49	0.30	0.49	0.10	0.12
1998 (El Niño)	0.54	0.26	0.54	0.07	0.18

2.4.6 Variability in inundation area and volume

The variability of the wetland inundation area and volume is valuable because it provides an overall representation of what the system has experienced historically in terms of inundation area patterns. Water level results from the hydrodynamic simulations were used in combination with a stage volume curve to estimate the corresponding inundated areas and volume for each day of the simulation period (one year). The stage volume was built from the available topography of the wetland (Figure 2-26).

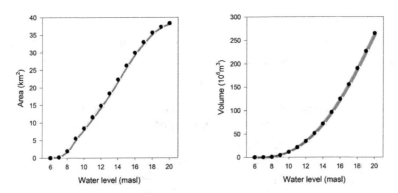

Figure 2-26 Stage-volume curve for Abras de Mantequilla wetland

For the analysis of the variability in inundation area, three approaches were performed. A first approach was to have a general overview of the inundation variability. Daily inundation area results from the different hydrological conditions simulated were pooled together for this purpose. The pooled simulation results show that, historically, the wetland has experienced changes in inundation area from 5–27 km² (Figure 2-27), given the set up of the hydrodynamic model.

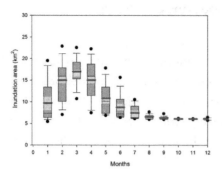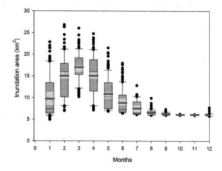

Figure 2-27 Boxplot of wetland inundation area (km²). Built with a stage volume curve and Delft3D-FLOW simulations results (one year simulation), from: dry year (1990); extreme wet year (El Niño-1998). Each month built with the daily values of each simulation. Blue line (mean). Left panel: upper and lower black dots show the 5th and 95th percentiles. Right panel: black dots show each outlier. Months display from left to right: January (1), to December (12).

A second approach was to calculate the monthly average of the inundation area for each year/condition from daily-simulated results. Results of this approach indicated that historically the wetland has experienced an average monthly flooding from 5–23 km². A high variability between the different simulations is evident during the wet season period (January to April). During the dry season, there was no variation among the simulations, since all reached a value of 5 km² (Figure 2-28). Nevertheless, the exception was the maximum historical since this time series includes the entire set of extreme wet conditions for a long historical period (1962-2010), and therefore includes also the isolated high events that occurred during this long period (Figure 2-28).

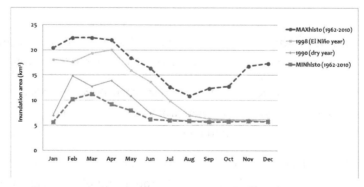

Figure 2-28 Monthly average of wetland inundation area (km²). Calculated from a stage volume curve and Delft3D-FLOW simulations results (one year simulation), from: dry year (1990); extreme wet year (El Niño-1998); Minimum and Maximum historical (period 1962-2010).

2.5 Water balance and relative contributions of inflows

2.5.1 Water balance

The calculation of the water balance was performed in order to estimate the contribution of the inflows (Nuevo River, Upper Chojampe tributaries, precipitation) and outflows (Nuevo River, evaporation, evapotranspiration, infiltration) (Eq 2-6). The inflow values for Nuevo River and Chojampe tributaries were obtained from the model estimated boundary conditions, while the outflow of Nuevo River from the results of the simulation.

$$(Q_{\text{NR inflow}} + \sum_{i=1}^{4} Q_{\text{tributary } i} + P) - (Q_{\text{NR outflow}} + E + Ev + I) = \frac{\Delta s}{\Delta t} \qquad \text{(Eq 2-6)}$$

where:

$Q_{\text{NR inflow}}$ = discharge entering the wetland from Nuevo River

$Q_{\text{tributary } i}$ = discharge from the four tributaries from Chojampe subbasin

P = precipitation over the wetland

$Q_{\text{NR outflow}}$ = discharge in Nuevo River flowing outside the study area

E = evaporation from the wetland inundated area

Ev = evapotranspiration from area not inundated

I = infiltration

S = storage

For the estimation of the remaining parameters, the following methodologies were applied:

Precipitation (P) was calculated with the data from Pueblo Viejo and Vinces gauging stations; the area of influence was considered to be the total modelled area.

Evaporation (E) was calculated with the Penman Open Water Evaporation equation (Penman and Keen, 1948), with parameters from Table 2-11, and considered to occur at the open water of the wetland inundated area.

Evapotranspiration (Ev) was calculated with the Penman-Monteith equation, and considered to occur at not inundated areas. A crop height of 0.6 meters was used in combination with the parameters from Table 2-11. This height value was selected because the vegetation present in the study area corresponds mainly to short-term crops (rice, maize) that have this average value. Furthermore, forest cover is almost absent (3% of the total area) (Arias-Hidalgo, 2012).

Infiltration (I) was calculated with the assumption that occurs in the saturated part of the wetland inundated area. A saturated hydraulic conductivity was estimated as 10 mm/d, since the soil type in the area is loamy clay and inorganic clay (Arias-Hidalgo, 2012).

Storage (S) was calculated with the water level variation and the stage-volume curve built from the topography (Figure 2-26).

Table 2-11 Parameters for calculation of open water evaporation

PARAMETERS	VALUES
Temperature (°C)	25
Wind (m/s)	0.5
Sunshine (hours)	4
Relative humidity % (RH)	89

Results from the water balance analysis (Table 2-12) determined that overall the Nuevo River inflow accounted for 86% of the total inflow, and the four tributaries from Chojampe subbasin for 11% of the total inflow (average of all simulations) (Figure 2-29). This pattern is maintained for the different years simulated, with minor oscillations in the contribution percentages (Figure 2-29).The contribution of Nuevo River inflow ranged from 83% (1998-el Niño) to 89% (2006). During wettest years, the contribution of the total inflow of Chojampe was higher than for the rest of simulations, indicating an increase in run-off during this extreme wet condition. Precipitation contribution directly in the area had a minor contribution ranging from 2.3% (1990-dry year) to 3.6 % (1992-wet year). The overall assessment of the outflows determined the significance of Nuevo River, which accounted with percentages above 95% while evaporation open water, evapotranspiration, and infiltration accounted for 0.4%, 2%, and 0.8%, respectively. Evapotranspiration was slightly higher during dry conditions (1990-dry), with percentages around 3% (Figure 2-29). Regarding volumetric errors, the results showed that during wettest years (1992 and El Niño-1998) the errors were higher than during the average and dry year (Table 2-12).

Table 2-12 Estimated water balance for the different conditions/years simulated
(whole year simulation)

		Units	1990 (Dry)	2006 (Ave)	1992 (Wet)	1998 (El Niño)
Inflows	Upper Chojampe tributaries:					
	El Recuerdo	10^6m^3	189.2	240.1	299.9	542.4
	AdM T1	10^6m^3	2.0	5.8	17.6	21.8
	AdM T2	10^6m^3	3.6	10.2	30.9	38.1
	Abanico T1	10^6m^3	17.5	44.4	120.6	153.9
	Nuevo River	10^6m^3	1607.3	3025.3	3691.8	4443.1
	Precipitation	10^6m^3	42.6	78.4	155.5	177.0
Outflows	Evaporation open water	10^6m^3	-11.2	-13.2	-14.6	-16.0
	Evapotranspiration	10^6m^3	-63.1	-60.7	-59.0	-57.4
	Infiltration	10^6m^3	-20.6	-24.2	-26.8	-29.3
	Outflow Nuevo River	10^6m^3	-1786.9	-3252.3	-4080.7	-5184.3
	Total Inflow	10^6m^3	1862.1	3404.2	4316.2	5376.3
	Total outflow	10^6m^3	-1881.8	-3350.5	-4181.1	-5287.0
	Storage	10^6m^3	2.2	0.0	1.4	-69.0
	Volumetric error	10^6m^3	-21.9	53.7	133.7	158.4
	Simulation time	days	363.0	363.0	363.0	363.0
	Discharge error	(m^3/s)	-0.7	1.7	4.3	5.0

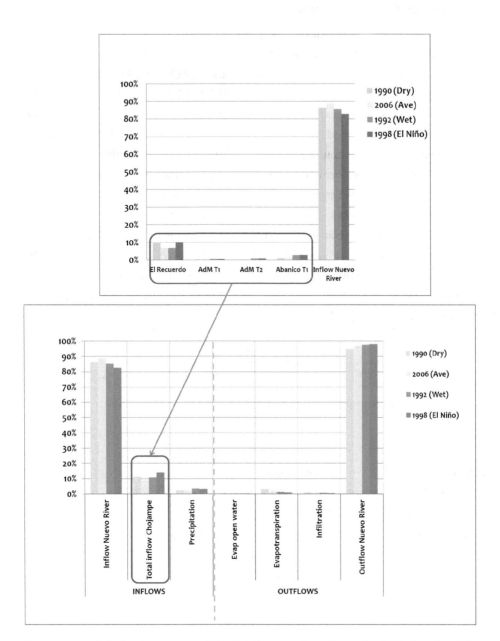

Figure 2-29   Relative importance (%) of inflows and outflows in AdM for the different conditions/years simulated. Upper panel: Inflows contribution; Lower panel: INFLOWS & OUTFLOWS (separated by dashed line)

2.5.2 Relative contributions of inflows

One of the objectives of this chapter was also to evaluate the contributions of the different inflows but at specific wetland locations. Thus, a tracer analysis was applied to evaluate the transport in the wetland using the WAQ module of the DELFT3D software suite, which was coupled with DELFT3D-FLOW. For this purpose, one conservative tracer was assigned to each boundary condition of the model, while an additional tracer was assigned to the initial condition. Results from five observation points located in the upper, middle and low wetland were used for this analysis. The simulations were also performed for each year condition. The model solves the advection diffusion-reaction equation (Eq 2-7) on a predefined computational grid for a wide range of substances. There are two different parts: (i) solving the equations for advective and diffusive transport of substances in the water body; and (ii) model the water quality kinetics of chemistry, biology, and physics that determines the behaviour of substances and organisms. The model does not compute the flow, so it needs to be connected/coupled to a hydrodynamic flow model (Deltares, 2013a; Deltares, 2013d; Deltares, 2014).

$$\frac{\partial C}{\partial t} = -u\frac{\partial C}{\partial x} - v\frac{\partial C}{\partial y} + \frac{\partial}{\partial x}(Dx\frac{\partial C}{\partial x}) + \frac{\partial}{\partial y}(Dy\frac{\partial}{\partial y}) + S + P \qquad \text{(Eq 2-7)}$$

Source: Adopted from Blauw et al., (2008); Smits and van Beek (2013)

Where:

C = concentration (gm^{-3})

u, v, w = components of the velocity vector (m s^{-1})

Dx, Dy, Dz = components of the dispersion tensor (m^2 s^{-1})

x, y, z = coordinates in three spatial dimensions (m)

S = sources and sinks of mass due to loads and boundaries (g m^{-3} s^{-1})

P = sources and sinks of mass due to processes (g m^{-3} s^{-1})

A spatial representation of the tracer analysis for the two main inflows is presented in Figure 2-30. The two main inflows in terms of water contribution are represented for the same days in order to visualize the contribution of each inflow at the same time.

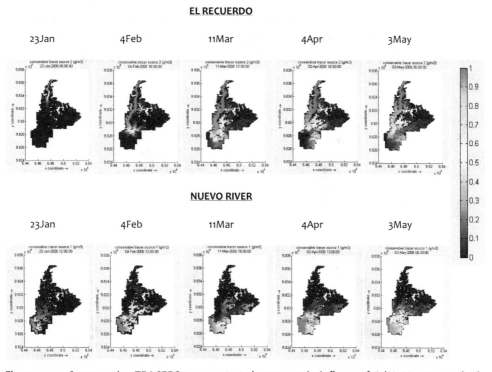

Figure 2-30 Conservative TRACERS transport at the two main inflows of AdM. Upper panel: El Recuerdo inflow; Lowe panel: Nuevo River inflow. One snapshot per month (Snapshots only for wet season).

The tracer analysis results showed that the upper wetland area (S1) is controlled by the Upper Chojampe (90-96%) influence that decreases in the middle area (S2) to (70-83%). At these two observation points, the Nuevo River inflow was not important (Figure 2-31). Results from AbanicoF, also located in the middle wetland area but in the Abanico branch, indicate that this point is dominated by the Abanico microbasin (40-78%) being higher during the wettest years. In this branch, the Nuevo River also showed an influence (4-28%) that decreased in the wettest years. Thus, during the wettest years, the Abanico branch is more influenced by the Abanico microbasin than by the Nuevo River (Figure 2-31).

At the lower wetland area (S3c and S7) the Nuevo River was the main source of water. At the intersection of the two branches (S7) the influence of the Nuevo River was between 19 and 52%, being higher during dry years. Upper Chojampe and Abanico also influenced this point: Upper Chojampe showed similar contributions over the years, while Abanico showed higher contributions during wettest years (up to 40%). At S3c, the closest wetland point to the inflow, the Nuevo River provides the dominant inflow (44-66%) being relatively constant, except for the El Niño year. The contribution of the Upper Chojampe subbasin at this point was between 23 and 39% (Figure 2-31).

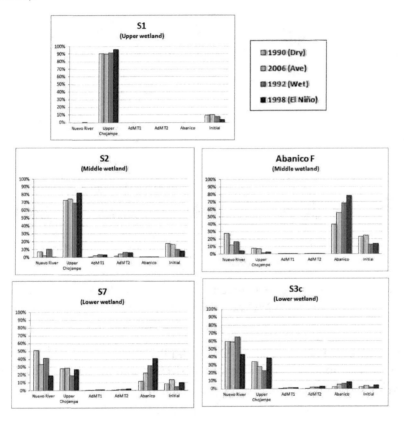

Figure 2-31 Relative contribution (%) of each tributary at observation points located at Upper, Middle and Low wetland areas, for the different conditions simulated.

2.5.3 Residence times

Several biogeochemical processes are driven by residence times. The residence times for various parts of the wetland were calculated using a passive tracer approach with prescribed initial conditions, in accordance with the procedure proposed by Takeoka (Takeoka, 1984). The procedure considers the remnant function, which is defined as the concentration of a compound at the initial time (C_0) and at time t $(C(t))$. From this concentration variation, the residence time was defined by equation (Eq 2-8) as:

$$t_r = \int_0^\infty \frac{C(t)}{C_0} dt = \int_0^\infty \chi(t) dt \qquad \text{(Eq 2-8)}$$

The residence time was calculated for the observation points used in the tracer analysis. The results indicate that the middle wetland areas (S2 and Abanico F) are the ones with higher residence times with averages of 20 and 30 days, respectively. Lowest residence times were found in areas dominated by one inflow (S1 and S3c), with average values of 12 and 5, respectively (Table 2-13) and (Figure 2-32).

Table 2-13 Residence time (days), at upper, middle and low wetland observation points.

YEARS	UPPER S1	MIDDLE S2	ABANICO F	LOWER S7	S3C
1990 (Dry)	14	26	34	12	4
2006 (Ave)	15	25	37	20	6
1992 (Wet)	12	15	19	7	3
1998 (El Niño)	8	16	30	22	9
AVERAGE	12	20	30	15	5

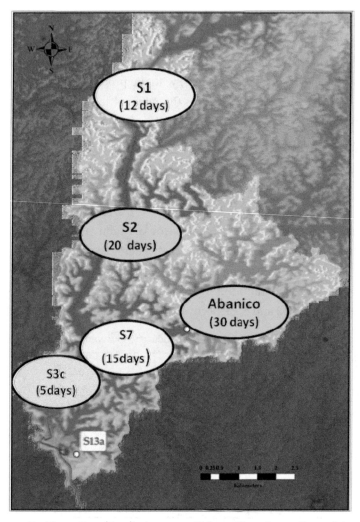

Figure 2-32 Residence time (days), at upper, middle and low observation points

2.6 Conditions during measurement campaigns 2011&2012

Once the hydrodynamic functioning of AdM wetland was determined, the next objective was to know which were the hydrological conditions during the years when the measurement campaigns were carried out (2011 and 2012), Figure 2-33 summarizes the precipitation conditions at those times (February 2011, January and March 2012).

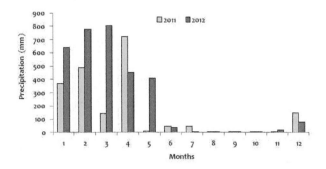

Figure 2-33   Monthly precipitation in Pichilingue station, for the sampling years 2011 & 2012. Months display from left to right: January (1), to December (12)

2.6.1 Boundary conditions

The six boundary conditions (Nuevo River inflow and outflow; El Recuerdo; AdMT1; AdMT2; AbanicoT1) for the years 2011 and 2012 were calculated following the same procedure as for the conditions presented earlier (section 2.4).

2.6.2 Initial conditions and water level

Initial water levels were determined for two areas: (i) inside the wetland and (ii) at the river inlet. For the wetland, the water levels were estimated from the limnometric station located inside the wetland. For the river inlet, the water levels were estimated based on the water levels downstream from the wetland, which were calculated with a rating curve. Initial velocities were set to zero. Table 2-14 presents the initial water levels for the simulated years.

Table 2-14 Hydrodynamic model-Initial conditions for water level (m)
for the sampling years 2011 & 2012

	WATER LEVEL (M)	
YEAR	WETLAND	INLET
2011 & 2012	9.35	8.95

The limnometric station only has recorded values until 2007, thus the years 2011 and 2012 have only simulated results and could not be compared with measured ones. However, the months where the maximum water levels occur follow the same patterns of the previous simulated conditions (Figure 2-34). Reference bed level in the wetland is 6 meters above sea level (m.a.s.l.).

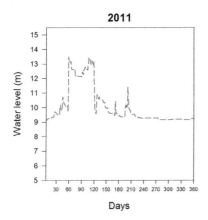
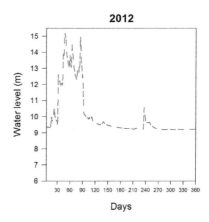

Figure 2-34 Simulated water levels at Abras de Mantequilla limnometric station 'El Recuerdo'. For the sampling years 2011 & 2012.

Relatively high water depths occurred during 2012, while lower water depths were observed during 2011. Overall, the wetland experienced fluctuations from 0.5–9.0 m for the two sampling years (Table 2-15).

Table 2-15 Simulated water depths results (m) of observations points located at the upper, middle, and lower wetland areas, for the sampling years 2011 & 2012

| | S1 | | | S2 | | | S7 | | |
| | (UPPER WETLAND) | | | (MIDDLE WETLAND) | | | (LOWER WETLAND) | | |
YEARS	MIN	MAX	AVERAGE	MIN	MAX	AVERAGE	MIN	MAX	AVERAGE
2011	1.3	5.8	2.4	0.5	5.5	2.5	0.9	7.4	3.3
2012	1.3	7.5	2.4	0.6	8.9	3.1	1.5	9.1	3.3

2.6.3 Flow velocities

Measured values from campaigns 2011 and 2012 indicated that velocities in the wetland area were up to 0.4 m/s and in the river inflow up to 0.8 m/s (Table 2-16). In the wetland area higher values were measured at observation point S7 (0.4 m/s) due to the proximity to the main inflow from the Nuevo River, which showed values between 0.6–0.8 m/s. Since the campaigns took place at the peak of the wet season (February 2011 and March 2012), these measured values represent typical velocities during this period of the year at these observation points. Measured and simulated velocities were in the same range (Table 2-16 &Table 2-17). Pooled velocities from sites located in the three wetland areas showed that average simulated velocities during the sampling years were up to 0.05 m/s, with maximum values up to 0.90 m/s (2012). In the Nuevo River average values were up to 0.26 m/s, and maximum ones up to 0.70 m/s (2012) (Table 2-17). Measured velocities were close to the maximum simulated velocities, confirming that the measurements were performed during the peak of the wet season.

Table 2-16 Measured velocities during campaigns 2011&2012

LOCATION	OBSERVATION POINTS	VELOCITY (m/s)
Wetland area	S1	0.3
	S2	0.2
	S7	0.4
Nuevo River inflow	S4	0.8
	S3a	0.6
Nuevo River outflow	S13a	0.8

Table 2-17 Simulated velocities (m/s) for the sampling years 2011 & 2012 (whole year simulation)

YEARS	WETLAND [A]			NUEVO RIVER [B]		
	MIN	MAX	AVERAGE	MIN	MAX	AVERAGE
2011	0.00	0.78	0.04	0.00	0.54	0.26
2012	0.00	0.90	0.05	0.00	0.70	0.24

[A] includes all simulated values from points: S1, S2, S7, S3c and Abanico F; [B] is represented by point S3a

Average velocities for years 2011 and 2012 were between 0.01 and 0.08 m/s (Table 2-18) while maximum ones were up 0.9 m/s (Table 2-19). Maximum velocities were evident in the lower area of the wetland (S7, S3c) due to the proximity to the Nuevo River inflow. At the middle area (S2) maximum values were up to 0.45 m/s, while at the upper wetland area maxima were only up 0.13 m/s (Table 2-19).

Table 2-18 Simulated average velocities (m/s) of each observation points located at the Lower, Middle and Upper wetland areas for the sampling years 2011 & 2012 (whole year simulation)

| | WETLAND AREAS | | | | |
| | LOWER | | MIDDLE | | UPPER |
YEARS	S7	S3c	S2	AbanicoF	S1
2011	0.06	0.07	0.03	0.00	0.02
2012	0.07	0.08	0.05	0.01	0.03

Table 2-19 Maximum simulated velocities (m/s) of observation points located at the Lower, Middle and Upper wetland areas for the sampling years 2011 & 2012 (whole year simulation)

| | WETLAND AREAS | | | | |
| | LOWER | | MIDDLE | | UPPER |
YEARS	S7	S3C	S2	ABANICO F	S1
2011	0.53	0.78	0.27	0.21	0.13
2012	0.45	0.90	0.45	0.28	0.13

2.6.4 Water balance

Once the water balance for the other conditions was established as well (Table 2-12), the magnitude of the inflows and outflows for the sampling years 2011 / 2012 were established on an annual basis (Table 2-20). Results from the water balance analysis indicated that the Nuevo River inflow importance was slightly higher during 2011 (90% compared with the other conditions), while 2012 had a similar importance as the wet year (1992). Thus, the overall importance of the Nuevo River inflow can reach up to 90%, which is close to the previously determined overall average of all simulations (86%). For the four tributaries from Chojampe subbasin, the year 2011 showed an 8% contribution, close to the an average year (9%), while 2012 (12%) was closer to the wet year 1992 (11%) (Figure 2-35).

Table 2-20 Estimated water balance for the sampling years 2011-2012 (whole year simulation)

		UNITS	2011	2012
INFLOWS	Upper Chojampe tributaries:			
	El Recuerdo	$10^6 m^3$	134.9	246.7
	AdM T1	$10^6 m^3$	5.4	9.0
	AdM T2	$10^6 m^3$	9.4	15.9
	Abanico T1	$10^6 m^3$	38.5	64.7
	Nuevo River	$10^6 m^3$	2154.1	2372.6
	Precipitation	$10^6 m^3$	59.4	102.2
OUTFLOWS	Evaporation open water	$10^6 m^3$	-11.5	-11.6
	Evapotranspiration	$10^6 m^3$	-62.7	-62.9
	Infiltration	$10^6 m^3$	-21.2	-21.2
	Outflow Nuevo River	$10^6 m^3$	-2236.6	-2612.0
	TOTAL INFLOW	$10^6 m^3$	2401.8	2811.1
	TOTAL OUTFLOW	$10^6 m^3$	-2332.0	-2707.6
	Storage	$10^6 m^3$	5.2	1.7
	Volumetric error	$10^6 m^3$	64.6	101.8
	Simulation time	days	363.0	363.0
	Discharge error	(m^3/s)	2.1	3.2

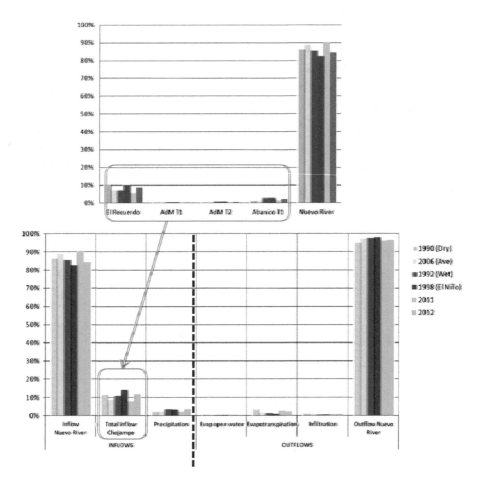

Figure 2-35 Relative importance (%) of inflows and outflows in AdM for the different conditions simulated. Upper panel: Inflows contribution; Lower panel: INFLOWS & OUTFLOWS (separated by dashed line)

2.6.5 Temporal inundation patterns

The inundation areas for the years 2011 and 2012 is presented with monthly averages calculated from daily-simulated results. The yearly trend of the inundation area for both years was in the overall range found for the other conditions; however, 2012 had some minimum values below the historical minimum (MINhisto) (Figure 2-36).

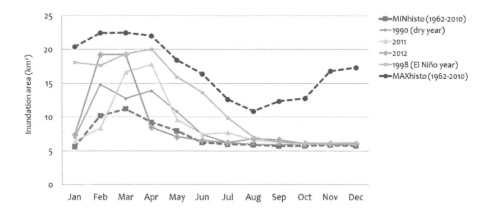

Figure 2-36 Monthly average of wetland inundation area (km²). Calculated from a stage volume curve and Delft3D-FLOW simulations results (one year simulation), from: dry year (1990); extreme wet year (El Niño-1998); MINIMUM and MAXIMUM historical (period 1962-2010), and sampling years 2011&2012.

2.6.6 Spatial inundation patterns

The spatial inundation patterns at the times of the monitoring campaigns 2011 and 2012 were obtained from the DELFT3D-FLOW model as illustrated in Figure 2-37. Even though both campaigns were carried out during the wet season, a clear difference (by about a factor of three) in inundation area is observed. These results are important for the evaluation of the water quality and ecological analysis as elaborated in the following chapters.

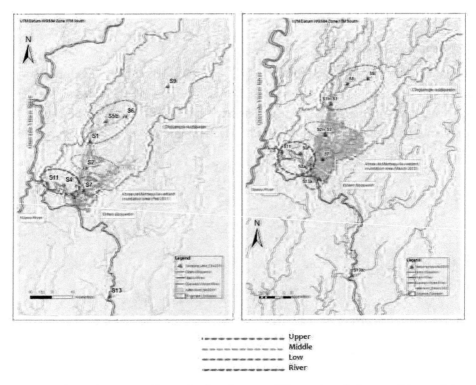

Upper
Middle
Low
River

Figure 2-37 Abras de Mantequilla wetland under two different inundation conditions. a) Low inundation conditions (LIC) during sampling of February 2011 (8.1 km²/810 hectares. b) High inundation conditions (HIC) during sampling of March 2012 (24.41 km² / 2441 hectares). Built based on DELFT3D-FLOW output. Circles with dotted lines represent the 4 wetland ZONES/areas.

2.7 Discussion

2.7.1 Natural variability of hydrodynamic conditions

The Abras de Mantequilla wetland is a river-wetland system that experiences extreme variations in hydrodynamic conditions: from low water depths and almost stagnant conditions during the dry season, to very dynamic ones during the rainy season. The Vinces River, located in the Quevedo-Vinces basin, diverts around 40% of the water to the Nuevo River, which constitutes the main inflow to the system. The wetland flooding starts in January and continues till April, with the exception of extreme wet years, when the rainy season lasts longer.

Temporal analysis of the boundary conditions showed that the timing of peak discharges varies from year to year, but occur usually during the months of February and March. Results from the hydrodynamic model indicated that historically, the wetland has experienced significant variations in water depth from 0.4 to 9.0 meters, in inundation areas from 5km^2 to 27 km^2, and in velocities from 0.0 to 0.9 m/s. The maximum discharges from the Nuevo River into the wetland system also showed high variations between the conditions simulated, increasing up to three times in magnitude from a dry condition year to an extreme wet one (up to 650 m^3/s). Overall, these extremes in hydrodynamic characteristics characterize this wetland system as being highly dynamic.

The evaluation of inundation patterns showed that during the rainy season the wetland has higher water depths (up to 9 meters) and inundated areas can reach maxima of 27km^2, depending on the year and condition evaluated. During the dry season, the inundation area is reduced to more or less constant values of around 5km^2.

2.7.2 Inflows assessment

The results from the hydrodynamic assessment revealed that the 'Nuevo River' is the main inflow to the wetland, with an overall contribution of about 86%. The four tributaries from the Chojampe subbasin contribute with around 11%. From these four tributaries, the inflow from the Upper Chojampe subbasin collected at the point 'El Recuerdo' is the most important, contributing up to 10% of this inflow. Direct precipitation on the system was not found important, contributing only with 3%. The Nuevo River is the main outflow of the wetland (95%), while evaporation, evapotranspiration and infiltration accounted for the rest.

2.7.3 Spatial analysis, inflows contribution and residence times

The tracer analysis indicated that the wetland could be divided into three main areas (upper, middle and low) based on the influence of the aforementioned inflows at the selected points. The 'upper part' of the wetland was dominated by the Upper Chojampe inflows where the contribution of the Nuevo River was almost nonexistent. Results from the model indicate the flow in this area is not exhibiting any dynamic effects in flooding and ebbing patterns. Overall, this area is characterized by low velocities (not higher than 0.2 m/s) and average residence times of about 12 days.

The 'middle part' of the wetland represents a transition area mainly influenced by the Upper Chojampe subbasin for the Abras branch (S2), and by the Abanico subbasin for Abanico F. Moreover, in these middle areas, the Nuevo River also showed an influence. However, since the Abanico branch has lower discharges than the Abras branch, the influence of the Nuevo River reaches further upstream (Galecio, 2013). The middle area (S2) presents a high variability in flow velocities, from values close to zero to values up to 0.6 m/s during extreme wet conditions. Model results at this point indicated that lower velocities occurred during periods of high influence of Nuevo River, while higher velocities during peak discharge from Upper Chojampe.

On the other hand, the Abanico branch is characterized by low velocities (not higher than 0.2 m/s) that did not show significant variation between dry and wet conditions. Average retention time in the middle areas is about 20 to 30 days for Abras (S2) and Abanico branch respectively, being the highest of the three-wetland areas. This is an important feature of the middle wetland area, since this may promote an increase in biological processes. It is well known, that both residence time and water depth support in describing the flow conditions of a system, and associated high flows with higher water depths and faster travel times (Van Breemen et al., 2002).

The 'lower part', located between the Nuevo River and the confluence of the two main branches of the wetland, is characterized by a strong influence from the Nuevo River. Nonetheless, this area experiences a mixing of the different tributaries, which is determined by the changes in magnitude of the Nuevo River. Thus, an increase in discharge of Nuevo River nearly removes the contribution from Upper Chojampe and Abanico subbasins, whereas a decrease in discharge promotes it. High velocities characterize this area and this pattern was observed during all conditions simulated. Due to the high dynamics of this area, retention times were lower than the ones of the middle area, with average values of about 12 days at the junction of the two branches, and 5 days close to the Nuevo River inlet point.

2.7.4 *Conditions during the sampling campaigns compared to historical conditions*

The hydrodynamic assessment performed in this chapter was useful also to position the sampling years of this research (2011 and 20120 in the historical spectrum. Temporal and spatial results of water depth, inundated areas and inflow contributions for both years were compared with the rest of the conditions simulated. As a result, it was found that the year 2011 resembles dry conditions, while the year 2012 is close to the wettest conditions, at least over the hydrological time span evaluated in this thesis.

3
ENVIRONMENTAL VARIABLES AND SPATIAL PATTERNS

3.1 Background

Tropical freshwaters often have clear hydrological seasonal inundation patterns, affecting flow velocity, water chemistry, and organism metabolic rates (Lewis, 2008). In the tropics, seasonality is based mainly on hydrology instead of hydrology and temperature, drivers typical for temperate latitudes (Lewis, 2008). In humid equatorial areas close to the sea, annual and daily fluctuations of water temperature are minor, and as a result tropical water systems are thermally quite stable (Lewis, 2008). Wetlands are also strongly influenced by the hydrology of their surrounding catchments, from which they receive nutrients and other dissolved and suspended material. Thus, wetlands act as a periodic or permanent sink of inorganic sediments, nutrients and organic carbon (Junk, 2002).

Hydrology is a major factor influencing chemical processes in the AdM wetland. The hydrologic conditions that occur (Chapter 2) influence biogeochemical processes and spatial variation of chemical substances (Mitsch and Gosselink, 2007). AdM is a seasonally flooded system with a strong influence of the Nuevo River inflow. The AdM study area is mainly surrounded by crops of rice and maize (Arias-Hidalgo et al., 2013), with a low human density population in the vicinity of the wetland water body. Organic matter produced by humans in the immediate surroundings is expected to be small. Concentrations in the AdM wetland are expected to be more influenced by external sources and inputs transported by the Nuevo River inflow.

This chapter focuses on the analysis of the environmental variables in the water body and sediments of the wetland. Spatial patterns and gradients of environmental variables are explored via multivariate analysis, which reduce the complexity of large datasets (Clarke and Warwick, 2001). This chapter identifies the overall range of water and sediments physico-chemical concentrations in the different wetland areas and in the river inflow. Two hypotheses are evaluated, the first is that the concentrations from the physico-chemical variables differ from the ones of the river inflow, and the second is that these concentrations may change depending on the inundation periods the wetland experiences. Thus, research questions in this chapter are:

- Do the spatial patterns of the environmental data differ from a low inundation period to a high inundation one?

- Which are the main environmental variables (physico-chemical, water depth, velocity) describing these environmental patters?

3.2 Field measurement campaigns

3.2.1 Selection of environmental variables

A suite of environmental variables to be collected in the water body and sediments was chosen to evaluate the trophic status and organic components of the AdM river-wetland (Table 3-1). This list of variables represents an overall synoptic survey of physical and chemical ambient conditions.

Table 3-1 Water and sediment variables selected for sampling

WATER VARIABLES	UNITS	SEDIMENT VARIABLES	UNITS
pH		Sulphides	mg/kg
Temperature	(°C)	Sand	%
Conductivity	µS/cm	Silt	%
Turbidity	NTU	Clay	%
Hardness	mg/CaCO$_3$/l	Nitrites (NO$_2$_N)	mg/kg
Alkalinity	mg/CaCO$_3$/l	Nitrates (NO$_3$_N)	mg/kg
Dissolved oxygen (DO)	mg/l	Ammonium (NH$_4$_N)	mg/kg
Biochemical oxygen demand (BOD)	mg/l	Dissolved inorganic nitrogen (DIN)	mg/kg
Chemical oxygen demand (COD)	mg/l	Organic nitrogen (N_organic)	mg/kg
Total suspended solids (TSS)	mg/l	Total nitrogen (N_Total)	mg/kg
Total solids (TS)	mg/l	Phosphates (PO4_P)	mg/kg
Nitrites (NO$_2$_N)	mg/l	Organic phosphorus (P_organic)	mg/kg
Nitrates (NO$_3$_N)	mg/l	Total phosphorus (P_Total)	mg/kg
Ammonium (NH$_4$_N)	mg/l	Organic matter	%
Dissolved inorganic nitrogen (DIN)	mg/l	Carbonates	%
Organic nitrogen (N_Organic)	mg/l	Organic carbon	%
Total nitrogen (N_Total)	mg/l	COD (Chemical oxygen demand)	mg/kg
Phosphates (PO4_P)	mg/l		
Organic phosphorus (P_Organic)	mg/l		
Total phosphorus (P_Total)	mg/l		
Silicates (SiO$_4$_Si)	mg/l		
Organic carbon	mg/l		
Chlorides	mg/l		
Sulphates	mg/l		
Sulphides	mg/l		
Chlorophyll_a	µg/l		
Secchi depth	m		
Water velocity	m/sec		

3.2.2 Identification of sampling sites and inundation conditions

The AdM wetland was sampled during two different inundation periods, both in the wet season (Figure 3-1). Low inundation conditions (LIC) occurred at the beginning of the wet season (Feb 2011) while high inundation conditions (HIC) were observed during the measurement campaign of March 2012. In terms of area, the wetland increases by about a factor of three from LIC to HIC (Chapter 2). In order to sample the wide range of abiotic and biotic variables, the campaigns were developed with the support of the National Institute of Fisheries of Ecuador (INP), University of Guayaquil, ESPOL Polytechnic University, and *Efficacitas*. INP provided technical staff, logistics, and equipment; University of Guayaquil and ESPOL staff and equipment, and *Efficacitas* equipment. The WETwin project (Chapter 1) provided financial support for the laboratory costs of analysing the samples.

First Measurement Campaign (LIC)

During the first campaign (February 2011), sampling sites were selected according to a spatial distribution representing the main hydrological features of the wetland and inflow river, using expert judgment and maps of the area. The sampling was performed over four days (February 8th till February 11th). The wetland was divided into three sections: (i) upper (S5, S6, S1); (ii) middle (S2); and (iii) lower (S7, S3b, S3c), which are lentic sites with flow velocities < 0.3 m/s (Table 3-2). To characterize the Nuevo river (main inflow to the wetland), sites were located along the Nuevo River (S11, S4, S3a) with velocities up to 0.8 m/s. S3a is located at the mouth of the wetland and S11 close to the Vinces River. S13 corresponds to the outflow (downstream section of Nuevo River). S9 is located at the North of the wetland representing the rainfall-runoff/drainage area of the Chojampe sub-basin (Figure 3-1a). A total of 39 physico-chemical variables were measured in the water column and sediments (Appendix C- Table C1). The variables were selected to cover a range of abiotic characteristics as a first effort for potential use for long term monitoring to support AdM wetland management (Alvarez-Mieles et al., 2013). Furthermore, an intermediate campaign was carried out during January 2012 at the start of the inundation season. This campaign focused on the sampling of physico-chemical variables (Appendix C-Table C2).

Second Measurement Campaign (HIC)

This campaign was performed during the peak of the wet season (March 2012) over 4 days from March 5th to March 8th. The same sites of February 2011 were sampled with the exception of S3b and S9 (Table 3-2). Furthermore, a new site S13a was included, located close to the S3b site of 2011, which corresponds to the immediate outflow of the wetland. This site represents better the immediate wetland outflow characteristics than S13 from 2011, which was located at a considerable distance from the wetland outflow. The latter was also sampled during the second campaign, but named S13b (Figure 3-1b). The same physico-chemical variables of February 2011 were sampled again during March 2012. Biotic sampling included phytoplankton, zooplankton, macroinvertebrates and fish (Chapter 4). Furthermore, two sites S1 and S2 (middle wetland) were sampled during the night to explore day/night variability (S1n and S2n in Figure 3-1b). For a complete list of variables see Appendix C-Table C3.

Table 3-2 Sampling sites

LOCATION	FIRST CAMPAIGN SITES	SECOND CAMPAIGN SITES
Upper wetland	S5	S5
	S6	S6
	S1	S1 (day & night) *
Middle wetland	S2	S2 (day & night) *
Low wetland	S7	S7 (day & night) *
	S3b	
	S3c	
River inflow	S3a	S3a
	S4	S4
	S11	S11
Outflow	S13	S13a
		S13b

*: Samples collected at surface and bottom level. The rest of samples at surface level

a) LIC   **b)** HIC

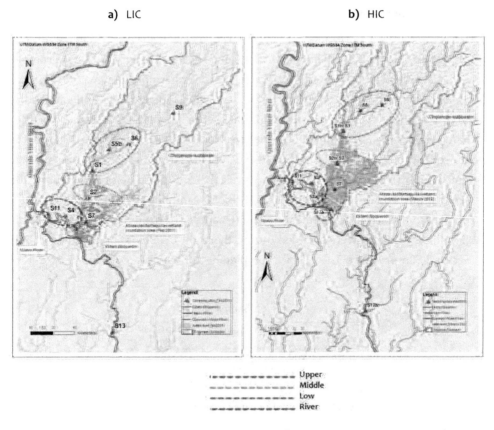

Figure 3-1 Abras de Mantequilla wetland under two different inundation conditions: a) Low inundation conditions (LIC) during sampling of February 2011 (8.1 km^2/810 hectares; b) High inundation conditions (HIC) during sampling of March 2012 (24.41 km^2 / 2441 hectares). Inundation maps are output of DELFT-FLOW (Chapter 2). Circles with dotted lines represent the four wetland ZONES/areas that were identified.

3.2.3 Sampling procedure for water body and sediment

Water samples were collected with a Niskin bottle at the water surface (upper layer of the water column), and then transferred to clean, pre-rinsed and labelled bottles. The bottles were maintained at 4 °C, until further analysis in the laboratory. The samples were pre-processed at the end of each sampling day following the standard procedure for each variable and maintained either refrigerated or frozen, according to standard procedure. At the end of the monitoring campaign, all the samples were transported to INP (National Institute of Fisheries in Guayaquil) for the culmination of the analysis. During the second campaign (March 2012), water samples were also collected from the bottom of the water column at S1, S2. S7. Sediments were collected with a Van Veen grab. Water and sediment samples were analyzed following Standard Methods (Eaton et al., 2005). Figure 3-2 and Figure 3-3 show the different components measured and sampled during the fieldwork campaigns.

Figure 3-2 Topography and water depth measures

Figure 3-3 Water quality sampling; upper panel (first campaign), lower panel (second campaign)

3.2.4 Data analysis of environmental variables

Multivariate techniques were applied to evaluate the environmental data using PRIMER (V6) statistical software (Clarke and Warwick, 2001). For the water body, 24 physico-chemical variables were evaluated, and for the bottom sediment 14 variables. A complete set of all collected data is provided in Appendix C. The same variables were analyzed for both inundation conditions to allow comparison. Environmental matrices were built for each condition and, subsequently, data was normalized to construct the resemblance matrix. A Principal Component Analysis (PCA) was performed to determine the main variables of the environmental data. PCA is a technique that creates an orthogonal basis that expresses the variability within the dataset. The PCA technique was applied separately for water and sediment variables for both low (LIC) and high (HIC) inundation conditions. The environmental data of both conditions was combined into one matrix to evaluate the patterns of the data in an integrated way, using cluster analysis and Multidimensional scaling (MDS) (Clarke and Warwick, 2001).

3.3 Sampling results within the water body

3.3.1 Low inundation conditions

During low inundation conditions (LIC), the water temperature was around 25 °C in the wetland sites, and ranged from 26 to 31 °C in the river sites (Table 3-3). Dissolved oxygen (DO) had a wide variation (1.2 to 5.5 mg/l) within the wetland sites, while the river values were higher (4.3 to 6 mg/l). Total suspended solids (TSS) ranged between 13 and 80 mg/l, with higher values in the river sites. DIN (sum of: NO_2-N; NH_4-N; NO_3-N) fluctuated between 0.02 and 0.52 mg N/l at wetland sites, and between 0.3 to 0.5 mg N/l at river sites. Nitrate was always the dominant fraction of the DIN in both wetland and river sites. At wetland sites, N:P ratio molar (inorganic fractions) ranged from 1 to 13, whereas at river sites increased to 35, and ranged from 9 to 53. Among wetland sites, upper sites (S5, S6) had higher ratios than middle and low sites (Appendix C- Table C1). Total nitrogen fluctuated between 0.3 and 0.8 mg/l at wetland sites, and between 0.7 and 0.9 at river sites. Organic nitrogen was the main constituent of total N. Total phosphorus was between 80-110 µg P/l in the wetland, and between 60-90 µg P/l at river sites. Silicate ranges were similar for wetland and river sites, with slightly higher maximum values at river sites. Sulphides showed similar values for both type of sites (0.1 - 0.2 mg/l). Chlorophyll-concentrations were higher at wetland sites with maximum values up to 8.3 µg/l. Secchi depth was also higher at wetland sites (0.6 and 1.4 m). BOD ranged between 0.1 and 2.2 mg/l at the wetland, and from 0.2 to 0.5 mg/l at the river. COD ranged from 17– 69 mg/l at wetland sites, and 17–51 mg/l at river sites (Table 3-3). BOD and COD variables were included to evaluate the possible organic pollution due to the remnants of the crops from the dry season. A complete list of sampling sites and results is contained in Appendix C- Table C1.

Table 3-3 Minimum and maximum values of abiotic variables in the water column (Wetland and river sites). Low inundation conditions (LIC) Feb 2011.

		WETLAND		RIVER	
VARIABLES	UNITS	MIN	MAX	MIN	MAX
pH		6.8	7.3	6.7	7.4
Temperature	(°C)	25.9	30.7	24.7	25.4
Conductivity	µS/cm	21	34	26	32
Turbidity	NTU	3	256	19	158
Hardness	mg/CaCO3/l	9.3	14.0	9.3	12.4
Alkalinity	mg/CaCO3/l	30.5	48.7	29.4	36.5
DO	mg/l	1.2	5.5	4.3	5.8
BOD	mg/l	0.1	2.2	0.2	0.5
COD	mg/l	17.1	68.6	17.1	51.4
TSS	mg/l	13	23	24	80
TS	mg/l	53	88	60	106
NO2_N	mg/l	0.003	0.013	0.005	0.008
NO3_N	mg/l	0.0002	0.47	0.25	0.46
NH4_N	mg/l	0.01	0.04	0.01	0.03
DIN	mg/l	0.02	0.52	0.27	0.48
N_Organic	mg/l	0.2	0.5	0.4	0.5
N_Total	mg/l	0.3	0.8	0.7	0.9
PO4_P	µg/l	40	100	20	80
P_Organic	µg/l	10	40	10	60
P_Total	µg/l	80	110	60	90
N:P					
(DIN:PO4)	ratio	0.9	13	9.4	53
SiO4	mg/l	24	38	29	42
SiO4_Si	mg/l	7.2	11.5	8.7	12.9
Sulphates	mg/l	0.8	4	1.1	10.5
Sulphides	mg/l	0.1	0.2	0.1	0.2
	µg/l				
Chlorophyll_a	(mg/m³)	0.8	8.3	0.3	1.9
Secchi depth	m	0.65	1.4	0.2	1

Spatial patterns

A PCA analysis conducted on 24 water variables accounted for 62.6% of the total variation in water samples ordination during low inundation conditions (LIC) (Figure 3-4). PC1 contributed 43.1% of the variation and was positively correlated (+ve) with temperature (0.75), conductivity (0.6), alkalinity (0.9), phosphates (0.7), total phosphorus (0.8), chlorophyll-a (0.6) and Secchi depth (0.92); and negatively correlated (-ve) with turbidity, total suspended solids, dissolved oxygen, nitrogen (inorganic, organic and total), N:P ratio (-0.95), sulphates (-0.9), sulphides. PC2 contributed 19.5% of the variation and correlated positively (+ve) with COD, inorganic nitrogen, phosphates, and negatively (-ve) with conductivity, hardness, organic phosphorus (-0.81), chlorophyll-a and velocity (Table 3-7). A clear division between river and wetland sites can be observed from Figure 3-4. Water samples from the upper wetland sites clustered well with the river sites, showing a river-type character. Sediments showed a different pattern and grouped with wetland sites, indicating similar characteristics (Figure 3-8).

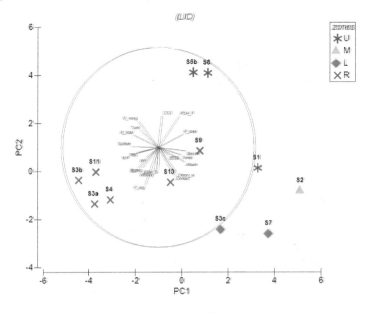

Figure 3-4 Principal component analysis of 24 WATER variables for LOW inundation conditions (LIC). 62.6% of the variation explained (PC1 : 43.1% and PC2 : 19.5%). Sampling sites with symbols indicating the wetland zones: U: upper, M: middle, L: low and R: river sites. Vectors represent the direction and gradient of the water variables. Longer vectors described higher correlations with the components (PC1 and PC2).

3.3.2 High inundation conditions

During high inundation conditions (HIC), the water temperature ranged from 27–32 °C in wetland sites, and from 27–29 °C in river sites. Minimum values of 0.7 mg/l of dissolved oxygen (DO) were recorded at the bottom of the water column in wetland sites, while at the river sites values ranged from 2.9–6.4 mg/l. Total suspended solids (TSS) ranged between 10–55 mg/l at wetland sites and were up to 97 mg/l in river sites. DIN values (the sum of NO_2-N; NH_4-N; NO_3-N) fluctuated between 0.05-0.32 mg N/l at wetland sites, and between 0.21–0.36 mg N/l at river sites. As in previous campaigns, nitrates were the dominant fraction of DIN in both wetland and river sites (Table 3-4). Total nitrogen fluctuated between 1.4–2.6 mg/l at wetland sites, and between 1.1–1.8 at river sites, having organic nitrogen as the main constituent of total N. Total phosphorus concentrations recorded were between 70-140 µgP/l at the wetland and between 60-80 µgP/l at river sites. At wetland sites, average N:P ratio (inorganic fractions) ranged between 2 and 24, whereas at river sites increased to 28, and ranged from 12 to 40. Upper sites (S5, S6) had similar ratios than middle (S2) and low (S7) sites (Appendix C- Table C3). Silicate ranges were similar for wetland and river sites, and similar to the previous campaign. Sulphides showed similar values for both type of sites during each sampling campaign. Chlorophyll-concentrations were higher at wetland sites with maximum value of 24 µg/l. Secchi depth was higher at wetland sites (0.5~1.0 m) whereas river sites had values of 0.2 m. BOD ranged between 0.10– 3.7 mg/l at the wetland, and between 0.3–1.2 mg/l at the river. COD ranged between 12–110 mg/l at wetland sites, and 6.7–68.3 mg/l at river sites (Table 3-4). A complete list of sampling sites and results is provided in Appendix C- Table C3.

Table 3-4 Minimum and maximum values of abiotic variables in the water column (Wetland and river sites). High inundation conditions (HIC) March 2012.

Variables	Units	WETLAND MIN	WETLAND MAX	RIVER MIN	RIVER MAX
pH		6.1	7.2	7.2	7.4
Temperature	°C	26.7	31.8	27.0	28.9
Conductivity	µS/cm	57	73	57	75
Turbidity	NTU	9	109	10	106
Hardness	mg/CaCO3/l	9.6	14.3	9.6	9.6
Alkalinity	mg/CaCO3/l	34.3	44.4	30.3	34.3
DO	mg/l	0.7	6.1	2.9	6.4
BOD	mg/l	0.2	3.7	0.3	1.2
COD	mg/l	11.7	110.0	6.7	68.3
TSS	mg/l	10	55	27	97
TS	mg/l	40	164	74	166
NO2_N	mg/l	0.004	0.009	0.004	0.006
NO3_N	mg/l	0.04	0.31	0.20	0.35
NH4_N	mg/l	0.002	0.01	0.002	0.005
DIN	mg/l	0.05	0.32	0.21	0.36
N_Organic	mg/l	1.1	2.5	0.8	1.4
N_Total	mg/l	1.4	2.6	1.1	1.8
PO4_P	µg/l	30	80	20	40
P_Organic	µg/l	10	110	30	60
P_Total	µg/l	70	140	60	80
N:P (DIN:PO4)	ratio	2.2	24	12	40
SiO4	mg/l	24	42	27	47
SiO4_Si	mg/l	7.2	12.8	8.1	14.2
Sulphates	mg/l	1.7	2.6	1.2	2.5
Sulphides	mg/l	0.4	1.0	0.4	0.8
Chlorophyll_a	µg/l	2	24	4	10
Secchi depth	m	0.5	1.0	0.2	0.2

Spatial patterns

The PCA analysis conducted on 24 water variables accounted for 52.3% of the total variation in the water samples ordination during high inundation conditions (HIC). Thus, the variation was less than during low inundation conditions. PC1 contributed 38.5% of the variation and was positively correlated with total suspended solids, turbidity, dissolved oxygen, inorganic nitrogen, N:P ratio, silicates and velocity; and negatively (-ve) with temperature, Secchi depth, alkalinity, organic nitrogen, phosphates and sulphates. PC2 contributed 13.8% of the variation, correlating negatively (-ve) with turbidity, chlorophyll-a, organic and total phosphorus (Table 3-7). When the analysis was conducted excluding S2 (bottom sample), the variation increased to 55. 9% (Figure 3-5). PC1 accounted for 43.2% of the variation and was correlated with the same variables when S2(b) was included in the analysis. A slight increase in correlation with organic phosphorus was observed. PC2 contributed 12.7% of the variation. However, the correlations with PC2 changed when S2(b) was included, and a strong negative correlation with organic phosphorus from the first approach did not occur. This probably occurred because S2(b) was mainly responsible for this correlation due to the higher concentration of organic phosphorus in this site. PC2 correlated negatively (-ve) with pH, temperature, BOD, and chlorophyll-a. During HIC, both water body (Figure 3-5) and bottom sediment samples (Figure 3-9) from upper wetland sites clustered well with the rest of wetland sites located at middle and low areas, showing a clear separation from the Nuevo River inflow sites. This suggests that once the wetland reaches its maximum inundation capacity, upper wetland sites resemble those in the main body of the wetland.

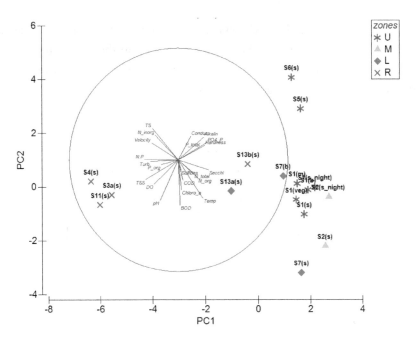

Figure 3-5   Principal component analysis of 24 WATER variables for HIGH inundation conditions (HIC). 55.9% of the variation explained (PC1 : 43.2% and PC2 : 12.7%). Sampling sites with symbols indicating the wetland zones: U: upper, M: middle, L: low and R: river sites. Vectors represent the direction and gradient of the water variables. Longer vectors described higher correlations with the components (PC1 and PC2).

3.3.3 Combined analysis

When both sets of samples (LIC and HIC) were combined, the cluster analysis revealed five significant groups (Figure 3-6) suggesting that a higher number of samples could increase the significant divisions. The initial split at 8.6 distance isolated site S2 (bottom sample) from all the other sites (SIMPROF test: π=0.48, p=0.1%). A second division at 7.6 divided wetland sites from river sites (SIMPROF test: π=0.47, p=0.1%). However, the wetland cluster included also two river sites. A third split at 6.9 divided wetland sites according to the inundation conditions (SIMPROF test: π=0.31, p=0.4%). The next division at 6.2 separated river sites also according to the inundation conditions (SIMPROF test: π=0.84, p=0.1%). The MDS ordination showed an agreement between both representations, clearly displaying the same cluster groups (stress: 0.13) (Figure 3-6). ANOSIM confirmed significant differences for all the tested factors: zones (R=0.5, P=0.001); location (R=0.52, P=0.001); and inundation conditions (R=0.56, P=0.001). This combined analysis revealed that for the water column, the factor 'location' (river or wetland) had more influence than 'inundation condition' since it was the initial factor dividing the samples. Thus, all wetland sites and all river sites clustered together independently of the inundation condition. Conversely, for sediment variables, the factor 'inundation condition' seemed to have more influence in the spatial representation of the samples. Thus, river and wetland sites of HIC formed one cluster, and river and wetland sites of LIC formed another two clusters.

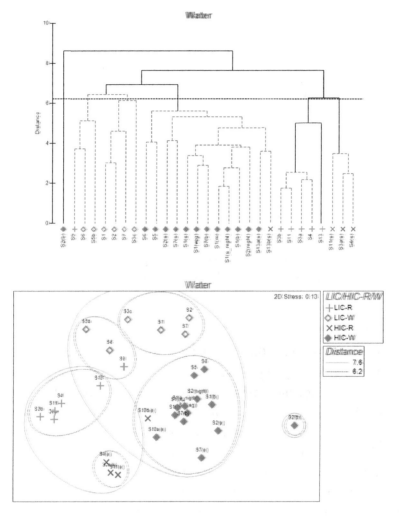

Figure 3-6 Group average cluster analysis (left) and MDS ordination (right) for WATER samples (both conditions integrated). Based on Euclidean distances resemblance matrices (built from normalized data). Continuous lines represent the divisions for which SIMPROF test (P< 0.05, p<5% in percentage) confirmed further subdivision structure to explore. Red dashed lines show the group structure with no evidence from the SIMPROF test. The x- axis shows sampling sites, symbols described the condition of sampling (LIC or HIC), and sites location (R: river, W: wetland).

3.3.4 *Measured concentrations and spatial distribution*

Measured concentrations of chlorophyll-a, dissolved oxygen, nitrates and phosphates for the monitoring campaigns (February 2011 and March 2012) are presented in Figure 3-7, and provide an overall view of the range of concentrations during the two monitoring campaigns performed for the study area. These variables are just a part of the complete set of variables measured in the field. The difference of inundated area is evident between left and right panels. Left panels correspond to the monitoring campaign developed in February 2011 during low inundation conditions (LIC), while right panels correspond to the campaign of March 2012 during high inundation conditions (HIC). Overall, the values measured in the AdM are within the range of concentrations measured in the surrounding river basin previously (Alvarez, 2007; INP, 1998; INP, 2012; Prado et al., 2012). This provides confidence in using the measured values for the development of a dynamic model.

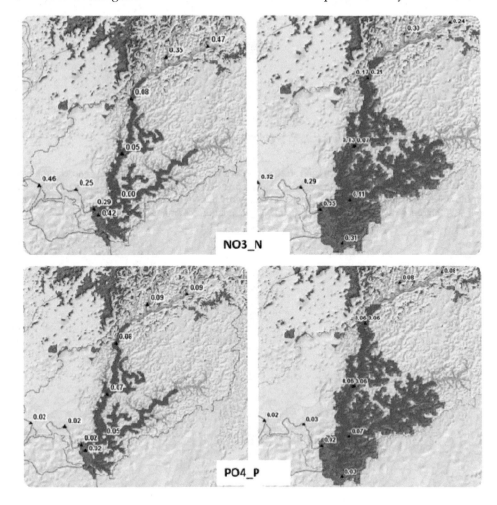

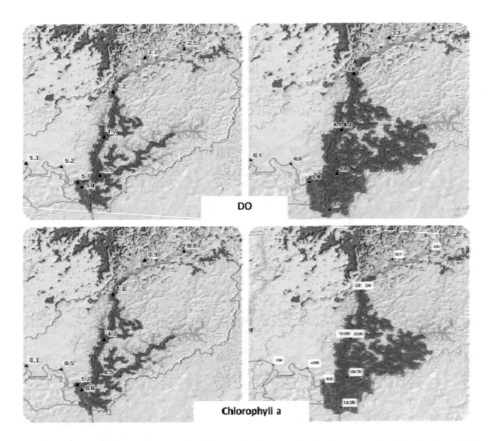

Figure 3-7 Measured concentrations of nitrates, phosphates, dissolved oxygen (mg/l) and chlorophyll-a (mg/m^3=µg/l) at observation points located at the upper, middle and low wetland areas; inflows and outflow. Left panels: Monitoring campaign February 2011 during low inundation conditions (LIC); Right panels: Monitoring campaign March 2012 during high inundation conditions (HIC).

3.4 Sampling results of bottom sediment

3.4.1 Low inundation conditions

The sediment texture differed from wetland to river sites. At wetland sites silt was found to be the dominant fraction, while at river sites sand was the main fraction. Ammonium was the main constituent of inorganic nitrogen (DIN) in the sediments. Total nitrogen and total phosphorus were mainly represented by their organic fraction. In general, concentrations of total nitrogen were similar for both wetland and river sites (up to 2.4 mg/kg). Total phosphorus was slightly higher in wetland sites (up to 0.6 mg/kg). Organic matter and organic carbon were also higher in wetland sites (Table 3-5).

Table 3-5 Minimum and maximum values of abiotic variables in sediments (Wetland and river sites, Feb 2011)

VARIABLES	UNITS	WETLAND MIN	WETLAND MAX	RIVER MIN	RIVER MAX
Sand	%	11.2	29.4	26	99.5
Silt	%	30.6	49	0.4	49.6
Clay	%	21.56	56.4	0.1	24.4
Sulphides	mg/kg	1.6	4.4	1.6	4
NO2_N	mg/kg	0.001	0.006	0.001	0.003
NO3_N	mg/kg	0.002	0.020	0.010	0.010
NH4_N	mg/kg	0.18	0.8	0.27	0.41
DIN	mg/kg	0.2	0.8	0.3	0.4
N_organic	mg/kg	0.3	1.8	1.6	2.1
N_Total	mg/kg	1.1	2.2	1.9	2.4
PO4_P	mg/kg	0.01	0.19	0.02	0.08
P_organic	mg/kg	0.15	0.37	0.05	0.17
P_Total	mg/kg	0.2	0.6	0.1	0.2
Carbonates	%	0.5	1.2	0.5	0.6
Organic matter	%	14	25	2	14
Organic carbon	%	5.6	7.6	0.7	5.8
COD	mg/kg	195	417	198	367

Spatial patterns

For low inundation conditions (LIC), the PCA analysis performed on 14 sediment variables accounted for 72.4% of the total variation in the sediment samples ordination during LIC (Figure 3-8). PC1 contributed 58% of the variation and was positively correlated (+ve) with sand (0.85), organic and total nitrogen; and negatively correlated (-ve) with silt, clay, inorganic nitrogen, phosphorus (inorganic, organic and total), organic matter and organic carbon. PC2 contributed 14.4% to the variation, showing moderate (+ve) correlations (r < 0.6) with sulphides, silt, total nitrogen, organic carbon and COD (Table 3-7).

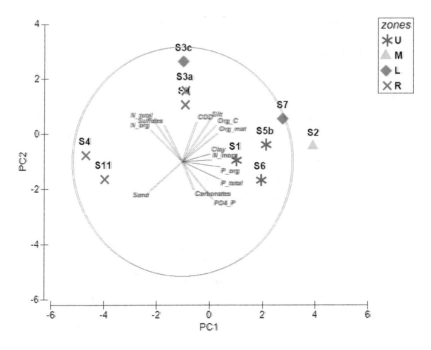

Figure 3-8 Principal component analysis of 14 SEDIMENT variables for LOW inundation conditions (LIC). 72.4% of the variation explained (PC1 : 58% and PC2 : 14.4%). Sampling sites with symbols indicating the wetland zones: U: upper, M: middle, L: low and R: river sites. Vectors represent the direction and gradient of the water variables. Longer vectors described higher correlations with the components (PC1 and PC2).

3.4.2 High inundation conditions

Sediment texture remained the same as for low inundation conditions. Again, ammonium was the main constituent of inorganic nitrogen (DIN) in the sediment. Also, total nitrogen and total phosphorus were mainly represented by their organic fraction. In general, concentrations of total nitrogen were in the same order of magnitude for both areas. Total phosphorus was slightly higher in wetland sites (up to 2.2 mg/kg). Organic matter and organic carbon (percentages) were also higher in wetland sites with values up to 29% and 9.6 % respectively (Table 3-6).

Table 3-6 Minimum and maximum values of abiotic variables in sediments (wetland and river sites, March 2012)

VARIABLES	UNITS	WETLAND MIN	WETLAND MAX	RIVER MIN	RIVER MAX
Sand	%	8.3	28.9	32.4	100.0
Silt	%	51.7	89.3	0.01	64.3
Clay	%	2	20	0.03	12.4
Sulphides	mg/kg	0.2	0.6	0.1	0.4
NO2_N	mg/kg	0.001	0.010	0.001	0.014
NO3_N	mg/kg	0.002	0.013	0.0001	0.008
NH4_N	mg/kg	0.23	0.51	0.13	0.28
DIN	mg/kg	0.2	0.5	0.1	0.3
N_organic	mg/kg	4.5	5.0	3.4	5.0
N_Total	mg/kg	5.0	5.4	3.6	5.3
PO4_P	mg/kg	0.09	0.14	0.04	0.14
P_organic	mg/kg	0.23	2.05	0.01	0.11
P_Total	mg/kg	0.3	2.2	0.1	0.2
Carbonates	%	0.8	1.8	0.9	1.1
Organic matter	%	17	29	4	14
Organic carbon	%	5.3	9.5	1.4	5.4
COD	mg/kg	267	490	228	298

Spatial patterns

A PCA analysis performed on 14 sediment variables accounted for 64% of the total variation in the sediment samples ordination during HIC (Figure 3-9). PC1 contributed 42% of the variation, correlating positively with sand and negatively with silt, clay, all fractions of nitrogen and phosphorus (inorganic, organic and total), COD, organic matter, organic carbon. Compared with LIC, an inverse relation to PC1 was observed for nitrogen (organic and total), probably influenced by the higher concentrations of these components in the wetland sites during HIC. PC2 accounted for 22.3% of the variation, being positively correlated with COD, carbonates, sulphides and inorganic nitrogen, and negatively with phosphorus (organic and total) (Table 3-7).

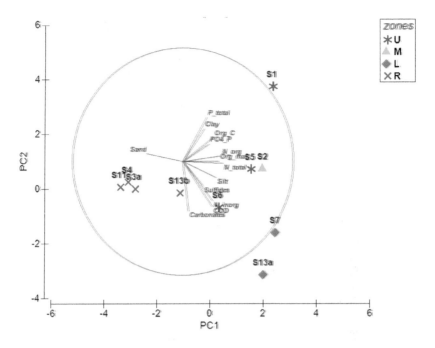

Figure 3-9 Principal component analysis of 14 SEDIMENT variables for HIGH inundation conditions (HIC). 64% of the variation explained (PC1 : 42% and PC2 : 22.3%). Sampling sites with symbols indicating the wetland zones: U: upper, M: middle, L: low and R: river sites. Vectors represent the direction and gradient of the water variables. Longer vectors described higher correlations with the components (PC1 and PC2).

3.4.3 Combined analysis

When both sets of sediment samples (LIC and HIC) were combined, a cluster analysis revealed four significant groups. The initial split at a distance 7.5 isolated site S1 (HIC). A second division at 6.1 separated two river sites of LIC (S4, S11) from all the rest. Both splits had (SIMPROF test: $\pi=0.41$, p=0.1%). A third split at 5.1 divided the sites according to the inundation conditions (SIMPROF test: $\pi=0.31$, p=0.2%). The last significant split at distance 4.4 divided the HIC sites into two groups: one containing the river sites and another the wetland sites (SIMPROF test: $\pi=0.34$, p=0.7%) (Figure 3-10). The MDS ordination confirmed the cluster groups (stress: 0.1) (Figure 3-10). Although the R ANOSIM was low to moderate, results were significant for all the tested factors: zones (R=0.3, P=0.006); location (R=0.26, P=0.009); and inundation conditions (R=0.47, P=0.001). However, pairwise test for zones, showed a strong correlation between river and middle areas (R=0.64, P=0.04), indicating there are strong differences between these two zones.

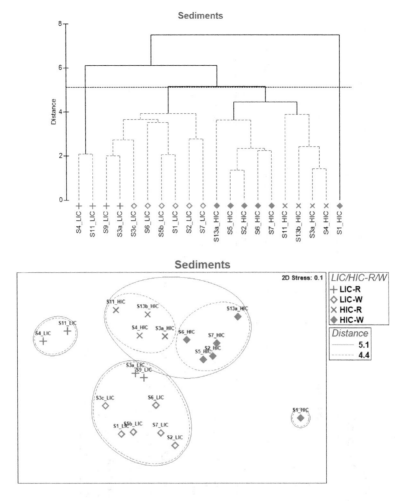

Figure 3-10 Group average cluster analysis (left) and MDS ordination (right) for SEDIMENT samples (both conditions integrated). Based on Euclidean distances resemblance matrices (built from normalized data). Continuous lines represent the divisions for which SIMPROF test (P< 0.05, p<5% in percentage) confirmed further subdivision structure to explore. Red dashed lines show the group structure with no evidence from the SIMPROF test. The x- axis shows sampling sites, symbols described the condition of sampling (LIC or HIC), and sites location (R: river, W: wetland).

Table 3-7 Component loadings (correlations) (using Pearson's r) between the PC components and the environmental variables. TV: total variation explained by PC1+PC2. LIC: low inundation conditions. HIC: high inundation conditions.

a) Water (LIC)
TV: 62.6%

	pH	Temp	Conduc	Turb	TSS	TS	Secchi	DO	BOD	COD	Hardness	Alkalin	N_inorg	N_org	N_total	PO4_P	P_org	P_total	N:P	SiO4_Si	Chloro_a	Sulfate	Sulphides	Velocity
PC1 (43.1%)	-0.41	0.75	0.54	-0.60	-0.70	-0.58	0.92	-0.68	0.34	0.12	0.70	0.88	-0.68	-0.54	-0.82	0.69	-0.44	0.82	-0.95	-0.01	0.59	-0.89	-0.78	-0.13
PC2 (19.5%)	-0.22	-0.16	-0.61	0.38	-0.10	-0.43	-0.08	-0.45	-0.17	0.69	-0.63	-0.32	0.58	-0.44	0.27	0.68	**-0.81**	0.28	-0.17	-0.48	-0.54	0.02	-0.19	-0.54

b) Water (HIC)
TV: 52.3%

	pH	Temp	Conduc	Turb	TSS	TS	Secchi	DO	BOD	COD	Hardness	Alkalin	N_inorg	N_org	N_total	PO4_P	P_org	P_total	N:P	SiO4-Si	Chloro_a	Sulfate	Sulphides	Velocity
PC1 (38.5%)	0.50	-0.66	-0.32	0.69	0.90	0.74	-0.85	0.72	-0.03	-0.18	-0.74	-0.72	0.68	-0.56	-0.44	-0.74	0.36	-0.29	0.95	0.73	-0.16	-0.71	-0.07	0.75
PC2 (13.8%)	-0.15	0.11	0.43	-0.63	-0.20	0.34	-0.15	0.28	0.05	-0.52	0.02	0.15	0.44	-0.32	-0.25	0.49	-0.85	-0.69	0.04	0.14	-0.49	0.20	-0.02	0.20

c) Water (HIC) excluding S2(b)
TV: 55.9%

	pH	Temp	Conduc	Turb	TSS	TS	Secchi	DO	BOD	COD	Hardness	Alkalin	N_inorg	N_org	N_total	PO4_P	P_org	P_total	N:P	SiO4-Si	Chloro_a	Sulfate	Sulphides	Velocity
PC1 (43.2%)	0.50	-0.68	-0.35	0.94	0.90	0.73	-0.85	0.74	-0.05	-0.14	-0.74	-0.72	0.69	-0.55	-0.44	-0.81	0.76	-0.35	0.95	0.72	-0.11	-0.71	-0.08	0.74
PC2 (12.7%)	-0.59	-0.57	0.35	-0.03	-0.29	0.47	-0.15	-0.40	-0.67	-0.32	0.21	0.34	0.39	-0.27	-0.20	0.26	-0.11	0.35	0.01	0.25	-0.55	0.28	-0.15	0.25

d) Sediments (LIC)
TV: 72.4%

	Sulphides	Sand	Silt	Clay	N_inorg	N_org	N_Total	PO4_P	P_org	P_Total	Org_mat	Org_C	COD	Carbonates
PC1 (58%)	0.48	0.85	-0.73	-0.72	-0.75	0.83	0.76	-0.79	-0.97	-0.96	-0.91	-0.87	-0.37	-0.30
PC2 (14.4%)	0.45	-0.37	0.53	0.09	0.02	0.39	0.52	-0.47	-0.06	-0.22	0.35	0.45	0.50	-0.35

e) Sediments (HIC)
TV: 64.1%

	Sulphides	Sand	Silt	Clay	N_inorg	N_org	N_total	PO4_P	P_org	P_total	Org_mat	Org_C	COD	Carbonates
PC1 (41.8%)	-0.45	0.79	-0.73	-0.47	-0.66	-0.85	-0.87	-0.58	-0.50	-0.52	-0.79	-0.67	-0.65	-0.12
PC2 (22.3%)	0.38	-0.12	0.25	-0.51	0.61	-0.09	0.03	-0.27	-0.69	-0.50	-0.01	-0.37	0.70	0.77

3.5 Concentrations, gradients and key variables

Overall, the water and sediment concentrations were found to have similar ranges compared with previous studies in the AdM wetland system and Vinces River Basin (Alvarez, 2007; INP, 2012; Prado et al., 2012). The AdM concentrations are also in the range of the ones collected in the surrounding Guayas River Basin (INP, 1998).

Average nitrate concentrations at wetland sites (0.2 mg NO_3-N/l) were in the range of the ones described for unpolluted rivers in the paper by Meybeck (1982) for tropical waters (0.016-0.24mg NO_3-N/l) (Meybeck, 1982), while river inflow concentrations with values around 300 µg NO_3-N/l were over this range. A maximum concentration up to 470 µg NO_3-N/l was measured at an upper wetland site (S6) during LIC, probably related to the inflow /run-off characteristics of this site.

Phosphates concentrations at wetland and inflow sites (20-100 µg PO_4-P/L) were both over the range described for unpolluted rivers in the tropics (1 to 24 µg PO_4-P/L) (Meybeck, 1982), thus indicating no P limitation in this system. High concentrations of total phosphorus also characterized other tropical systems (Talling, 1992), probably because chemical weathering of phosphorus seems to be more efficient at higher temperatures (Lewis, 1996). Furthermore, most of the phosphorus in rivers is particle bound, and since high precipitation and associated discharges increase erosion rates, resulting in higher fluxes of phosphorus from the landscape (Howarth et al., 2006). Such patterns are also observed in the Amazon basin linked to the Andes sediment weathering (McClain and Naiman, 2008). In the Guayas River Basin, a positive phosphorus balance was found, suggesting that phosphorus is retained in the soils (Borbor-Cordova et al., 2006). Furthermore, phosphorus concentrations in lakes depends of the equilibrium between internal and external loading, physico-chemical and biological processes in the water column (Jennings et al., 2003)

An extensive study of nutrients budgets of 10 sub-catchments belonging to the Guayas River Basin determined that N, and P budgets are mainly influenced by agricultural activities and associated nutrient inputs. However, only a small fraction of N (14%) and P (38%) inputs are leached to the rivers. Nitrogen river exports in this basin have been associated to land use and agricultural activities, whereas P appears to be driven by runoff and erosion process (Borbor-Cordova et al., 2006)

N:P ratios in AdM wetland had a great variation not only among river and wetland areas, but also within wetland and river areas, during both inundation periods. The established Redfield N:P ratio relation 16:1 (Redfield, 1934) developed for oceans, deviates constantly in this study. Inland waters exhibit a high degree of variability in stoichiometric ratios than oceans due to the variation in inputs that they receive from their watershed, and the relatively fast flushing (Sterner, 2009).

Lower N:P molar ratios (inorganic fraction) occurred at wetland sites, probably due to higher denitrification and phytoplankton uptake at wetland central areas with higher residences time. Low N:P ratios occur at high P, this suggests that N limitation increases at higher P concentrations (Downing and McCauley, 1992). Overall, AdM central areas can be described as nitrogen limited with an average N:P ratio of 7 for both inundation periods. Nitrogen has been found as the limiting nutrient in other tropical South-American lentic environments (Lewis, 1986). In tropical lakes, nitrogen rather than phosphorus limitation appear to occur, hence the ratio dissolved inorganic nitrogen (DIN) to soluble reactive phosphorus (SRP) is usually lower than in temperate waters (Lewis, 1996).

During LIC, N:P ratios at upper wetland sites were higher than the ones of middle and low wetland areas, whereas, during HIC they were similar suggesting a system homogenization during HIC. River inflow indicated phosphorus limitation with N:P ratios >16, average around 30, and maximum up to 53, thus differing from the low ratios occurring at disturbed catchments (Downing and McCauley, 1992; Saunders and Kalff, 2001).

An increase of total nitrogen during HIC was evident, and is attributed to the increase in the organic fraction, since inorganic fractions remained similar during both inundation periods. These higher organic fractions are probably related to an increase of N in the living fraction (phytoplankton) and non-living one (detritus), when the wetland is at its maximum inundation capacity.

Silicates concentrations were uniform for both wetland and inflow sites. Silicates uniformity is characteristic of river waters and change in discharge rates apparently not affects concentrations (Wetzel, 2001c). Concentrations of silicates were from 7 to 14 $mgSiO_4$-S/l for both periods, and no difference was observed between wetland sites and river inflow. Diel variation was minor, since night concentrations increased only 1-2 mg/l compared to day ones. Therefore, AdM waters have no limitation for this nutrient and has concentrations in the range of the world average for drainage natural waters (13 $gSiO_4$-S/l) (Wetzel, 2001c). Constant availability of silicates allowed the proliferation of those specific diatoms species with high silicates requirements as *Melosira granulata* (Kilham, 1971) (Chapter 4).

Chlorophyll-a concentrations were from 1 to 24 µg/l during both periods, indicating high inflow dynamics, and possible seasonal effects related to run off. Higher concentrations were evident during HIC in wetland sites (up to 24 µg/l). During LIC, river inflow sites and upper wetland sites had similar concentrations (up to 2 µg/l). On the other hand, during HIC, upper sites concentrations increased up to 6 mg/l, being similar to the ones of middle wetland areas, suggesting a stabilization of upper areas once the wetland is at its maximum inundation capacity. Overall, average concentrations in wetland area were from 4 to 9 µg/l for LIC and HIC, respectively, and in the river inflow sites from 1 to 6 µg/l for LIC and HIC, respectively. Based on overall average concentrations, AdM wetland river-wetland system could be described as a mesotrophic system (Huszar et al., 2006; Wetzel, 2001d). This trophic status is also described with the nutrient and total organic carbon concentrations measured in the system. However, due to the dynamics of the system, a fixed trophic status should be considered with caution.

Suspended solids concentrations had similar ranges for the river inflow during both inundation periods (24-97 mg/l). Wetland concentrations were from 13-55 mg/l, and this maximum was recorded during HIC only in a low wetland site (55 mg/l), indicating higher intrusion of the Nuevo River towards the wetland main area.

PCA analyses on water and sediment variables summarized the environmental patterns and provided the key variables. The associated variables for the AdM wetland-river system were the ones that had higher correlations (>0.7) with the PC components. For the water column: temperature, total suspended solids, DO, secchi, alkalinity, nitrogen and phosphorus (organic and inorganic), N:P ratio, were key variables during both conditions. Furthermore, during HIC, silicates and velocity were also influencing. For sediments: sand and silt, nitrogen and phosphorus content (inorganic and organic), organic matter and organic carbon were found to be the most influencing due to their higher correlations with the PC components.

Environmental gradients were observed inside the AdM wetland-river system, indicating the wetland is clearly divided between river sites with higher concentrations of DO, TSS, organic phosphorus, higher N:P ratios and velocities, and wetland sites with higher concentrations of organic nitrogen, alkalinity, chlorophyll-a, secchi. Still, overall the water and sediment concentrations in the AdM wetland system are quite similar.

4

COMMUNITY STRUCTURE OF BIOTIC ASSEMBLAGES

4.1 Background

Water systems that have not been studied previously can benefit from a initial survey of environmental and biological data (Field et al., 1982). Patterns, trends or relationships can be extracted from what can be complex data sets. The number of species obtained are usually large and the patterns 0f community structure often not obvious by simple inspection of data (Clarke and Warwick, 2001). The reduction of this complexity by graphical representation of the relationships between biota in the different samples is often referred to as 'representation of communities' (Clarke and Warwick, 2001). Several approaches have emerged over the last decades to visualize and analyse the structure of biotic assemblages in exploratory studies. Among them multivariate analysis and non-parametric techniques help to evaluate biological and other environmental data.

Previous data on biota in the AdM wetland have been collected separately in specific studies for macroinvertebrates (Van den Bossche, 2009), and plankton (Prado et al., 2012). This chapter reports on the first attempt in assess community structure of different biotic groups, and their relation with environmental data. Biotic communities of the AdM wetland were sampled during two different inundation periods of 2011 and 2012. The biotic community structure was evaluated across zones, including factors such as diel variation and sampling efforts. Main research questions in this chapter include:

- What are the patterns of the biotic communities sampled during both inundation periods?

- Do biotic communities show different responses depending on different flow regimes?

- What is the spatial distribution of the biotic communities along a river-wetland gradient?

- Do the biotic communities structure between the different wetland zones differed?

- Are there differences in structure of these communities from day to night?

- Do communities from littoral differ from those of pelagic areas?

- Which species/taxa typified the AdM wetland?

- Are changes in biotic communities composition between both inundation periods?

- Are environmental variables associated with the biotic community patterns?

4.2 Field measurement campaigns

4.2.1 Sampling methods and inundation conditions

The AdM wetland was sampled for biota simultaneous with other environmental variables during the periods described in Chapter 3. While the previous chapter assessed chemical substances and physical variables, this chapter focuses on biotic sampling of (i) phytoplankton; (ii) zooplankton; (iii) macro invertebrates; and (iv) fish (Figure 4-1). During the second field measurement campaign (March 2012) plankton communities were also sampled with vertical hauls in addition to horizontal tows. Furthermore, two sites S1 and S2 (middle wetland) were also sampled during the night to explore day/night variability (Figure 3-1). Phytoplankton and zooplankton samples were collected with nets of 55 and 200 μm mesh sizes, respectively. Horizontal tows of 5 minutes at a speed 2 knots (1 knot=1.852 km/h) collected phytoplankton and zooplankton. The samples were preserved in a 4% formalin solution. Phytoplankton was also collected from the water column with the Niskin Bottle and fixed with lugol's iodine. Macro-invertebrates were collected with a hand net of mesh size 500 μm. In the river section, sampling was performed along the banks. Inside the wetland, the habitats sampled comprised mainly floating and emerging vegetation. Each sample was preserved in 70% alcohol for further analysis in the laboratory. Fish were sampled with a seine net along the banks of the wetland, where vegetation was present. The net was placed approximately 2 meters from the shore and then a confinement was performed, walking towards the shore while making an 'U' with the net. The net had weights that reached the bottom, so as to prevent the escape of specimens. The specimens were washed out from the net and the type of vegetation collected within the net was recorded. The samples were preserved in 4% formalin solution for further analysis and identification in the laboratory (Figure 4-2). During the campaign of March 2012 (HIC), the seine net used was smaller ($6m^2$) than the one used during the campaign of February 2011 ($30m^2$). A conversion of the values of 2011 was performed on the values of 2011, only for the analysis that combined both campaigns.

Figure 4-1 Sampling biotic communities

4.2.2 Identification of biotic communities

Identification of phytoplankton was based on a standard keys and guilds (Bourrelly, 1966; Bourrelly, 1968; Bourrelly, 1970; Desikachary, 1959; Komárek and Anagnostidis, 2005; Parra, 1982; Prescott, 1982). Zooplankton was identified with several keys (Alonso, 1996; Brues et al., 1954; Edmondson, 1966; Pennak, 1989). Phytoplankton was fixed in lugol's iodine and quantified with the drip technique (Semina, 1978). This technique count all cells present in one drop of sample, and the result is extrapolated to the total volume collected. Zooplankton were counted using sub-samples of 25 ml in a Dolfus chamber (Boltovskoy, 1981). Macroinvertebrates samples were analyzed following the procedure recommended by De Pauw and Vanhooren (1983), consisting of four main steps: sieving, sorting, preservation and identification. Each sample was poured and rinsed with water over a tower of standard sieves with different mesh sizes (2000, 1000, 710 and 500 microns respectively) in order to separate the organisms by size and remove excess sand and silt (De Pauw and Vanhooren, 1983). Organic material (leaves, macrophytes) were carefully rinsed, inspected to remove attached animals and discarded. The content of each sieve was sorted in a white tray. All specimens were separated as far as possible, by order and preserved in small bottles with 70% alcohol. Later identification occurred under the stereomicroscopic was performed. Specimens were identified to family level with the keys of Roldán (Roldán, 1996; Roldán, 2003), with the exception of Anelida and Arachnida that were identified to class and suborder level, respectively. Fish identification was performed using these taxonomic keys (Eigenmann and Myers, 1927; Gery, 1977; Laaz et al., 2009; Maldonado-Ocampo et al., 2005)(Figure 4-2).

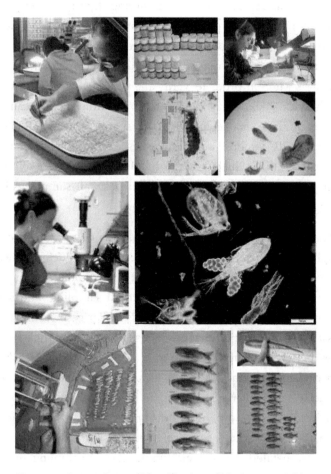

Figure 4-2 Processing and identification of biotic communities

4.2.3 *Data analysis of biotic communities*

The community structure of each biotic assemblage was explored by different multivariate techniques using PRIMER (V6) statistical software (Clarke and Warwick, 2001). For all methodologies, each biotic community (phytoplankton, zooplankton, macroinvertebrates, and fish) was analyzed independently for both low inundation (LIC 2011) and high inundation conditions (HIC 2012). Biological matrices for each community were built with abundance data. The matrices were square root transformed to construct Bray-Curtis similarity matrices for the different multivariate analyses. This moderate type of transformation was selected in order to retain all the collected information but at the same time minimize the influence of dominant species, which is more suitable for initial exploratory studies (Clarke and Warwick, 2001). Significance level was set at $P < 0.05$ for all routines.

CLUSTERING and MDS. Hierarchical clustering (group average linking) was applied to the Bray-Curtis similarity matrix including the SIMPROF (similarity profile) routine (Clarke et al., 2008) to test if there were significant differences among clusters. Cluster analysis is recommended to be used in combination with ordination techniques such as MDS in order to verify the consistency of both representations (Clarke and Warwick, 2001). Thus, non-metric multidimensional scaling (MDS) was applied to the resemblance matrix to construct a spatial representation of community composition similarity among sampling sites. A combined analysis of both inundation conditions (LIC+HIC) was performed for each biotic assemblages in order to visualize general spatial patterns. For this purpose, all samples collected during both monitoring campaigns were pooled together.

ANOSIM. Analysis of similarities (ANOSIM), a non-parametric permutation procedure applied to the (rank) similarity matrix, was conducted to test hypotheses about differences in community spatial distribution, sampling technique (horizontal tows or vertical hauls), pelagic (limnetic) or littoral sampling, and sampling time. This method uses Bray–Curtis distance as a dissimilarity measure and is suitable when there is *a priori* information that can be used to test the null hypothesis (H0) of 'no differences between groups' (Clarke et al., 2008; Clarke and Warwick, 2001). Thus, *a priori* divisions or 'factors' were set up for the data analysis. For spatial distribution, the wetland-river system was divided into four zones (upper, middle, low and river inflow). For sampling effort, *a priori* divisions were horizontal tows and vertical hauls. Furthermore, vertical hauls also included two *a priori* subdivisions: littoral-vegetated and pelagic (limnetic). For sampling time, the *a priori* factors were day and night. Macroinvertebrate and fish communities were sampled in littoral areas. For these two groups spatial distribution (zones) and time of sampling (day and night) were the *a priori* divisions tested. The aim was to explore if there were statistically significant differences in community composition regarding all these *a priori* selected factors. The criterion 'spatial distribution' was analyzed for both conditions: low inundation conditions (LIC) and high inundation conditions (HIC) while the other criteria (sampling effort, sampling time, pelagic- limnetic or littoral) were introduced in the second monitoring campaign (HIC). To summarize the spatial distribution, the criteria 'location' (river/wetland) was also tested. Finally, to test possible differences in the communities between LIC and HIC, the factor 'condition' was examined.

SIMPER. If differences among *a priori* defined 'factors' were shown by ANOSIM, a following step was to determine which species/taxa were responsible for those differences. For this purpose, the SIMPER 'similarity percentages' routine was implemented. This technique, frequently used to complement ANOSIM and MDS, indicates which species are responsible either for an observed clustering pattern or for differences among sets of samples. More specifically, it determines the percentage contribution of each taxa to the average similarity 'within' a group or to the average dissimilarity 'between' *a priori* defined groups (Clarke and Warwick, 2001). Similarities within wetland zones were analyzed to determine which taxa of each biotic assemblage typified each zone of the river-wetland system. In order to determine possible differences between both inundation conditions (LIC & HIC) SIMPER results from both campaigns were analyzed separately. The major taxa contributing to these within/intra areas similarities for each biotic assemblage and for each condition are detailed in Appendix D-Table D1. The similarities for each biotic assemblage per zone and for both inundation conditions are presented. Dissimilarities between the different *a priori* factors followed the same approach.

BEST and LINKTREE. BEST routines from PRIMER (BVSTEP and BIOENV algorithms) were applied to investigate which environmental variables best explain the patterns of the biotic communities. The test operates by permutation and utilizes only rank dissimilarities by selecting a subset of variables in one matrix 'environmental' which best matches the multivariate pattern of samples in a different 'biotic' matrix. It explains a biotic assemblage structure with a subset of environmental variables (Clarke and Gorley, 2005). The null hypothesis tested was that there was no relationship between environmental and the biotic community patterns of the investigated biotic assemblages phytoplankton, zooplankton, macroinvertebrates and fish. Each biotic assemblage (phytoplankton, zooplankton, macroinvertebrates, and fish) was analysed independently. As a first step, a resemblance matrix (Bray-Curtis) for each biotic assemblage was created with square root transformed data. This is the 'fixed resemblance matrix' that describes the biotic relationships among samples, and in the context of the BEST analysis is referred to as the 'response matrix' (Clarke et al., 2008). Secondly, for each biotic community a match with the environmental data was performed, since e.g. different water levels were sampled at different locations (littoral and pelagic). Thus, for each biotic community a specific environmental matrix explanatory or driver matrix' was built with normalized environmental data. Diversity indices of each biotic community were included in the environmental matrix to explore the relations between each biotic assemblage and the biotic indices of the other assemblages. These indices were previously calculated in the DIVERSE routine. Spearman rank correlation (ρ) was used to compare both matrices, and high rank correlations between variables in both matrices were searched. The level to which a particular subset of environmental variables captures the pattern of the biotic community samples is measured by correlating the matching entries of the two matrices. Then, the BEST procedure selects the combination of environmental variables that maximizes (ϱ) and therefore 'best explains' the biotic structure (Clarke et al., 2008). BVSTEP performs a stepwise search over the tested variables, fitting first the environmental variable with the strongest relationship, subsequently adding the variable with the next strongest relationship. BVSTEP searches in a hierarchical way, adding (forward stepping) and deleting (backward elimination) variables one at a time. On the other hand, BIOENV tests all possible combinations of variables, from each environmental variable separately through to all at the same time (Clarke and Warwick, 2001).

Owing to the large number of environmental variables available to explore, BVSTEP was more suitable to apply in this study, since BIOENV can become prohibitive and time consuming when there are more than 15 variables. As first step, several runs with BVSTEP were performed for each biotic community before selecting the best-fit ones. Subsequently, to confirm the BVSTEP results of those selected runs, BIOENV was applied to validate the results, and generally, it was found that the same subset of variables was selected. In order to evaluate the level of significance of the rank correlation results between the two matrices evaluated, the routine was run with 999 permutations. The subset, which maximizes the value of the Spearman rank correlation coefficient (ϱ), is the one with collective properties that best capture, in quantitative terms, the subjective *a priori* habitat distinctions as represented by the model matrix. The significance of P (< 0.05) was ascertained by 999 random permutation tests. As a next step, the subset of variables selected by the BVSTEP routine was included as explicative variables in the LINKTREE routine. The procedure is a constrained type of cluster analysis based on the biotic resemblance matrix that involves a divisive partition of the biotic samples. Each of the divisions is explained in thresholds of the environmental variables. This routine utilizes the ranks of the resemblances and calculates the ANOSIM R values between the two groups formed at each division, which provides a measure of the degree of separation (R is the difference of the average rank dissimilarities between and within groups and reaches its maximum at 1 when all dissimilarities between the two groups exceed any dissimilarity within either group). Orthogonal to that is B (%), which describes how well separated the two groups of samples are in the current split, in relation to the maximum separation of the first split. B% is an absolute measure of group differences at that level.

LINKTREE was run in conjunction with the SIMPROF test (with a criterion of P <0.05) to provide stopping rules for the tree divisions, and also to be consistent with the criterion applied for the cluster analysis. This P level is recommended for exploratory analysis. The SIMPROF test stops unwarranted subdivisions when there is no significant multivariate structure among the remaining biotic samples, and samples below that point are considered as homogeneous (Clarke et al., 2008). As a result, LINKTREE constructs a hierarchical tree that shows how the biotic samples are successively split into groups according to the environmental variable(s) maximizing the separation of these samples in a multidimensional space (Bauman et al., 2013). The LINKTREE routine was run with the same biotic resemblance matrix previously built (for Cluster, MDS and BVSTEP routines), and the selected subset of normalized environmental data. Each biotic community was analyzed independently with their respective subset of environmental variables and under the two inundation conditions. The RELATE routine was applied after elucidating the variables selected by BVSTEP to compare the biotic matrix with the environmental matrix that was applied. A matrix with the variables selected by the BVSTEP routine was built for both LINKTREE and RELATE routines.

4.3 Phytoplankton

4.3.1 Sampling with Niskin Bottle

During LIC, 38 species belonging to 22 families and 9 classes were identified in the wetland and river sites (pooled results) for phytoplankton collected with a Niskin bottle. Wetland sites contained almost double the number of species collected at river sites (32 and 17, respectively). Both, river and wetland shared some species. Phytoplankton densities ranged from 30000 to 860000 cell/l. Lower densities were observed at river sites and maximum at S7 (Figure 4-3a). Dominance of Cryptophyceae (*Cryptomonas sp.*) was observed at S7, with 500000 cell/l (58% of the total phytoplankton community). Cryptophyceae was present at four sites and especially abundant at the lower wetland sites S3c and S7. Chlorophyceae and Euglenophyceae were present at five and four sites, respectively, and also abundant in the lower wetland. Fragilariophyceae had high relative abundances at river sites (Figure 4-3b). Pooled wetland samples indicated the dominant species as *Cryptomonas sp.* (39%) *Ankistrodesmus acicularis* (9%), *Trachelomonas* (7%), *Phacus* (6%), *Euglena sp* (4%). The rest of the species (27) comprised < 4% of the total and collectively 36%. At river sites, 8 of the 17 species identified comprised 74% of the total abundance: *Synedra sp* (29%), *Fragilaria sp.* (11%), *Nitzschia acicularis*, *Alaucoseria granulata*, *Closterium acerosum*, *Oscillatoria sp*, *Pseudoanabaena sp*, *Ulnaria ulna* (6% each). The other 9 species were present in percentages lower than 4%.

During HIC, a total of 57 species belonging to 20 families and 9 classes were identified in the wetland and river sites. A substantially higher number of species were found at wetland sites (54), compared with river inflow sites (12). Densities ranged from 50000 to 370000 cell/l. Lower densities were observed at river sites (Figure 4-3b). Results from vertical profiles showed that densities at surface level double those at the bottom level in S1 and S2, and triple at S7 (Figure 4-3c). Diel variation was observed at the two sites sampled during day and night (S1 and S2 at surface level). Thus, densities at S1 (s_night), were half the day densities S1(s). The same occurred for S2 (s_night), but at this site densities were three times lower than day ones S2(s) (Figure 4-3c). Furthermore, a reduction in the number of species collected at night was observed for both sites, decreasing from 11 species during day to 6 at night.

Regarding spatial distribution of the classes during HIC, Cryptophyceae, was present in all wetland sites except S1 (bottom) and S2 (vegetation). At S7 densities at surface level were five times higher than at the bottom. Bacillariophyceae and Coscinodiscophyceae were the second and third groups in importance but their distribution along the wetland varied. Chlorophyceae showed similar densities at surface and bottom samples (S1, S7), and at S2 was only present in surface samples. Coscinodiscophyceae were recorded at similar densities at both depths (S2, S7); however, at S1 (m) higher densities were found at 3 meters (middle water column) compared with the surface (s) (Figure 4-3c). The classes Bacillariophyceae, Chlorophyceae, Coscinodiscophyceae and Fragilariophyceae had generally higher relative abundances in bottom samples than surface ones. Cryptophyceae and Euglenophyceae showed the opposite (Figure 4-3d). Thus, although densities of Bacillariophyceae were similar at surface and bottom samples (Figure 4-3d), this class showed a higher contribution to the total abundance in middle and bottom samples when Cryptophyceae was absent (Figure 4-3d). Pooled wetland samples indicated that dominant species were *Cryptomonas* sp. (37%); *Melosira granulata* (11%) and *Cyclotella comta* (4%). The rest of the species (51) were present in percentages below 4% contributing together 48%.

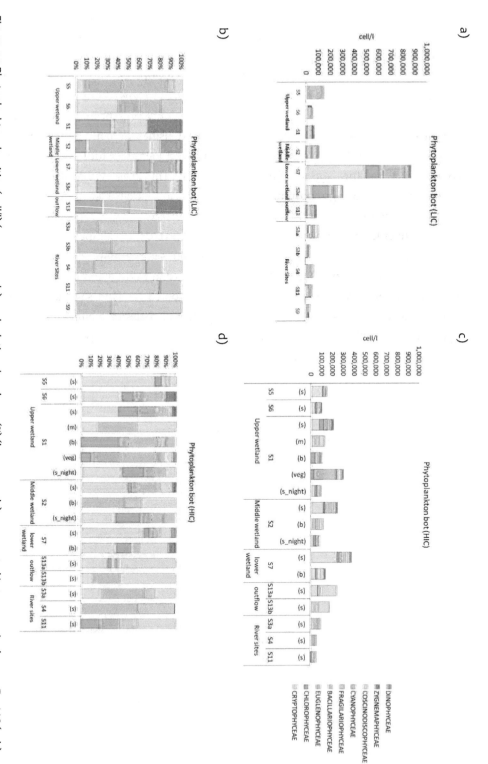

Figure 4-3 Phytoplankton densities (cell/l) (upper panels); and relative abundance (%) (lower panels) represented in taxonomic classes. For LIC (a -b); for HIC (c-d). Samples collected with Niskin bottle.

Spatial patterns

For LIC, Cluster and SIMPROF analysis revealed three groups for phytoplankton sampled with Niskin bottle (19% similarity level; SIMPROF test: π= 4.31, p=0.5%). A similarity threshold of 25% divided the samples in 5 groups. The superimposition of these two sets of groups on the MDS ordination confirmed the agreement between both representations (stress: 0.1). A clear division was observed between river sites (S3b, S4, S11, S9) from wetland sites (S5, S6, S3c, S7) (Figure 4-4). For HIC, Phytoplankton (collected with Niskin bottle), did not show evidence of significant splits in community structure (SIMPROF test: p >0.05 for all the subdivisions). The arbitrary 20% threshold divided the sites into four groups. The 2D ordination exhibits a wide spread of the samples, where separation between river and wetland sites is not clear. Only sites located at the lower area (S7, S13a) cluster tightly at 40% similarity (stress: 0.19) (Figure 4-4). A combined analysis of both inundation conditions suggested a division between LIC and HIC samples (Figure 4-5).

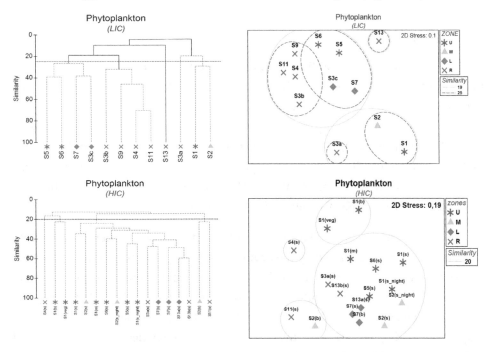

Figure 4-4 Group average cluster analysis (left column) and MDS ordinations (right column) for phytoplankton collected with Niskin bottle during LIC (low inundation conditions/upper panel) and HIC (high inundation conditions/lower panel). Based on Bray Curtis similarity matrices (built from square root transformed abundance data). Continuous lines represent the divisions for which SIMPROF test (P< 0.05, p<5% in percentage) confirmed further subdivision structure to explore. Dashed lines show the group structure with no evidence from the SIMPROF test. The x- axis shows sampling sites, symbols represent wetland areas (U: upper; M: middle; L: low; R: river). The y-axis represents Bray-Curtis similarity (%).

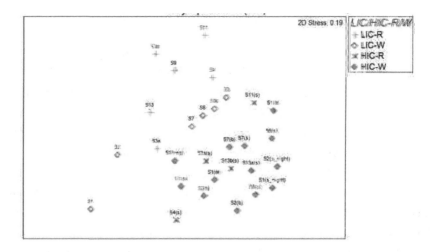

Figure 4-5 MDS ordination for phytoplankton collected with Niskin bottle. Pooled samples of both sampling conditions (LIC: low inundation conditions; HIC: high inundation conditions). R: river sites; W: wetland sites. Based on Bray Curtis similarity matrices.

Similarities/Dissimilarities

During low inundation conditions (LIC), the upper area was characterized by *Cryptomonas, Trachelomonas* and *Synedra*; the middle by *Oscillatoria, Synedra* and *Cryptomonas*; and the river area by *Synedra*. The lower area had more species for the intragroup similarity than upper and river areas where only 3 species dominated (Table D1 - Appendix D). Overall, average similarities were from 13 to 22. During high inundation conditions (HIC), *Cryptomonas* was the typical species for the wetland areas with contributions up to 48%, and *Melosira granulata* was the main species in the river area (34%) (Table D1 - Appendix D).

ANOSIM test for the '*river and wetland*' factor did not show differences for both inundation periods, confirmed by the ordinations (Figure 4-4), where river sites were grouped with wetland sites. When the samples of both inundation periods were combined (LIC+HIC), minor differences were observed (Global R=0.27, P=0.005). For the factor '*wetland zones*' a low R (R= 0.125, P= 0.048) was obtained (Table D2 - Appendix D). *Cryptomonas* was the top discriminator species between river and wetland areas for LIC, HIC and also for the combined conditions analysis; due to its higher abundances in the wetland area. During LIC, *Synedra sp.* was the second species differentiating the two areas due to its higher abundance in the river, while for HIC and combined conditions, *Melosira granulata* distinguished both areas, being more abundant in the river.

For combined conditions (LIC+HIC), Synedra sp. confirmed its dominance in the river (Table D13 - Appendix D). *Cryptomonas* sp and *M. granulata* were the differentiating species of inundation conditions, because of their higher abundances during HIC. Some species were more abundant during LIC, notably *Trachelomonas sp.*, *Oscillatoria sp.*, and *Pseudanabaena sp.*, indicating that these species are more important during the initial periods of wetland inundation (Table D17 - Appendix D).

4.3.2 Sampling by horizontal tows

During LIC, phytoplankton collected with horizontal tows had lower densities in the middle and in the lower wetland areas compared with the upper wetland and river areas. A maximum of 778,440 cell/m^3 was observed at S5 (upper wetland), attributed to the high densities of *Fragilaria longissima*. Site S1 and the inflow site S3a showed similar densities (Figure 4-6a). Fragilariophyceae was present at river and wetland sites with *Fragilaria longissima* as dominant species, with a major contribution at S5. Bacillariophyceae were also common, but with no dominance of one species. Higher densities of Bacillariophyceae were observed at the upper wetland sites. Other classes were present only in low densities (Figure 4-6b). Considering all wetland samples of LIC together, four species dominated the community sampled by horizontal tows: *Fragilaria longissima* (52%), *Gomphonema gracile* (9%), *Navicula sp.*(9%), *Nitzschia palea* (7%), comprising together 77%. The other 38 species were present in percentages lower than 4%, comprising together 23%. At river sites, *Fragilaria longissima* was the dominant species (58%), followed by *Fragilaria sp* (6%), *Gomphonema sp* (6%), *Nitzschia pale*a (34%), *Synedra goulardii* (3,4%), comprising together 77% of the total community at river sites. The other 27 species were present in percentages lower than 3%, comprising together 23%.

During HIC, results from horizontal tows showed that wetland sites had twice the number of species collected at river sites (41 compared with 18). Densities were between 5848 and 239768 cell/m^3, thus lower than during LIC. Inflow sites (S3a) showed similar densities as the upper wetland (S6). Lower densities were observed at S1 and S2. No clear trend was observed for day-night sampling, with lower densities at night, but higher at S2 (Figure 4-6c). A higher contribution of Coscinodiscophyceae (mainly represented by *Melosira granulata*) in the area of the wetland outflow (S13a) was found. A decreasing gradient of this class was observed towards the middle wetland area. On the other hand, Fragilariophyceae (represented by *F. longissima*) was dominant in the upper wetland area. Bacillariophyceae were present in all sites but with lower contribution (Figure 4-6d). Considering all wetland samples of HIC together, two species dominated the samples: *M. granulata* (57%), and *F. longissima* (17%). The other 39 species were present in percentages below 2%, comprising together 26%. At river sites, from the 18 species identified, main species were *M. granulata* (27%), *F. longissima* (24%), *Fragilaria capucina, Fragilaria sp, Frustulia saxonica, Synedra sp* (6% each). The other 12 species were present in percentages lower than 3%, comprising together (29%).

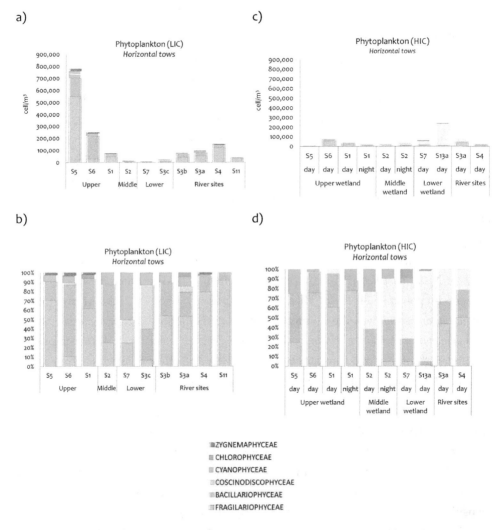

Figure 4-6 Phytoplankton densities (cell/m³) (upper panels); and relative abundance (%)(lower panels) represented in taxonomic classes. For LIC (a -b); for HIC (c-d). Samples collected with horizontal tows. Same scales (a, c) to allow comparison.

Spatial patterns

During LIC, phytoplankton collected with horizontal tows clustered in two major groups at 10% similarity level (SIMPROF test: $\pi= 4.48$, p=0.1%). The low wetland sites (S7 and S3c) were highly similar (55% similarity) and were segregated from the rest of the sites at 28% similarity. However, subgroups formed above 28 % similarity were not significant according to SIMPROF test (P > 0.05) (Figure 4-7). During HIC, the SIMPROF test applied to the cluster analysis did not indicate significant evidence of group substructure (Figure 4-7). A combined analysis exhibited a distribution according to the inundation condition since samples of LIC were observed at the upper part of the 2D ordination, and samples of HIC at the lower part (Figure 4-8).

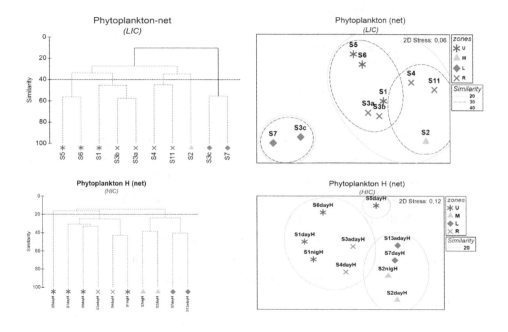

Figure 4-7   Group average cluster analysis (left column) and MDS ordinations (right column) for phytoplankton collected with horizontal tows during LIC (low inundation conditions/upper panel) and HIC (high inundation conditions/lower panel). Based on Bray Curtis similarity matrices (built from square root transformed abundance data). Continuous lines represent the divisions for which SIMPROF test (P< 0.05, p<5% in percentage) confirmed further subdivision structure to explore. Dashed lines show the group structure with no evidence from the SIMPROF test. The x- axis shows sampling sites, symbols represent wetland areas (U: upper; M: middle; L: low; R: river). The y-axis represents Bray-Curtis similarity (%).

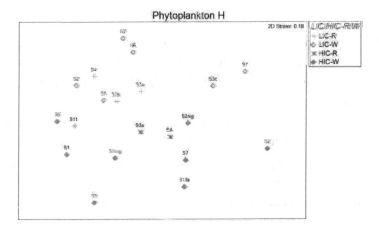

Figure 4-8 MDS ordination for Phytoplankton (collected with horizontal tows). Pooled samples of both sampling conditions (LIC: low inundation conditions; HIC: high inundation conditions). a) R: river sites; W: wetland sites. b) ZONES (U: upper; M: middle; L: low; R: river). Based on Bray Curtis similarity matrices (built from square root transformed abundance data)

Community similarities/dissimilarities

Fragilaria longissima was the typical species in the upper wetland and river areas during both LIC and HIC, due to its high contribution to the intragroup similarity and high SIM/SD ratios (Table D6 - Appendix D). The lower area was characterized by *Microcystis* and *M. granulata* during LIC, while during HIC, the middle and lower areas were typified by *M. granulata* (with over 50% contributing to the average similarities in both areas). During HIC, the average similarity within groups was lower (33%) than during LIC (45%). Fewer species contributed to the area similarity in the upper wetland and river areas in HIC compared with LIC (Table D1 - Appendix D), likely due to the higher flow conditions in HIC. The testing for differences in structure among different factors found that for the factors 'river and wetland' no differences of structure were evident during both conditions (Figure 4-7) and river sites grouped with wetland sites. For the factor 'wetland zones' during both LIC and HIC, phytoplankton (horizontal tows) showed a separation among zones (R=0.70, P=0.002) and (R=0.49, P= 0.028), respectively (Table D2 - Appendix D) (Figure 4-7). Combined analysis of both inundation conditions also indicated that some differences occurred for the 'wetland zones' factor (R= 0.32, P=0.003) with significant pairwise differences (Table D2 - Appendix D) as displayed in Figure 4-8. The factor 'inundation condition' exhibited relatively low values (R=0.24, P=0.005), although a separation is displayed in Figure 4-8.

4.3.3 Sampling by vertical hauls

Vertical hauls had noticeably higher densities of phytoplankton per m³ than horizontal tows (Figure 4-9a). Higher densities were collected at the littoral zones with more vegetation than in open water (pelagic), with the exception of S5 (upper wetland). A maximum density of 4964 x10³ cell/m³ at S1 was observed during the night (Figure 4-9a), probably due to a low grazing activity (Figure 4-16a). On the other hand, at S2 a very low density of phytoplankton was observed at night (Figure 4-9a). Diel variation was observed at S1, where night samples for both pelagic and vegetated sites showed higher densities than day samples. However, at S2 the opposite occurred, with higher densities in vegetation during day time (Figure 4-9a), thus not consistent pattern was observed. Possible causes could be related to growth rates, grazing effect or just and effect of spatial variation.

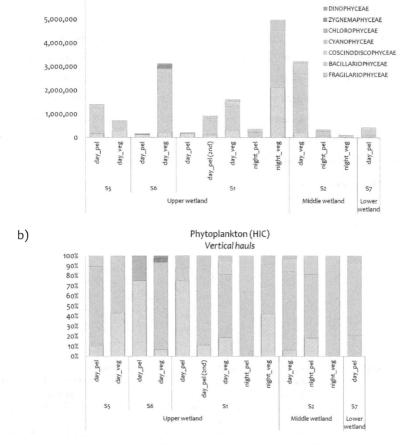

Figure 4-9 Phytoplankton densities (cell/m³) (a), and relative abundance(%) (b). Represented in taxonomic classes. Samples collected with vertical hauls.

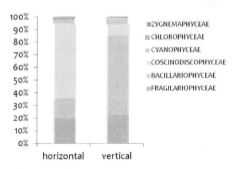

Figure 4-10 Comparison of the contribution (%) of each taxonomic class of phytoplankton for horizontal tows and vertical hauls. Both during HIC (Pooled results of wetland sites).

Spatial patterns

The spatial patterns obtained from phytoplankton collected with vertical hauls exhibit two significant groups (SIMPROF: π=1.83, p=3.1%) that isolate Site 2 (collected at night in vegetation) from the rest of sites. However, this splitting occurred at a low similarity level (less than 10%). A 20% similarity threshold divided the samples into six groups that are also clearly display in the 2D ordination (stress 0.12), with the central group conformed mainly by the samples collected in vegetation (Figure 4-11). A combined analysis of horizontal and vertical samples did not show a separation according to inundation (Figure 4-12a), but led to separation on the basis of littoral compared with pelagic-limnetic (Figure 4-12b).

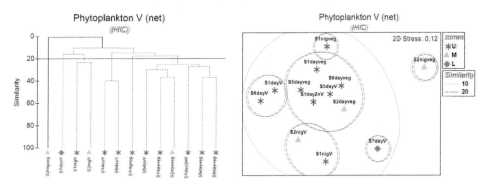

Figure 4-11 Group average cluster analysis (left column) and MDS ordinations (right column) for phytoplankton collected with vertical hauls during HIC (high inundation conditions) Based on Bray Curtis similarity matrices (built from square root transformed abundance data). Continuous lines represent the divisions for which SIMPROF test (P< 0.05, p<5% in percentage) confirmed further subdivision structure to explore. Dashed lines show the group structure with no evidence from the SIMPROF test. The x- axis shows sampling sites, symbols represent wetland areas (U: upper; M: middle; L: low). The y-axis represents Bray-Curtis similarity (%).

a)

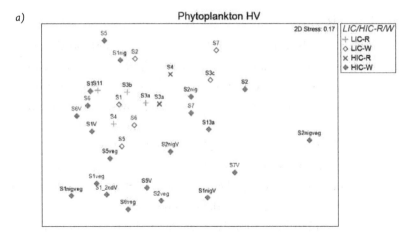

b)

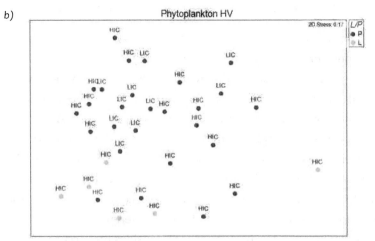

Figure 4-12 MDS ordination for phytoplankton. Pooled samples of both sampling conditions (LIC: low inundation conditions; HIC: high inundation conditions), horizontal tows and vertical samples pooled (HV). a) R: river sites; W: wetland sites. b) L: littoral; P: pelagic- limnetic. Based on Bray Curtis similarity matrices (built from square root transformed abundance data)

Community similarities

Community similarities based on vertical tows of phytoplankton during HIC also identified *F. longissima* as the main species for the Upper area, contributing 49% to the similarity. More species were responsible for the intra group similarity in the Upper Wetland area compared with the horizontal tows, indicating that vertical hauls collected more species than horizontal tows in this area. The middle area was characterized by *M. granulata* and *F. longissima* (both contributing 50% each to the similarity within this area) (Table D1 - Appendix D). The higher contribution of *F. longissima* in the middle area could be attributed to the higher sinking that this species can experience with higher residence times, and therefore can be captured due to the vertical sampling effort. Lastly, average similarities within the haul groups were lower than the ones of horizontal tows (Table D6 - Appendix D).

When testing for differences in structure, the factor 'wetland zones' exhibited minor differences among zones during HIC (R=0.36, P=0.05), (Table D2 - Appendix D). For the factor 'diel variation' moderate differences in structure between day and night were observed for vertical hauls (R=0.29, P=0.05), while the rest of the biotic communities did not show significant differences (Table D2 - Appendix D). The factor 'littoral/pelagic' phytoplankton collected during HIC showed a moderate division between littoral hauls in vegetation and pelagic samples (R= 0.35, P=0.002), where pelagic samples included both horizontal tows and vertical hauls (HV) (Table D2 - Appendix D). However, when the analysis compared only vertical hauls (pelagic and littoral) results were not significantly different. Thus, these results were apparently more influenced by the difference in densities that occur when horizontal tows were added in the analysis. For combined conditions (LIC+HIC), only phytoplankton net (HV) was significant for the 'littoral-pelagic factor' (R= 0.35, P=0.003) (Table D2 - Appendix D), and illustrated in its ordination (Figure 4-12b). Dissimilarities between sampling efforts (horizontal tows and vertical hauls) were analyzed with HIC samples. The 50 % of the dissimilarities was explained by 12 species, with *F. longissima*, *M. granulata*, *N. recta* and *Melosira varians* as the top ones, while 40 species explained 90% of the dissimilarities (Table D12 - Appendix D). For the littoral-pelagic factor the dissimilarity was attributed to the higher abundance collected with vertical hauls, not to a presence/absence. Only *Nitzschia amphibia* and *Neidium affine* were collected with vertical hauls. *N. amphibia* was only observed in the littoral area, possibly indicating an association to vegetation.

4.4 Zooplankton

4.4.1 Sampling by horizontal tows

During LIC, 53 species belonging to 28 families, 16 orders and 11 classes of zooplankton were collected in the wetland and river sites using horizontal tows, however, few species did dominated the samples. Zooplankton densities at wetland sites ranged between 2884 to 14880 org/m³. River sites showed a decreasing pattern from the wetland inlet (5340 org/m³) to 6 org/m³ in the outermost site (S11) located closer to the Vinces River. Main groups during this sampling period were Rotifera and Rhizopoda; Copepoda was important in two sites (S1, S7); Cladocera was present at all sites but with low densities (Figure 4-13a). During HIC, horizontal tows at wetland sites during daytime sampling had densities between 19 and 1410 org/m³ (Figure 4-13c). Site S13a had densities at daytime similar to the ones collected at night in S1 and S2. In these sites, densities increased around three folds compared with day sampling. Low densities were observed at river sites, similar to the ones in the upper sites S5 and S6 (Figure 4-13c). Dominant groups during HIC changed with respect to LIC (Figure 4-13b), with greater dominance of Cladocera and Copepoda , while Rotifera was frequent but in very low densities (Figure 4-13d).

Spatial patterns

During LIC, the zooplankton collected with horizontal tows showed two significant groups, that implied the isolation of a single river site (S11) from the rest of sites at 17% similarity level (SIMPROF: π=4.48, p=2.1%). As for phytoplankton, five subgroups were displayed in the MDS ordination when an arbitrary 40% similarity threshold was set up. Subgroups were also clearly represented in the MDS (stress: 0.08), corresponding to a strong ordination (Figure 4-14). During HIC, the community was also clustered in two significant groups at 21% similarity (SIMPROF: π= 8.04, p= 0.1%). A tight cluster integrated by wetland sites (S1, S2, S7, S13) was observed at 50% similarity, which is clearly separated from river inflow sites (S3a, S4) and from the two upper wetland sites (S5, S6). The latter four sites shared similar hydrodynamic conditions, since both have higher velocities and behave as inflows to the wetland. An arbitrary threshold of 40% divided the sites into four groups which is in agreement with the 2D ordination and provides an excellent representation since the stress value is < 0.05 (Figure 4-14). The combined representation of both conditions displayed a clear separation with samples of LIC shown in the upper section of the 2D ordination (Figure 4-15).

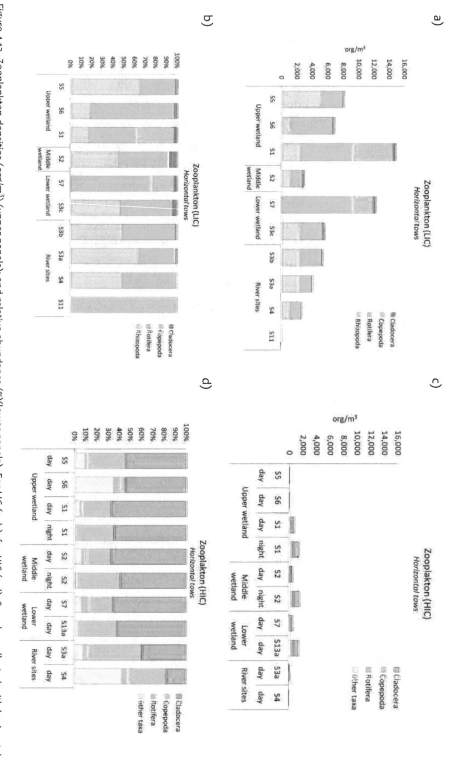

Figure 4-13 Zooplankton densities (org/m³) (upper panels); and relative abundance (%)(lower panels). For LIC (a -b); for HIC (c-d). Samples collected with horizontal tows. Same scales (a, c) to allow comparison. Main groups represented

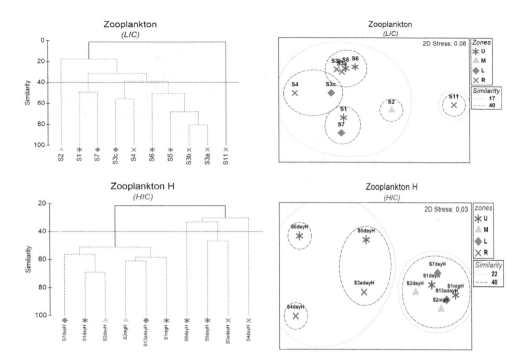

Figure 4-14 Group average cluster analysis (left column) and MDS ordinations (right column) for zooplankton collected with horizontal tows during LIC (low inundation conditions/upper panel) and HIC (high inundation conditions/lower panel). Based on Bray Curtis similarity matrices (built from square root transformed abundance data). Continuous lines represent the divisions for which SIMPROF test (P< 0.05, p<5% in percentage) confirmed further subdivision structure to explore. Dashed lines show the group structure with no evidence from the SIMPROF test. The x- axis shows sampling sites, symbols represent wetland areas (U: upper; M: middle; L: low; R: river). The y-axis represents Bray-Curtis similarity (%).

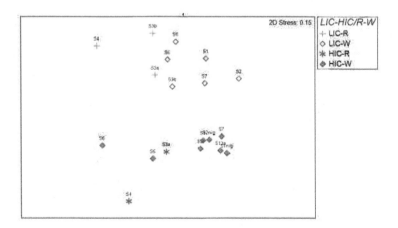

Figure 4-15 MDS ordination for Zooplankton. Pooled samples of both sampling conditions (LIC: low inundation conditions; HIC: high inundation conditions). R: river sites; W: wetland sites. a) Collected only with horizontal tows (H);   b and c) horizontal and vertical samples pooled (HV). Symbols at c) Sampling effort (H: horizontal; V: vertical). Based on Bray Curtis similarity matrices.

4.4.2 *Sampling by vertical hauls*

During HIC, vertical hauls provided greater estimated densities (Figure 4-16a) than horizontal tows (Figure 4-13c). Vertical hauls taken in littoral zones (with vegetation) had higher densities than in pelagic zones. Diel variation densities at S2 increase twofold from day (56575 org/m³) to night (93050 org/m³). Upper sites (S5, S6) had lower densities compared with the other sites, but also with higher densities in vegetation. A high density was observed in the lower wetland site S7 (29774 org/m³) despite being a pelagic sample (Figure 4-16a). Dominant groups for vertical hauls were Cladocera and Copepoda (Figure 4-16b).

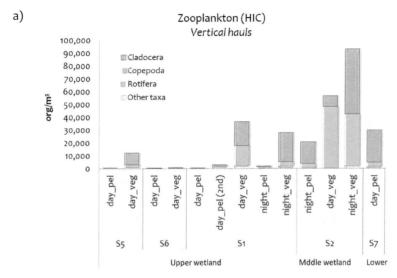

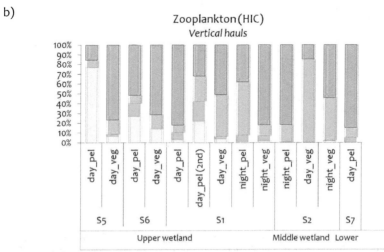

Figure 4-16 Zooplankton densities (cell/m³) (a), and relative abundance (%) (b). Main groups represented. Samples collected with vertical hauls. (pel: pelagic; veg: littoral vegetation).

Spatial patterns

Zooplankton collected with vertical hauls also split into two significant clusters at 18% similarity (SIMPROF: π=4.24, p= 0.1%). A 40% similarity threshold divided the sites into five subgroups, also clearly observed at the 2D ordination (stress: 0.08). One of the subgroups includes only samples collected in littoral vegetated areas (veg) (Figure 4-17).

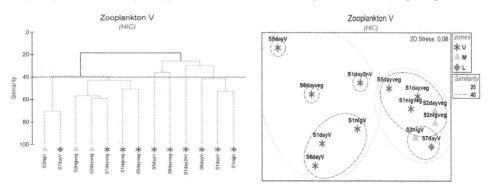

Figure 4-17 Group average cluster analysis (left column) and MDS ordinations (right column) for zooplankton collected with vertical hauls during HIC (high inundation conditions) Based on Bray Curtis similarity matrices (built from square root transformed abundance data). Continuous lines represent the divisions for which SIMPROF test (P< 0.05, p<5% in percentage) confirmed further subdivision structure to explore. Dashed lines show the group structure with no evidence from the SIMPROF test. The x- axis shows sampling sites, symbols represent wetland areas (U: upper; M: middle; L: low). The y-axis represents Bray-Curtis similarity (%).

Community similarities

Vertical hauls of zooplankton during HIC identified *C. sphaericus* and *M. venezolanus* as the main species contributing to the intragroup similarities in the upper area, and *Mesocyclops* (copepodite) in the middle area (Table D1 - Appendix D) and (Table D7 - Appendix D).

When testing for differences in community structure the spatial distribution factor 'wetland zones' was almost significant for vertical hauls (R=0.36, P=0.05). The factor 'littoral/pelagic', showed a moderate division between littoral (vertical hauls in vegetation) and pelagic samples (R= 0.25, P=0.024), when pelagic samples included both horizontal tows and vertical hauls. For the factor 'sampling effort' during HIC, zooplankton showed differences in structure between horizontal and vertical samples (R=0.35, P=0.001) (Table D2 - Appendix D). Thus, sampling effort influences the structure, but related to difference in abundance, since both sampling efforts collected the same species. A visual representation of this pattern is depicted in Figure 4-18. This pattern was also confirmed when samples of both conditions (LIC+HIC) were combined for the analysis (R=0.34, P=0.001) (Table D2 - Appendix D).

The division by 'inundation condition' was observed when vertical haul samples were combined with horizontal tow samples (R=0.65, P=0.001) (Table D2 - Appendix D), and depicted in Figure 4-19. *Moina micrura* discriminated middle and lower from the upper wetland, due to the higher abundance in the middle area. In the lower area, higher densities of *M. micrura* were estimated from vertical hauls compared with horizontal tows (Table D12 - Appendix D). For 'littoral/pelagic' factor, five species were the main discriminators between littoral and pelagic areas, contributing 50% to the total dissimilarity, due to their higher abundance in the littoral area. A total of 22 species explained 90% of the total dissimilarity. A key discriminator species of the littoral area was *C. sphaericus* due to its higher abundance in the littoral and high Diss/SD ratio.

When horizontal and vertical hauls were pooled to account for pelagic samples, the same five species were the main discriminators between littoral and pelagic areas (Table D15 - Appendix D). For 'sampling effort', 3 species contributed 30% to the total average dissimilarity between both sampling efforts: *M. micrura*, *Mesocyclops* (copepodite) and *M. venezolanus*. The dissimilarity was also related to the higher densities collected with vertical hauls, since all species were present in both sampling efforts. A total of 24 species explained 90% of the dissimilarity (Table D12-Appendix D).

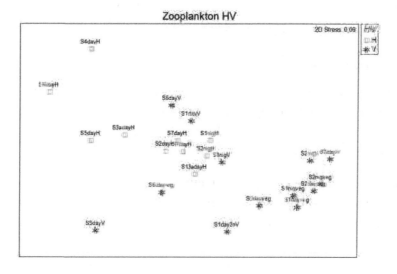

Figure 4-18 MDS ordination for zooplankton during high inundation conditions (HIC). Symbols (H: horizontal, V: vertical). Based on Bray Curtis similarity matrices (built from square root transformed abundance data).

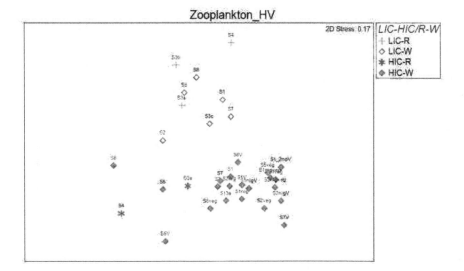

Figure 4-19 MDS ordination for Zooplankton. Pooled samples of both sampling conditions (LIC: low inundation conditions; HIC: high inundation conditions). R: river sites; W: wetland sites. Horizontal and vertical samples pooled (HV). LIC sites above the line; HIC sites below the line.

4.5 Macroinvertebrates

During LIC, 51 families belonging to 13 orders were collected at wetland and inflow sites. In wetland sites Amphipoda, Gastropoda, Diptera, Coleoptera, comprised 74% of the total community. In river sites, Diptera was the dominant group (40%), followed by Ephemeroptera (18%) and Coleoptera (17%). Wetland Sites S1 and S5 had higher densities than river sites S4, S9, S11, S13, that were characterized by higher flow velocities, suspended solids and less aquatic vegetation (Figure 4-20 a & b). During HIC, 64 families belonging to 20 orders, were collected at both wetland and inflow sites. Class Insecta dominated the community (78%), with the Orders Ephemeroptera, Diptera and Coleoptera as most representatives (30, 24 and 12% each). Class Gastropoda contributed 7% of the total and was mainly represented by the family Planorbidae. Densities ranged from 65 specimens (inflow site S4b) to 1072 specimens (wetland site S1 during night sampling) (Figure 4-20 c & d).

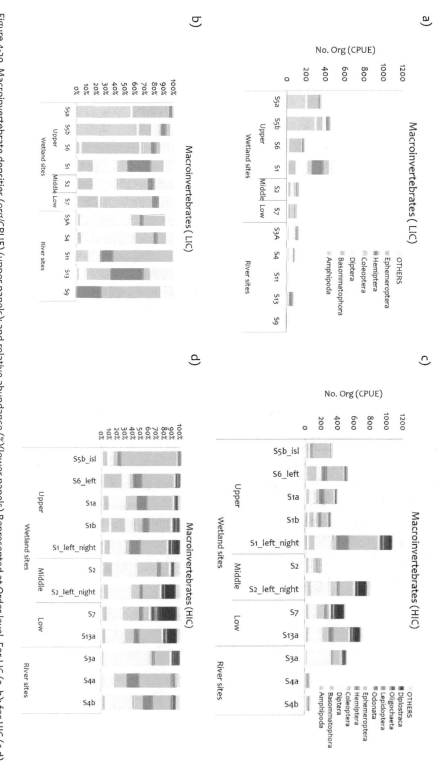

Figure 4-20 Macroinvertebrate densities (org/CPUE) (upper panels); and relative abundance (%)(lower panels) Represented at Order level. For LIC (a -b); for HIC (c-d). Samples collected at littoral vegetated areas. Same scales (a, c) to allow comparison.

4.5.1 Spatial patterns

For LIC, a higher number of distinct grouping represented by continuous lines in the cluster dendrogram were recognized (SIMPROF test: P < 0.05). River sites (S9, S11) were clearly differentiated from the rest of sites. At 40% similarity, four subgroups were clearly depicted in the 2D ordination (stress: 0.05). The dendrogram showed the higher similarity (74%) for river sites (S3a, S4), followed by two other pairs of sites (S2, S7) (S9, S11) both with similarities around 60% (Figure 4-21). For HIC, the first significant division was at 36% similarity (SIMPROF: π=5.36, p=0.1%) separating the two river sites (S4a-b) from the rest of the sites. The following significant splits at around 60% similarity formed four clusters, confirmed by the 2D ordination (stress: 0.08). Upper wetland sites (S1, S5, S6) were grouped in one cluster and middle and lower sites (S2, S7, S13) in another. Site S1 defined as upper (*a priori*), sometimes cluster with middle sites and other times with upper sites (Figure 4-21). Combined analysis of conditions illustrated a division between river and wetland sites, with river sites distributed at the left side and wetland at the right side of the ordination (Figure 4-22).

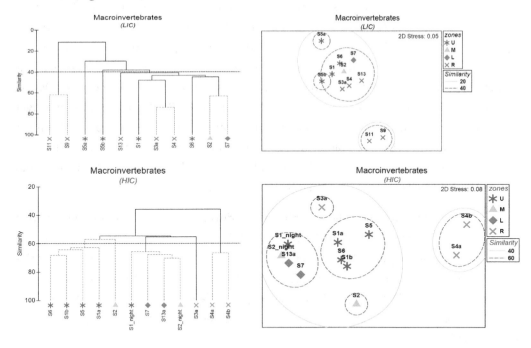

Figure 4-21 Group average cluster analysis (left column) and MDS ordinations (right column) for macroinvertebrates during LIC (low inundation conditions/upper panel) and HIC (high inundation conditions/lower panel). Based on Bray Curtis similarity matrices (built from square root transformed abundance data). Continuous lines represent the divisions for which SIMPROF test (P< 0.05, p<5% in percentage) confirmed further subdivision structure to explore. Dashed lines show the group structure with no evidence from the SIMPROF test. The x- axis shows sampling sites, symbols represent wetland areas (U: upper; M: middle; L: low). The y-axis represents Bray-Curtis similarity (%).

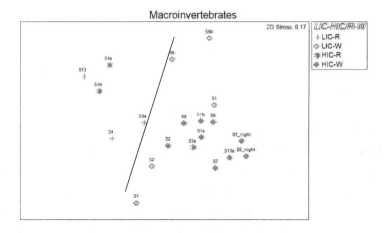

Figure 4-22 MDS ordination for Macroinvertebrates. Pooled samples of both sampling conditions (LIC: low inundation conditions; HIC: high inundation conditions). a) R: river sites; W: wetland sites..Based on Bray Curtis similarity matrices (built from square root transformed abundance data).

4.5.2 Similarities/dissimilarities

During LIC, the main contributors for intragroup similarity were: Planorbidae at the upper (43%) and middle areas (14%); Hydrophilidae and Noteridae at the lower (15% each); and Baetidae at the river (54%). Results also indicated that more species play a role in the total contribution at the middle compared with upper and river areas, where just one taxa was the main contributor to within group similarity (Table D1 and D8 -Appendix D). During HIC, Baetidae was the initial contributor to the intragroup similarity at the upper and river area. Planorbidae was also present in the upper area but with a lower contribution compared with LIC. Chironomidae increases its contribution at the upper area during HIC. At middle areas, Planorbidae and Chironomidae had similar contributions during both inundation conditions (Table D1- Appendix D).

The factor 'river and wetland', showed differences during both inundation periods with a high R ANOSIM during HIC (R=0.75, P=0.009) (Table D3 - Appendix D), evident in the visual ordinations (Figure 4-21 & Figure 4-22). Baetidae was the leading species distinguishing both areas, followed by Chironomidae and Planorbidae, all more abundant in the wetland (Table D16 - Appendix D). During LIC, Planorbidae was the leading species distinguishing river from wetland areas, and during HIC was among the top three, with higher abundances in the wetland area.

The factor 'zones' revealed significant differences during high inundation conditions (HIC) (Global R=0.49, P=0.005), with pairwise test showing differences between the upper area compared with the lower and river areas (Table D3 - Appendix D) (Figure 4-21). Baetidae, Chironomidae and Lynceidae were the main responsible taxa for the dissimilarity between upper and middle areas, although present in both areas, Baetidae had higher abundance in the upper area, while Chironomidae and Lynceidae in the middle. The contribution of the species to the average dissimilarity was homogeneous. Lynceidae and Scirtidae distinguished the lower area from the rest of the areas, due to their higher abundance in the lower area (Table D10 - Appendix D).

For the 'inundation condition" an intermediate R was found (R=0.37, P=0.004) (Table D3-Appendix D), hence a less evident division (Figure 4-22). Baetidae and Chironomidae were the first discriminating taxa for this factor, although present in both conditions, they had higher average abundances during HIC. The rest of the taxa contributing up to 50 % of the dissimilarity were also more abundant during HIC, with the exception of Hyalellidae (higher abundance in LIC) (Table D19 -Appendix D).

4.6 Fish

The fish assemblage was represented by 22 species belonging to 11 families and 5 orders during LIC, with a similar number of taxa collected during HIC: 21 species belonging to 10 families and 5 orders. Littoral zones were sampled during both inundation periods, four sites during LIC: S1, S2, S7 (wetland), and S3a (area of the river inflow), and 16 sites during HIC. During both periods, the most representative family was Characidae comprising 89% and 87% of the total abundance for LIC and HIC, respectively. *Astyanax festae* was the dominant species for both periods comprising 44 and 39 % of the total number of specimens collected, for LIC and HIC, respectively. The second dominant species was *Rhoadsia altipinna* (28%) during LIC and *Landonia latidens* (20%) during HIC. Densities ranged from 395 to 746 CPUE during LIC, with the maximum in the lower wetland site (S7) (Figure 4-23a). *Astyanax festae* was especially abundant in wetland sites while *Rhoadsia altipinna* was in the river inflow site. During HIC, densities ranged from 12 to 329 specimens. The highest density was collected in the middle wetland area (S2) during night sampling. Lower densities were collected in the upper wetland sites S5 and S6, and river inflow area (Figure 4-23b). During the campaign in March 2012 (HIC), the seine net used was smaller (6m^2) than the one used during February 2011 (30m^2) (LIC). Hence, the lower values of HIC compared with LIC.

a)

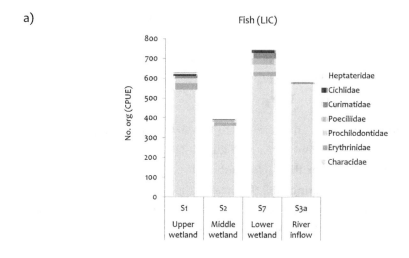

b)

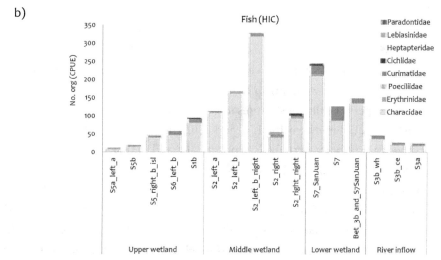

Figure 4-23   Fish densities (org/CPUE) (upper panels); and relative abundance (%)(lower panels). Represented at Family level. Samples collected at littoral vegetated areas. LIC (a); HIC (b). Families with abundance below 1% not represented.

4.6.1 Spatial patterns

The SIMPROF test did not provide evidence of significant clusters of fish communities during HIC. Six groups were formed when an arbitrary threshold of 50% was applied. The 2D ordination shows that the sites were grouped mainly in two central clusters (stress: 0.15), one of them including river and upper wetland sites and the other middle and lower sites (similar to the macroinvertebrate assemblage) (Figure 4-24). Combined analysis showed a separation between conditions with HIC samples located at the low part of the representation. Furthermore, inflows sites appear to group closer to upper wetland sites (S5 and S6) (Figure 4-25).

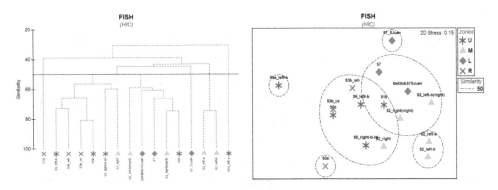

Figure 4-24 Group average cluster analysis (left column) and MDS ordinations (right column) for littoral fish during HIC (high inundation conditions). Based on Bray Curtis similarity matrices (built from square root transformed abundance data). Continuous lines represent the divisions for which SIMPROF test (P< 0.05, p<5% in percentage) confirmed further subdivision structure to explore. Dashed lines show the group structure with no evidence from the SIMPROF test. The x- axis shows sampling sites, symbols represent wetland areas (U: upper; M: middle; L: low). The y-axis represents Bray-Curtis similarity (%).

4.6.2 Similarities/dissimilarities

During HIC, *Astyanax festae* was the major species for upper and middle areas with contributions of over 30%. *Landonia latidens* was the second contributor for the middle area, and *Hemibrycon polyodon* typified the lower area with a high contribution of 28% (Table D1- Appendix D), and high Sim/SD ratio (Table D9 - Appendix D).

For the factor 'river and wetland' no clear differences were found during HIC (R=0.25, P= 0.05) (Table D3- Appendix D). During LIC, *Rhoadsia altipinna* was the top discriminator due to its high abundance in the river area, followed by *A. festae* due to its higher abundance in the wetland. During HIC and combined conditions *Astyanax festae* was the top species followed by *Landonia latidens*, since both species more abundant in the wetland area than in the river.

For the factor 'zones' significant differences among zones during HIC were found (R=0.37, P=0.003), substantiated with the spatial representations (Figure 4-24). Pairwise tests revealed that the middle area differed from all other areas (Table D3 - Appendix D). During HIC, *L. latidens* was the main contributor to the dissimilarity between the middle and the rest of the areas, because of its higher abundance in the middle (Table D11 - Appendix D). During LIC, *A. festae* was the main contributor to the dissimilarity between the upper and middle area. *Rhoadsia altipinna* typified the river area, distinguishing it from the rest of the areas. A decreasing gradient in abundance was observed for *Rhoadsia altipinna* from the river towards the upper area (Table D11 - Appendix D).

For the 'inundation condition' factor, an intermediate value for R was found (R=0.42, P= 0.012) (Table D3 - Appendix D) as depicted in Figure 4-25. *Rhoadsia altipinna* was the distinguishing species differentiating both conditions, being 4 times higher during LIC (Table D20 - Appendix D). *A.festae*, a common species in the wetland had lower abundances during HIC; probably attributed to an increase in predation during the periods of maximum inundation. *H. polyodon* was only collected during HIC, while other species with lower contributions to the dissimilarity were only collected during LIC. Overall, the majority of species were present during both conditions, with some variation in abundances. (Table D20 - Appendix D).

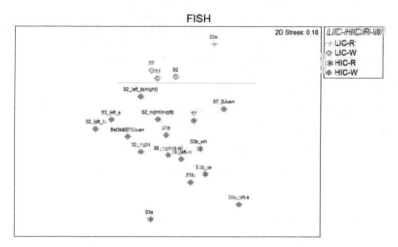

Figure 4-25 MDS ordination for littoral Fish. Pooled samples of both sampling conditions (LIC: low inundation conditions; HIC: high inundation conditions). a) R: river sites; W: wetland sites. Based on Bray Curtis similarity matrices (built from square root transformed abundance data).

An overall higher abundance of the species was observed during LIC compared to HIC (Table D16 a & b - Appendix D). However, this higher abundance was due to the use of a larger net during HIC ($30m^2$) compared with the one of HIC ($6m^2$). This was evident with *A. festae* that during LIC had an average abundance of 17.1 compared to 5.9 during HIC. However, if a division by 5 is made to LIC abundances (to make both conditions comparable), the resulting average abundance will be 3.4 which is then lower than the one of HIC; indicating that higher abundances of this species can be caught during HIC conditions (Table D16 - Appendix D).

4.7 Summary of similarities

A summary of biotic community similarities for each wetland zone showed higher values of similarities during HIC for three communities: phytoplankton bottle, zooplankton and macroinvertebrates as detailed in column SIM (Ave) in Table D1 - Appendix D, and Figure 4-26. This finding suggests a possible homogenization of the habitat conditions when the wetland is at its maximum inundation capacity. Phytoplankton collected with 60 μm net was the only assemblage showing lower similarities during HIC compared with LIC. Overall, the analysis revealed an increase in the number of species indicative of intragroup similarity from LIC to HIC for zooplankton and macroinvertebrates communities. For both phytoplankton communities, there was no discernible pattern, since some areas showed an increase and others a decrease in the number of taxa (Table D1 in Appendix D). Figure 4-26 presents an overall range of similarities for each thropic group, showing an increasing trend from lower to higher trophic groups, and thus suggesting that higher trophic groups were more similar in their community structure than lower ones. The explanatory results of this analysis are presented in the Appendix D- Table D1 and Tables D5 to D9.

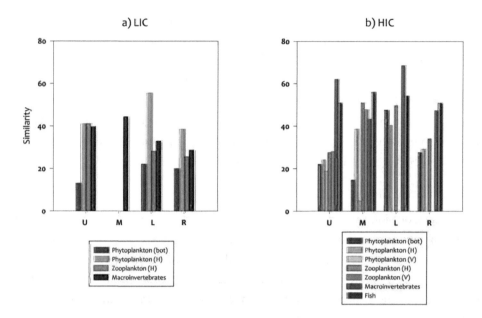

Figure 4-26 Within area/zones similarities for each biotic assemblage, from SIMPER results based on abundance data (square root transformed). a) LIC: low inundation conditions; b) HIC: High inundation conditions. ZONES (U: upper, M: middle, L: lower, R: river)

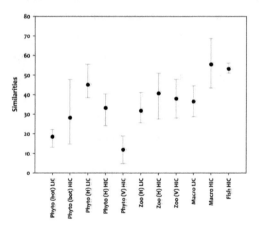

Figure 4-27 Ranges of within area similarities for each biotic assemblage (Range AvSIM) including all areas/ZONES. From SIMPER results based on abundance data (square root transformed). Black dots represent the average similarities (SIM Ave). H: horizontal tows; V: vertical hauls. LIC: low inundation conditions; HIC: high inundation conditions.

4.8 Summary of dissimilarities

Overall, average dissimilarities between areas were lower during HIC than during LIC for all biotic groups but fish (Table 4-1 see DISS (Ave)). The analysis revealed that the number of species responsible for the dissimilarities between areas increase in number from LIC to HIC, for all biotic groups but fish. For higher trophic groups, a small number of species explained the dissimilarities (Table 4-1). A decreasing trend of dissimilarities from lower to higher biotic groups was observed (Figure 4-28), thus, dissimilarities decrease with increasing trophic level.

Table 4-1: SIMPER results for dissimilarities (DISS) between wetland areas defined *a priori*. Biotic groups that were significant for the ANOSIM test (factor ZONES) are in bold. H: horizontal tows, V: vertical hauls. LIC: Low inundation conditions; HIC: high inundation conditions. DISS (Ave) is the average of the dissimilarities of all areas. DISS (Range): describes the minimum and maximum dissimilarity for all areas.

	Average dissimilarity between wetland areas								Sps responsible for dissimilarity (range (C=% contribution to DISS)	
BIOTIC GROUP	U & M	U & L	M & L	U & R	M & R	L & R	DISS (Ave)	DISS (Range)	C 50%	C 90%
Phyto (bot) LIC	86.5	82.9	90.7	87.6	92.1	82.7	87.1	83-92	5-8	13-23
Phyto (bot) HIC	80.7	75.6	69.6	82.3	82.8	71.7	77.1	70-83	6-12	18-35
Phyto (H) LIC	82.3	88.5	93.9	65.8	71.2	89.3	81.8	66-94	5-10	10-36
Phyto(H) HIC	92.7	88.2	76.0	75.3	78.5	72.9	80.6	73-93	5-8	21-27
Phyto (V) HIC	87.0	96.3	86.2				89.8	86-96	5-11	16-35
Zoo (H) LIC	78.0	59.4	73.7	68.0	93.6	70.7	73.9	59-94	3-5	10-21
Zoo (H) HIC	61.3	63.2	44.5	72.3	73.0	79.5	65.6	45-79	5-9	23-30
Zoo (V) HIC	74.7	79.2	48.5				67.5	48-79	3-4	15-18
Macro LIC	60.3	65.0	36.7	74.5	65.6	75.0	62.9	37-75	6-9	19-33
Macro HIC	44.5	42.2	39.7	56.0	61.3	59.7	50.5	40-61	9-14	36-43
Fish LIC	22.9	26.5	36.7	56.8	46.9	54.0	40.6	23-57	3-5	12-14
Fish HIC	55.9	53.7	54.2	44.4	62.5	56.6	54.6	44-63	4	10-11

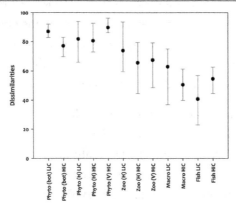

Figure 4-28 Dissimilarities (Range) for each biotic group during LIC and HIC. H: horizontal tows; V: vertical hauls. Black dots represent the DISS (Ave) as in Table 4-1.

4.9 Linking biotic assemblages with environmental variables

Although environmental conditions of a particular system cannot predict which species will dominate (Huisman et al., 2001), they can at least narrow down the probabilities that a particular 'functional group' can be present in a given habitat condition (Reynolds et al., 2002). The understanding of why certain species are more prevalent than others under a certain assemblage, provides the basis of making probabilistic predictions of community structures (Reynolds et al., 2002).

4.9.1 Low inundation conditions

For *phytoplankton* collected with the Niskin bottle, the complete set of variables (including water, sediment and biotic indices) could not explain the phytoplankton community structure, since results were not significant. When inorganic nutrients were forced (a special function in the software) in the BVSTEP routine, the contribution of these variables was minimal ($\varrho= 0.32$). Thus, nutrients alone could not explain either the structure of this community. When the analysis was performed only considering water variables, but including also two additional river sites, results were marginally significant ($\varrho=0.55$; P=0.07) selecting as main driving variables: Alkalinity, pH, DO (Table 4-2). This subset was selected as input for the LINKTREE analysis (Figure 4-29a).

For *zooplankton* assemblage collected with horizontal tows, the variables selected by the BVSTEP routine showed organic matter in sediments, followed by the inorganic fraction of nitrogen in water, diversity of phytoplankton (collected by horizontal tows) and total nitrogen in water, as possible variables structuring this community. Although the correlation was high ($\varrho= 0.76$), results were not significant (P= 0.18) (Table 4-2). The inclusion of these variables in the LINKTREE routine generated one significant split (A) that separated river site S11 from all the others on the basis of its lower phytoplankton diversity and organic matter in sediment but higher total nitrogen. These three alternative descriptors defined the same split, with a high R=0.99 displayed on the y-axis scale at B%=99 (Figure 4-29b).

For the *macroinvertebrate* assemblage, four variables were selected as main drivers: diversity of phytoplankton (bot), richness of phytoplankton (net), total suspended solids (TSS), and NH4_N in sediments. Diversity of phytoplankton-bot revealed a strong relationship, attaining alone a high correlation ($\varrho= 0.623$), compared to the overall one ($\varrho=0.77$; P=0.049) (Table 4-2). When BVSTEP was run only with water and sediment variables, results were not significant, indicating that abiotic variables alone could not explain the patterns of the macroinvertebrate community. The inclusion of the selected drivers in the LINKTREE routine generated an initial split (A) that defined phytoplankton diversity as the top variable discriminating the groupings (Figure 4-29c), split that separated river sites (S9, S11) from all other sites due to its low diversity (<0.9) with a strong ANOSIM R= 1, displayed at B%=100. Sites with phyto diversity (> 1.8) continued to a second split (B) determined by the concentrations of ammonium in sediments that separated the lower site S7 (>0.8 NH4_N mg/kg) from the remaining sites (< 0.75 NH4_N mg/kg). The third split (C) was determined again by phyto diversity. Sites with phyto diversity < 2.6 were divided based on the concentration of TSS, which clearly separated river from wetland sites (split D). With this assemblage, an explicative step-by-step subdivision was observed (Figure 4-29c).

Table 4-2 BVSTEP AND RELATE ROUTINES. a) BVSTEP routine results: Environmental driving variables for each biotic assemblage during low inundation conditions (LIC). Resemblance measure based on Euclidean distances. Driving variables appeared in order of importance (the first variable has the higher weight). b) RELATE routine tested matching between Biotic matrix (based on Bray Curtis similarity) and Environmental matrix (built with variables selected by BVSTEP routine and based on Euclidean distances). Both routines (a & b) based on Spearman Rank correlation method (ρ); Significance level (P) and 999 permutations.

Biotic Group		a) BVSTEP		b) RELATE	
	Driving variables	(ρ)	P	(ρ)	P
Phytoplankton (bot) (all sites only water variables)	Alkalin, pH, DO	0.55	0.07*	0.55	0.001
Zooplankton(H)	Org_mat_sed, N_Inorg, Div_phytonet, N_total	0.76	0.18*	0.76	0.002
Macroinvertebrates	Div_phytobot, Rich_phytonet, TSS, NH4_N(sed)	0.77	0.049	0.77	0.001

*: Spearman Rank Correlation coefficient (ρ) no significant when P >0.05, thus, no significant rank correlation between the selected driving variables and the response biotic matrix. **bot:** collected with Niskin bottle, **H:** horizontal tows. **Div:** diversity; **Rich:** richness

a)

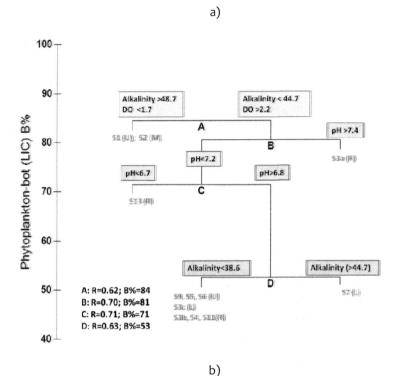

b)

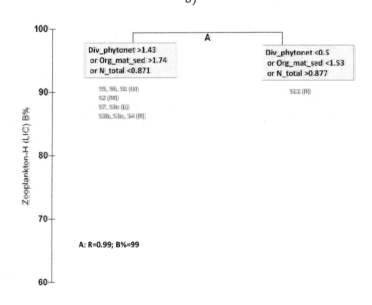

c)

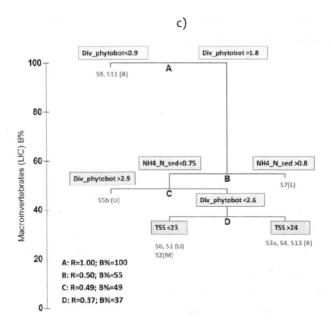

Figure 4-29 Linkage tree analysis (LINKTREE) for: a) Phytoplankton (collected with Niskin bottle); b) Zooplankton H (horizontal tows) and macroinvertebrates during low inundation conditions (LIC). The divisive clustering of sampling sites are driven by the explained thresholds of the environmental variables (with SIMPROF test P< 0.05). R: ANOSIM R statistic provides a measure of the degree of separation between 2 groups. B%: is the absolute subgroup separation, in relation to the maximum separation of the first split.

4.9.2 High inundation conditions

When the analysis for *phytoplankton* collected with Niskin bottle was run with the full set of environmental variables, results were not significant (ϱ=0.54; P=0.11), and forcing inorganic nutrients gave a minimal contribution (ϱ= 0.15), on this overall correlation. Thus, nutrients alone showed no indication of having influence on the structure of this community. When only water variables were included in the analysis, results were significant (ϱ=0.53; P=0.04), selecting turbidity, temperature, TSS, COD, Sulphides and Silicates (Table D4 - Appendix D).

For phytoplankton (vertical hauls), water and sediment variables did not explain the community pattern, even when nutrient variables were forced in the analysis. When biotic indices were gradually added to the analysis, results approached marginal significance (ϱ=0.65; P=0.08), and a subset integrated by abundance of zooplankton and macroinvertebrates, silt, and depth was selected (Table D4 - Appendix D). The inclusion of the four variables selected by BVSTEP in the LINKTREE routine provided with only one significant split (A) defined by zooplankton abundance (R=0.86; B% =94). Site S2 (sampled at night and in vegetation) separated from all the other sites due to its maximum zooplankton abundance (> 93050 org/m³); while zooplankton densities < 56575 org/ m³ described the rest of sites (Figure D1a - Appendix D).

For zooplankton collected with *horizontal* tows, *when* all variables were selected for the analysis, resulting potential drivers were total solids, conductivity, organic nitrogen, sand content and three macroinvertebrates indices (ϱ=0.90; P=0.001). When the analysis considered only water variables, resulting drivers were total solids, total nitrogen, pH, and the organic forms of nitrogen and phosphorus (ϱ=0.82; P=0.015) (Table D4 - Appendix D). In both cases, total solids was the top discriminating variable that best groups the sites. The LINKTREE was built based on the first set of selected BVSTEP variables, producing one significant split explained by thresholds of total solids and macroinvertebrates indices. The sites were separated in two groups with a high ANOSIM R=0.91. The left group included wetland sites, and the right the river sites plus two upper wetland sites (probably grouped together also due to the similar type of inflow pattern) (Figure D1b - Appendix D).

For *zooplankton* collected with vertical hauls, temperature was the single abiotic variable that best group the sites (ϱ=0.43) in combination with silicates, water depth, phosphates, organic nitrogen, richness of phytoplankton (ϱ=0.62; P=0.05) (Table D4 - Appendix D). The input of these variables in the LINKTREE produced two significant splits that selected temperature and depth as explicative variables. The first one (A) separated the upper wetland sites (S5, S6) from the rest of sites based on their lower temperatures (< 28.3). (R=0.71; B%=89). A second split (B) divided the rest of the sites based on depth thresholds (<6 and >7m), but with a lower ANOSIM (R=0.41; B%=56) (Figure D1c -Appendix D).

For the *macroinvertebrate* community, two different approaches were analyzed: one including river and wetland sites, and another only wetland sites. In the first approach, water depth was selected as a key driver (ϱ=0.63); contributing to the overall optimum attained in conjunction with zooplankton diversity, organic nitrogen in sediments and water (ϱ=0.8; P=0.02) (Table D4 - Appendix D). The selection of depth was attributed to the difference in depth values of the river site S4 compared to the rest of the sites. The second approach excluded the river site, selecting total phosphorus as top environmental driver (ϱ=0.45), and BOD as the second. Organic nitrogen and zooplankton diversity were selected as in the first approach, whereas organic nitrogen (in sediments) was replaced by total nitrogen in sediments (ϱ=0.78; P=0.02) (Table D4 - Appendix D). A first LINKTREE used as input the variables selected in the first BVSTEP approach, resulting in two significant divisions. A first division (A) distinguished river sites (S4a, S4b) from the rest of sites (all wetland sites) due to their higher depth, zooplankton diversity values but lower organic nitrogen (R=0.98; B%=99). The second division (B) separated the wetland sites based on their organic nitrogen content in water (R=0.58; B%=54). Higher concentrations (> 1.95 mg/l) grouped middle and lower wetland sites, together with one upper site (Split B-right). Lower organic content described mainly the upper sites (Split B-left) (Figure D2a - Appendix D). A second LINKTREE with the variables selected by the second approach (only wetland sites), resulted in one significant split explained only by the content of organic nitrogen in water and wetland sites were divided exactly as in split (B) of the first approach.

For the *littoral fish* community, the full set of environmental variables was tested as a first step. From this analysis, an intermediate correlation was found (ϱ=0.56), with a driving subset integrated by zooplankton and macroinvertebrates indices, BOD, organic matter and temperature in sediments. However, results could not be confirmed as significant (P=0.07). Since sediment variables appeared as potential drivers, a second step was to perform the analysis only considering sediment variables, and results were significant (ϱ=0.46; P=0.03) confirming again the selection of temperature and organic matter in sediments. A test including only water variables was also performed, but results were not significant (Table D4 - Appendix D). Selected variables of the first and second approach were introduced in the LINKTREE routine separately, but in both cases, the runs resulted in empty trees. Nevertheless, since fish assemblage has a higher position in the trophic chain, it was consider important to illustrate the 'potential' divisions that this selected subset of variables produce in grouping this assemblage, thus a LINKTREE without SIMPROF test was run (D2b -Appendix D). The first split (A) was described by zooplankton richness, isolating one of the upper wetland sites (S5) from the rest of sites based on its lower value (<5). Macroinvertebrates indices determined the second split (B), separating upper and river sites based on their lower values (B-right), compared with sites located in middle and lower wetland areas (B-left). A third division (C) was driven by BOD and macroinvertebrate richness; grouping middle sites due to their higher BOD measures (>3.6) but lower macro richness (<28) (C-left). The rest of the sites (C-right) were grouped on the basis of lower BOD but higher macroinvertebrate richness. Split (D) showed four alternative descriptors. with temperature and organic matter in sediments separating middle wetland sites based on their higher values of temperature (>25.2) and organic matter (>28.9%) (D-left), from the lower wetland sites (D-right) (Figure D2b - Appendix D).

4.10 Discussions

4.10.1 Spatial patterns

In AdM wetland, the distribution patterns of the biotic communities observed from the clustering and ordination analysis showed that river and wetland sites usually clustered separately. However, biotic patterns were not as obvious as those of the environmental variables (Chapter 3). During LIC, similarity levels that produce these two main clusters (river, wetland) started generally around 20% for all communities; with significant splits and low stress values, indicating good separation of the samples. The more specific factor 'zones' grouped upper wetland sites with river inflow sites; although not as clear as observed for the environmental variables during LIC. During HIC, the separation river-wetland was clear for zooplankton and macroinvertebrates, both communities with significant splits and low stress values. Separation was less evident for phytoplankton and fish communities, suggesting an increase in the homogeneity of the system when the wetland has a larger inundated area (at least for these two assemblages).

Inspection of pelagic-littoral patterns revealed that the similarities at which initial splits occurred for planktonic pelagic (limnetic) communities were always lower than those of littoral communities. Thus, initial clustering divisions for planktonic pelagic communities were generally between 10 and 20% similarity, while for littoral communities were between 20 and 40% similarity. This suggests that littoral communities are more similar than pelagic ones, probably due to their more specific zonation. On the other hand, pelagic communities are driven by the flow and therefore experience more mixing. Overall, from all the communities evaluated, macroinvertebrates had more significant divisions during both inundation conditions; but particularly during LIC, suggesting stronger zonation associated with low inundation area. ANOSIM test confirmed some of these ordination patterns and additionally suggested other differentiations that were not obvious with the clustering and ordination. During LIC, phytoplankton (horizontal) showed differences between 'zones', and macroinvertebrates between 'location' (river/wetland). During HIC, both plankton communities were differentiated for factors littoral/pelagic, and sampling effort, but with a moderate correlation. Furthermore, phytoplankton was significant for 'zones' and zooplankton for 'location' (river/wetland). Littoral communities (macroinvertebrates and fish), showed differences for 'zones' with a moderate R. Macroinvertebrates were also differentiated by 'location' (river/wetland) with a high correlation.

The combination of both sets of samples (LIC + HIC) for the ordination analysis showed that phytoplankton collected with bottle and horizontal tows, zooplankton, and fish showed a separation between inundation conditions, while macroinvertebrates separated according to the 'location' river/wetland. ANOSIM also indicated differences when LIC+HIC samples were combined. Phytoplankton (bottle) and phytoplankton (horizontal) showed moderate and low differences for the factor 'condition', suggesting overlap and thus a relatively constant structure for this assemblage despite inundation conditions. The other communities showed also moderate differences for 'condition', being more important for zooplankton. Phytoplankton collected with the Niskin bottle showed differences for the factor 'location' (river/wetland), probably due to an increase in the number of samples, since they did not appear when conditions were analyzed separately. Zooplankton and fish showed low differences and macroinvertebrates moderate ones. Therefore, a division river/wetland was more evident for some groups. Overall, all the communities showed overlap. The factor 'zones' was low for phytoplankton (horizontal) and fish, and intermediate for macroinvertebrates (all R < 0.4); while zooplankton did not show differences regarding zonation. Overall, cluster and ordination techniques were useful to identify general patterns of the biotic communities, while ANOSIM was useful for those factors that were not visibly with the previous techniques.

4.10.2 Typical species and ecological traits

Typical species and ecological traits from SIMPER analysis showed that Cryptophyceae (Cryptomonas), Euglenophyceae (*Trachelomonas*), Cyanophyceae (*Oscillatoria*) and Fragilariophyceae (*Synedra*) were the main taxa of the phytoplankton community collected with Niskin bottle during LIC. During HIC, Cryptomonas was the main species for all AdM wetland areas. At inflow locations, few species dominated compared with the other wetland areas where a higher number of species was observed, probably due to the more stable characteristics (e.g. lower velocities) of these areas. Cryptomonad algae are large nanoplankton flagellates well adapted to live in a wide-range of habitats, usually small-enriched lakes, able to grow at low intensities and tolerate wide variations in phosphorus concentrations (Ilmavirta, 1988; Reynolds, 2006; Reynolds et al., 2002; Wetzel, 2001d). In temperate areas, higher densities have been reported 2 m below the oxycline where oxygen levels are low (< 1mg/l), and light was limited, while nitrogen and phosphorus were highly available (Gervais, 1998; Sandgren, 1988). In AdM, *Cryptomonas* was also present at the bottom, but densities at surface level where five folds higher that in the bottom. The dominance of *Cryptomonas* in AdM could be attributed to the continuous nutrients contribution from the main inflow (Nuevo River), as in other freshwater systems where an increase of nitrates in circumneutral pH values have been related to higher cryptophytes biomass (Ilmavirta, 1988; Jones and Ilmavirta, 1988). Cryptomonad species are known to have a high nutritional quality and are highly susceptible to grazers (cladocerans, calanoid copepods and some rotifers) (Reynolds, 1995; Wetzel, 2001d). However, the predation rates on them are not high enough to diminish their population significantly (Gervais, 1998; Pedrós-Alió et al., 1995). The high wetland hydrodynamics at the lower areas of AdM is probably limiting this predation efficiency. In addition, the capacity of Cryptomonas to have high turnover rates (Reynolds, 2006; Wetzel, 2001d) could be another factor for its dominance in AdM. The presence throughout the wetland of *Trachelomonas* during LIC is probably related to the higher organic input into the system during the initial inundation periods of the wetland, since this species characterize shallow mesotrophic systems rich in organic matter (Reynolds et al., 2002; Wetzel, 2001d). *Synedra* was important at the wetland inflows during LIC. This genus is typical of shallow, enriched turbid waters, including rivers and are sensitive to nutrient depletion (Reynolds, 2006), but tolerant to nutrient and organic pollution (Bellinger and Sigee, 2010). Thus, the presence of *Synedra* at the wetland inflow could be associated to the constant flushing that occurs in this area and promotes the constant input of nutrients. *Nitzchia* also important in AdM together with *Synedra* are small celled and fast growing diatoms, characteristics of shallow, well-ventilated, enriched and turbid waters including rivers with tolerance to flushing, but sensitive to nutrient depletion (Reynolds et al., 2002).

For phytoplankton (net, algae community > 60 microns), a pennate diatoms, *Fragilaria longissima* was the typical species for the wetland inflows during both conditions, with higher average abundances during LIC. Although *Fragilaria* species are intermediate in maximal reproductive rates (Sommer, 1989). The epipelic community of several shallow lowland lakes has been found to be dominated by *Fragilaria* species (Bellinger and Sigee, 2010). Their prevalence in AdM could be attributed to their lower vulnerability to grazing compared to smaller taxa, their wide tolerance to nutrient concentrations, ability to grow fast with high nutrient availability, and capacity to avoid sinking when in colonies with more than 8 cells (Bellinger and Sigee, 2010; Padisák et al., 2003; Ptacnik et al., 2008; Salmaso et al., 2003).

Melosira granulata (*Aulacoseira granulata*) characterized the middle and lower AdM wetland areas during HIC, while during LIC *Melosira* and *Microcystis* typified the lower area. *Microcystis spp* is a slow growing, strongly K-selected, biomass-conserving S-strategist species characteristic of low latitude eutrophic lakes and tolerant to high insolation (Reynolds, 2006; Reynolds et al., 2002). However, the importance of *Microcystis spp* at the lower wetland area (subject to constant inflows and dynamics), contrasts with their functional ecology which has been described as sensitive to flushing (Reynolds et al., 2002). The other dominant species *Melosira granulata* is a widely distributed diatom, reported in high and low latitudes. *A. granulata* is frequent throughout the year in one of the lakes in Ireland (Salmaso et al., 2003) and has been described as a characteristic species of nutrient- enriched lakes with less diverse diatom assemblages (Leira et al., 2009).

Aulacoseira spp has been classified together with *Fragilaria spp* as a 'P-group' species, describing the eutrophic epilimnia of freshwater systems. Both species are present in shallow systems and large mesotrophic lakes at low latitudes, are tolerant to mild light and sensitive to silica depletion and stratification, thus, they have a strong dependence to physical mixing, requiring a continuous mixed layer of 2-3 m (Huszar et al., 2000; Reynolds, 2006; Reynolds et al., 2002). Early studies described *M. granulata* as consistently found in water with average silica content of 13 mg/l (Kilham, 1971), concentrations typical in AdM system. In temperate lakes, both species have been reported during summer and their growth has been attributed to the deeper mixing that promotes renewed phases of diatom abundance, given that silica remains available (Reynolds, 2006; Sommer, 1986). In other tropical systems, diatoms such as *Nitzschia acicularis*, also reported at AdM inflows, and other species of the genus *Aulacoseira* are abundant during periods of water column mixing in tropical Lake Victoria (Lung'Ayia et al., 2000). The prevalence of *F. longissima* and *A. granulata* in AdM confirms that this wetland is subject to continuous water mixing and not silicate-limited. Furthermore, the inter-annual recurrence of algae species in AdM suggests a high level of inter-annual constancy with environmental conditions recapitulating somehow each year (Reynolds, 2006). Nevertheless, the ability to predict the winners of multispecies competition 'dominant species' has not been possible yet (Huisman et al., 2001).

A shift in zooplankton community was evident between both inundation periods. Rotifers (*Lecane sp.* and *Platyias quadricornis*), and Protozoa (Rhizopods: *Arcella sp.* and *Difflugia sp.*) typified LIC, while macro-zooplankters (cladocerans and copepods) dominated HIC. The dominance of smaller zooplankters as rotifers and protozoans during LIC suggests that the larger specimens of macrozooplankton were probably subject to high predation during the first inundation periods of the wetland, although there was no independence evidence of higher predation rates, such as fish gut analysis, or higher densities of fish. Several studies associate predation and trophic status to the prevalence of these smaller zooplankters. High relative biomass of protozoan grazers and small zooplankton over larger zooplankton with high fish predation has been reported in temperate (Auer et al., 2004) and subtropical lakes (Havens et al., 2007). Auer et al., (2004) found that protozoans biomass increases with trophic status, similarly, a shift in dominance from macro to microzooplankton with increasing trophic state has been found in subtropical lakes. Although total zooplankton biomass showed a significant relationship with trophic status in these subtropical lakes, rotifers and nauplii showed the strongest relation to trophic status (Chlorophyll-a and total phosphorus), whereas cladocerans and copepods a weaker one (Havens et al., 2007). In contrast, a study of 81 shallow lakes in Europe determined that although the total zooplankton biomass was positively related to total phosphorus, not all taxa respond equal: rotifers biomass did not respond to changes in total phosphorus, while large cladocerans and cyclopoids responded positively (Gyllström et al., 2005). The AdM appears to be a system with higher productivity at the initial inundation periods (LIC) supporting dominance of protozoans and rotifers, but as the wet season (HIC) progresses, higher dominance of macrozooplankters was evident. However, this pattern could not be confirmed by the N, P and chlorophyll-a concentrations measured during both conditions.

Concentrations of total phosphorus were similar during both conditions but total nitrogen and chlorophyll-a were slightly higher during HIC. Therefore, monthly and seasonal sampling may clarify the occurrence or not of this shift in the components of zooplankton community.

During HIC, *Chydorus sphaericus* was the dominant species at the three AdM wetland areas, whilst rotifers and protozoans were rare. *C. sphaericus* is a small cladoceran common in temperate lakes, with a faster development than other chydorids, abundant in littoral areas, able to develop across a range of temperatures, and able to feed on a variety of food items (de Eyto and Irvine, 2001; de Eyto et al., 2002; Havens, 1991). In the AdM, *C. sphaericus* was found common in pelagic zones and littoral vegetated areas, but especially abundant in the littoral areas associated with macrophytes. The commonness of chydorids assemblages has been associated with trophic status, predation, and climate. The association of *C. sphaericus* to eutrophication related to increasing chlorophyll-a has been found in several studies, with a shift from diverse chydorids assemblages at littoral areas to monospecific assemblages of *C. sphaericus* in the open water (De Eyto et al., 2003). In AdM, although *Chydorus* dominated, it was found in association with other species of

Chydorids (e.g. *Leydigia cf leydigii, Alona sp.*) as well as other cladocerans, thus suggesting that high eutrophic conditions are not occurring in AdM yet.

A number of studies determined that high fish predation pressure leads to a shift to smaller zooplankton species (Auer et al., 2004; Hansson et al., 2004; Irvine et al., 1989). Studies on zooplankton community structure in shallow lakes from different climate zones associated them to the littoral fish community and the role of aquatic plants. For instance, subtropical lakes in Uruguay were characterized by small-bodied zooplankters, and temperate lakes in Denmark by large-bodied ones. This prevalence of smaller size zooplankton in lower latitudes was most likely a consequence of the stronger predation that planktivorous fish and invertebrates exert over zooplankton in subtropical lakes, supporting the hypothesis that higher predation pressure in warmer lakes is the main factor shaping the composition of cladocerans communities (Meerhoff et al., 2007a; Meerhoff et al., 2007b; Teixeira-de Mello et al., 2009). Similarly, in a study of subtropical lakes in Florida, it was also found that fish predation limited the crustacean zooplankton biomass (Havens et al., 2009).The stronger predation by fish in the subtropics has been attributed to the higher fish densities, richness, trophic diversity with predominance of omnivorous present in warmer lakes, and more frequent reproduction, compared with temperate ones (Teixeira-de Mello et al., 2009).

The typifying taxa of macroinvertebrates during LIC were Planorbidae, Hydrophilidae, and Noteridae at the wetland areas, and Baetidae at the river inflow. During HIC, Baetidae and Chironomidae increased and Planorbidae decreased in the upper AdM wetland areas, probably due to the higher velocities occurring during HIC in the upper area. Planorbidae populations have been described as affected by flooding associated with heavy rainfall, experiencing fluctuations in their populations due to changes in water flows and water levels (Standley et al., 2012; Woolhouse, 1992; Woolhouse and Chandiwana, 1990). At middle areas of AdM, the similar contributions of Planorbidae and Chironomidae during both inundation conditions suggested more stable conditions in this area despite the inundation condition. Chironomidae and Baetidae are common and abundant, and found to dominate tropical systems (Jacobsen, 2008; Kasangaki et al., 2008; Kibichii et al., 2007; Masese et al., 2009), in a previous research in the AdM (Van den Bossche, 2009); the Chaguana watershed (Dominguez-Granda et al., 2011); Quevedo-Vinces River (Alvarez, 2007). Planorbidae, a cosmopolitan snail family present in the tropics (Jacobsen, 2008).

During both conditions (LIC and HIC), the macroinvertebrate assemblage of middle area had a higher number of taxa responsible for the intragroup similarity than the upper and river areas, reflecting more stable conditions (biotic and abiotic) in this area; probably related to higher residence times. During HIC, this assemblage also had a higher number species contributing to the within area similarities compared to LIC, since during LIC one taxa group alone has more influence in the intragroup similarity suggesting dominance during low inundation periods (Appendix D -Table D8).

Littoral fish assemblages of the AdM wetland consist predominantly by small sized fish from the Characidae family (Alvarez-Mieles et al., 2013). The species of Characids are mostly small and abundant in rivers and associated habitats throughout the Neotropical region (Reis et al., 2003). In South America, Characiformes (tetras and their allies) comprise 1700 species, Siluriformes (1915 species), and Gymnotiformes (212 species) representing the 74% of all fishes in the continent (Reis et al., 2016). The north-west Pacific basins from Colombia, Ecuador, and north of Perú have a very high annual rainfall, and their ichthyofauna exhibits a high degree of endemism (Reis et al., 2016). Characiforms (tetras and related fishes), were found as the nine top ranked species in overall abundance, reflecting their prominent representation in the South American ichthyofauna (Winemiller, 1996).

Characids are an important source of food for higher trophic levels (top fish predators that have a value for local communities) and important seed dispersers in Neotropical floodplains. *Astyanax festae* (Boulenger, 1898) was the dominant species of Characids group in AdM. *A. festae* is a small (maximum 6.9 cm), omnivorous tetra known as 'Cachuela' (Laaz et al., 2009). *A. festae* is common in the Guayas River basin, and rivers from north-western Ecuador: Chone, Portoviejo, Santiago, Esmeraldas (Barriga, 1994; Eigenmann, 1922; Glodek, 1978). *Astyanax* species in southern Brazil were also described as omnivorous, feeding mainly of vegetal matter and insects, and also associated with macrophytes that provide autochthonous food and serve as habitat for organisms preyed by *Astyanax spp* (Vilella et al., 2002). A study of feeding behaviour of four species of characids, including one species of *Astyanax spp*, evaluated the importance of both autochthonous and allochthonous sources of food depending on the species. Furthermore, a high overlap in food items consumed was found, suggesting that competition was not regulating these species (Moraes et al., 2013).

Previous studies in the AdM wetland and associated 'Guayas River basin' reported the presence of the Characidae family (Florencio, 1993; INP, 2012; Laaz et al., 2009; Prado, 2009; Prado et al., 2012). Although, the different areas of the wetland share similar fish species from the Characidae family, there were species that seem to typify middle areas where higher residence times occur. The littoral fish assemblage included both common and endemic species. At the middle and lower wetland areas endemic species like *Phenacobrycon henni*, *Landonia latidens*, *Iotabrycon praecox*, *Hyphessobrycon ecuadoriensis* were collected. Upper and river areas showed similar abundances of the major taxa which is probably related to the similar hydrodynamics that both areas share. Other species typified the lower area, which is more influenced by the river inflow, while other species characterized the river inflow.

4.10.3 Explanatory variables

Results from this study revealed that the phytoplankton community collected with Niskin bottle was associated with pH and alkalinity during low inundation conditions (LIC). Both variables are essential factors affecting phytoplankton dynamics and composition (Ptacnik et al., 2008). During high inundation conditions (HIC), the community was influenced by another set of variables that include turbidity, temperature, TSS, COD, sulphides and silicates. The influence of inorganic nutrients was not evident during both conditions, since nutrients alone could not explain the structure of this community. For phytoplankton collected with vertical hauls, water variables alone did not explain either the structure, and a gradual inclusion of sediment and biotic indices provided a better explanation of the pattern, including abundances of zooplankton and macroinvertebrates, silt and depth as potential drivers. The inclusion of biotic indices of higher trophic levels could be suggesting a top-down control mechanism in this system.

The zooplankton assemblage collected with horizontal tows during LIC revealed as potential explanatory variables a subset that included organic matter in sediments, inorganic and total nitrogen in water and phytoplankton diversity collected with horizontal tows, but was not considered as significant. During HIC, results were more conclusive and defined total solids as main explanatory variable, with a clear threshold that separated wetland sites (< 88mg/l) from upper wetland and river sites (>116 mg/l). During both conditions, total nitrogen was also selected. In the wetland, the organic fraction is the main constituent of total nitrogen, especially during HIC that organic nitrogen represented 80% of the total nitrogen, which was homogeneous along the wetland sites suggesting a stabilization of the system in terms of productivity during HIC. The association of this community to total nitrogen might be suggesting an association with phytoplankton that contains particulate organic nitrogen. For zooplankton collected with vertical hauls, temperature was the primary explanatory variable structuring the community followed by depth, which is evident from the LINKTREE outcome. Inorganic nutrients (silicates, phosphates) also played a role but to a minor extent. Association with biotic indices (richness of phytoplankton and fish) was also determined from horizontal tows, with organic nitrogen integrated into the subset of explicative variables.

Macroinvertebrate community during LIC was best described by diversity of phytoplankton (bot). The farthest located river sites were clearly separated from the wetland and closer river sites. Ammonium in sediments and total suspended solids also played a role in defining the groupings. During HIC, depth was the top descriptor for this community. Diversity of zooplankton and the organic fractions of nitrogen in water and sediment also contributed in the separation of the groups. For the fish community, biotic indices of macroinvertebrates, zooplankton combined with sediment variables (temperature and organic matter) appeared as potential drivers in structuring this assemblage, probably due to the relation that macroinvertebrates have in the processing of organic matter (Tank et al., 2010).

"In nature we never see anything isolated, but everything in connection with something else which is before it, beside it, under it and over it"

Johann Wolfgang von Goethe

5

EVALUATION OF WATER QUALITY AND PRIMARY PRODUCTION DYNAMICS

5.1 Background and scope

The use of ecological models for environmental management started back in the seventies with the development of river eutrophication models. However, during that period the knowledge of ecosystems and ecological processes was limited (Jorgensen et al., 1991). The capabilities of computer-based environmental modelling using hydroinformatics techniques were outlined by Mynett (2002), Mynett (2003). Studies in the tropics show considerable variability in these conditions. Thus, there is still plenty to learn about the functioning of tropical systems, especially regarding their energy sources, ecological processes and interactions, including the dynamics and structure of food webs (e.g. consumers supported by them) (Boyero et al., 2009). Furthermore, tropical rivers and associated floodplains and wetlands are characterized by hydrographic seasonality that drives their variability in water depths, velocities, and water chemistry. This seasonality depends mainly on hydrology alone, instead of hydrology in combination with temperature, as occurs at temperate latitudes (Lewis, 2008).

Nutrients and energy supplies for primary production development and other physical and chemical mechanisms vary according to latitude (Lewis, 1987). Many of these mechanisms have been studied in temperate environments and compared with warmer latitudes to understand the role of climate, physical-chemical variables and food web interactions (Gyllström et al., 2005; Jeppesen et al., 2005), nutrient reduction fomenting top-down control (Jeppesen et al., 2007), and climate change impacts on biotic communities and eutrophication (Jeppesen et al., 2014; Jeppesen et al., 2010; Moss et al., 2011).

Concentrations of nitrogen, phosphorus and carbon in tropical systems are important due to their regulatory role in aquatic ecosystems (Lewis, 2008) with nitrogen usually being recognized as a more limiting factor in the tropics than in temperate systems (Lewis, 1996). These nutrients sustain the growth of primary producers that in turn support consumer's populations. Main production sources for higher consumers in river-floodplain food webs are phytoplankton, periphyton and fine particulate organic matter derived from algae (Winemiller, 2004), with higher standing crops of phytoplankton reported in floodplains than in their associated rivers (Davies and Walker, 2013).

Environmental problems from oxygen balance, eutrophication, pesticides and heavy metal pollution, habitats of endangered species etc. are often being investigated through the application of numerical models (Mynett and Chen, 2004). In this chapter, the application of the ecological model DELFT-WAQ-ECO (Deltares, 2014) in Abras de Mantequilla river-wetland system is presented. The model DELFT-WAQ-ECO has been widely applied in several aquatic environments, from coastal waters (Blauw et al., 2008; Li et al., 2010; Spiteri et al., 2011) to rivers (Chen et al., 2012), tropical lagoons (Velez, 2006), and reservoirs (Smits, 2007). In this chapter, the evaluation of key variables of water chemistry, primary production and zooplankton components is described for different hydrological conditions to describe the natural variability of this river-wetland system. This chapter aims to answer the following questions:

- How important is the role of hydrodynamics in defining temporal and spatial variability of nutrients and primary production?
- What is the relation between the inflows and the different wetland areas?
- To what extent does the inflow variability play a role in defining these concentrations?
- Do the concentrations follow an inflow pattern?
- Do the concentrations of these variables change when different hydrological conditions are evaluated?

5.2 Model set up

5.2.1 Motivation for eco-model implementation

The results of the different sampling campaigns performed during this study provide an overview at specific times during the wet season. To evaluate the full seasonal dynamics of this wetland, several sampling campaigns along a complete year would be needed. Furthermore, each year will have different hydrological conditions that generate different flow magnitudes and inundation patterns (Chapter 2), thus even one entire year of sampling cannot provide a complete assessment of the natural variability of the wetland both in terms of temporal and spatial variability. Here is when ECO modelling tools become essential to evaluate the long term natural variability of a system. The model results of the wetland hydrodynamics (Chapter 2) provided the basis to build the ECO model elaborated in this chapter.

5.2.2 Model description

The ECO model is based on the open source water quality-modelling framework DELFT3D-WAQ-ECO, a mathematical model for water quality and ecology that allows the selection of a wide range of substances and processes from a quite extensive process library. ECO is basically a eutrophication model that simulates concentrations of nutrients (N, P, Si), organic matter, dissolved oxygen, and quite a few additional substances, including growth of different groups of algae (Smits, 2007). An input file including: substances, processes, coefficients, is created to define the system to be modelled including substances, processes and coefficients, grid, initial conditions, flows, and meteorological forcing (Smits and van Beek, 2013). The model solves the advection-diffusion-reaction equation on a predefined computational grid for a wide range of substances. It has two different parts working together: (i) it solves the equations for advective and diffusive transport of substances in a water body and (ii) models the water quality kinetics of chemistry, biology, and physics that determines the behaviour of substances and organisms. The model does not compute the flow itself, so it need to be connected/coupled to a hydrodynamic flow model like DELFT3D-FLOW or TELEMAC (Deltares, 2013a; Deltares, 2013d; Deltares, 2014).

The ECO model for the Abras de Mantequilla (AdM) wetland was built by using the GUI (graphical user interface) of DELFT3D-WAQ-ECO that produces the input file. The flow input was derived from the 2D-FLOW model built in Chapter 2 for several hydrological conditions. Table 5-1 presents the general characteristics of the model-set up. The simulation period was set up for one year (January-December) for all the hydrological conditions. The substance groups refer to the 4 main groups described in the GUI. Several substances and associated processes are included in these four groups. The numerical scheme '15' (a fully implicit iterative method) was selected to solve the advection-diffusion equation. The method is an iterative solver: it makes an estimate of the concentration vector (e.g. the concentration at the previous time step), and subsequently checks if this estimate is correct or not. This scheme is implicit (both in vertical and in horizontal direction) which implies that it is computationally efficient if large time steps are used (one hour in the present study) (Deltares, 2014). Initial and boundary conditions for the different substances modelled were estimated both from field measurements and from literature as explained in the following sections. The observation points correspond to the ones of the hydrodynamic model (Chapter 2) and are located at upper, middle and low wetland areas, where inflows and outflows occur. The computational time step was set at one hour, and the output time series and maps were created on a daily basis.

Table 5-1 AdM ECO Model set up

Setting	Value/description
Simulation period	One year (for all the hydrological conditions evaluated)
Substance groups*	General. Oxygen-BOD. Eutrophication, Age and fraction
Integration method	Numerical scheme 15: Fully implicit iterative method
Initial conditions	Value were estimated from field measurements and literature
Boundary conditions	Values were estimated from field measurements and literature
Observation points	Same sites as the ones of the hydrodynamic model (upper, middle, and low wetland areas), inflow and outflow
Time step	One hour
Output	Time-series: 1 day, Maps: 1 day

*: this is the general division of substances in the GUI interface of DELFT-ECO.
Each of the four divisions include several state variables (explained in the section 5.2.3)

5.2.3 Substances included in AdM eco model

The following substances / state variables were considered for the set up of the AdM ECO mode: (i) four dissolved inorganic nutrients (nitrates, phosphates, ammonium, and silicates); (ii) dead organic matter (DETRITUS) including (a) particulate fractions of organic carbon, nitrogen, phosphorus (POC, PON, POP) and (b) dissolved fractions (DOC, DON, DOP); (iii) detritus in sediments (N, P); (iv) Phytoplankton biomass (C, N, P) for four algae groups, and (v) zooplankton biomass as a forcing function. The initial and boundary conditions for these substances are presented in Table 5-2. Variables obtained as output are: total nitrogen, total phosphorus, total organic carbon, total organic nitrogen, total organic phosphorus, total phytoplankton biomass, chlorophyll-a.

The phytoplankton biomass of the four algae groups (Diatoms, Green, Blue-green and Flagellates) was simulated with the BLOOM module which is part of DELFT-WAQ. BLOOM is a multi-species algae model, based on an optimization technique that distributes the available resources in terms of nutrients and light among algae species. The model optimizes the species composition to obtain the maximum growth rate under the given conditions (Deltares, 2014). Linear programming is used as an optimization technique to determine the species composition that is best adapted to prevailing environmental conditions, in which the species compete within the constraints for available nutrients (N, P, Si), available light (energy), the maximum growth rate and the maximum mortality rate (Smits and van Beek, 2013).

Major processes in BLOOM include the production and mortality of algae biomass, the uptake and excretion of the nutrients and the production and consumption of dissolved oxygen and carbon dioxide. One of the advantages of BLOOM is that it has been validated for a wide range of both freshwater and marine environments, resulting in a database with coefficients for 13 algal groups and species for which calibration is almost not required (Deltares, 2014). Process fluxes in BLOOM are calculated based on daily averaged meteorological forcing; thus a simulation time step of 24 hours was set up for phytoplankton processes.

5.2.4 *Processes included in AdM eco model*

The processes included in the AdM ECO model were selected from the Process Library and include among others: reaeration, nitrification, denitrification, decomposition of organic matter (detritus), consumption of electron acceptors, light extinction, settling and resuspension of particulates, phytoplankton processes (growth-respiration-mortality-settling), and grazing by zooplankton (Figure 5-1). The grazing module CONSBL was activated to include 'primary consumption' in the simulations. This module models grazing of phytoplankton and detritus based on grazer biomass forcing.

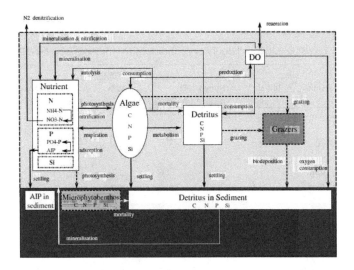

Figure 5-1 Schematic overview of variables and processes included in the ECO model. Note: Microphytobentos and AIP in sediment (not considered in the AdM ECO model set up). Source: Adopted from Blauw et al., (2008)

5.2.5 Initial conditions, boundary conditions and observation points

The initial and boundary conditions were determined based on field measurements of the two monitoring campaigns developed in the wetland during 2011 and 2012 as described in chapters 3 and 4. The initial conditions for the wetland were determined based on the field measurements that were developed inside the wetland area at sites S1, S2, S7, S3c. The values used for the boundary conditions come from the sites related to the inflows and outflows. For this purpose, field measurements of S4, S11, S3a were analyzed and used as input value for Nuevo River, the values of S5 and S6 for the Upper Chojampe inflows (El Recuerdo, AdMT1, AdMT2, Abanico T1), and the values of S13 for the wetland outflow (Figure 5-2). The ranges of these measurements were analyzed and an average value was selected as a reference input (Table 5-2). Extra nitrogen loads from agricultural activities in the area were also estimated (section 5.2.6 and nitrogen loads)were estimated and it was assumed that the nitrogen load contributes directly to the nitrate load, hence, it is added to the boundary condition for nitrate. Since there was no information for three of the small tributaries (AdM T1, AdM T2 and Abanico T1), their boundary conditions were considered to be the same as the one estimated for El Recuerdo. The observation points for the AdM ECO model correspond to the ones of the hydrodynamic model and to the sampling sites of the monitoring campaigns. At these observation points, output results for all the variables selected can be extracted from .his (time series) and .map (spatial) output files (Figure 5-2).

Table 5-2 Initial and Boundary conditions for the AdM ECO-Model

| | Units | Initial conditions | Inflows | | | | | Outflow |
			Nuevo River	El Recuerdo	AdM T1	AdM T2	Abanico T1	Nuevo River
Continuity	g/m³	1	1	1	1	1	1	1
Water Temperature	°C	28.2	26	27.9	27.9	27.9	27.9	26.5
Dissolved Oxygen	g/m³	3.0	5.6	2.6	2.6	2.6	2.6	3.9
TSS	g/m³	20.6	72.7	15.7	15.7	15.7	15.7	25.2
NUTRIENTS								
Nitrate	gN/ m³	0.11	0.32	0.34	0.34	0.34	0.34	0.34
Ammonium	gN/ m³	0.015	0.01	0.02	0.02	0.02	0.02	0.03
Phosphate	gP/ m³	0.07	0.02	0.09	0.09	0.09	0.09	0.03
Dissolved Silica	gSi/ m³	10.2	11.5	9.7	9.7	9.7	9.7	10.25
ALGAE								
Blue-green	gC/ m³	0.005	0.003	0.003	0.003	0.003	0.003	0.005
Diatoms	gC/ m³	0.08	0.06	0.06	0.06	0.06	0.06	0.17
Flagellates	gC/ m³	0.12	0.01	0.07	0.07	0.07	0.07	0.03
Greens	gC/ m³	0.033	0.001	0.01	0.01	0.01	0.01	0.03
DETRITUS								
POC1	gC/ m³	1.72	2.71	1.83	1.83	1.83	1.83	1.58
POC2	gC/ m³	1.72	2.71	1.83	1.83	1.83	1.83	1.58
DOC	gC/ m³	10.3	16.3	11.0	11.0	11.0	11.0	9.5
PON1	gN/ m³	0.18	0.09	0.11	0.11	0.11	0.11	0.15
PON2	gN/ m³	0.18	0.09	0.11	0.11	0.11	0.11	0.15
DON	gN/ m³	1.1	0.57	0.64	0.64	0.64	0.64	0.88
POP1	gP/ m³	0.012	0.021	0.002	0.002	0.002	0.002	0.012
POP2	gP/ m³	0.012	0.021	0.002	0.002	0.002	0.002	0.012
DOP	gP/ m³	0.006	0.009	0.006	0.006	0.006	0.006	0.005

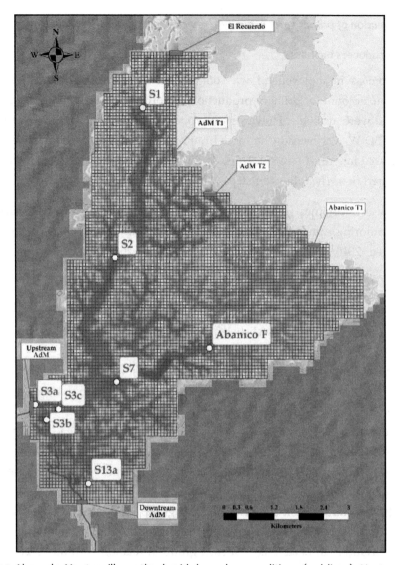

Figure 5-2 Abras de Mantequilla wetland grid, boundary conditions (red lines): Upstream AdM (Nuevo river-main inflow to the wetland), upper Chojampe (El Recuerdo, AdmT1, AdMT2, AbanicoT1), downstream AdM (wetland outflow). Observation points (white dots), for both hydrodynamic and ecological model. The observation points correspond to the sites of the sampling campaigns 2011 and 2012.

5.2.6 Estimation of primary producers, primary consumers, detritus and nitrogen loads

Primary producers biomass

Algae biomass in the wetland was one of the state variables of interest to be modelled to account for primary production. Chlorophyll-a concentrations measured in the field were available. However, chlorophyll-a cannot be used directly as input in DELFT3D-WAQ. Instead, the model requires that phytoplankton concentrations are expressed in terms of gC/m^3. Thus, an estimation of carbon concentration in algae was required first. The following steps were performed for this purpose:

1. The cell volumes (μm^3) of the different algae species collected during the monitoring campaigns were identified with different sources (Bellinger and Sigee, 2010; Jorgensen et al., 1991; Reynolds, 2006; Sicko-Goad et al., 1984; Snoeijs et al., 2002).

2. Using these identified volumes, the carbon content per cell (pgC/cell) for each species was calculated using the equations proposed by Menden-Deuer and Lessard (2000), who established a relation that differentiates diatoms and non-diatoms (Menden-Deuer and Lessard, 2000).

3. The resulting carbon content per cell and per species was related to the total abundance per species (cell/l) to compute the total carbon concentration for each species (gC/m^3) at each sampling site.

4. Subsequently, these results were classified in to the four algae groups that can be modelled in DELFTWAQ with the BLOOM module (Bluegreen, diatoms, flagellates and green algae), and the total carbon concentration for each group was obtained.

5. The resulting carbon concentration per group was related to the total carbon in algae (sum of the carbon concentrations of the four groups), to obtain the contribution in percentages of each group in relation with the total carbon in algae.

6. However, since cell densities could be underestimated, chlorophyll-a was introduced in the calculation because it is considered a better indicator of biomass (H. Los, personal comment). The percentage of each algae group (calculated in step 5), was multiplied by the chlorophyll-a concentration and then this result divided by the chlorophyll-a-carbon ratios available in the bloom.spe file of the DELFTWAQ.

7. This procedure was performed for each sampling site and for both sampling campaigns.

8. Finally, carbon concentrations for bluegreen (cyanobacteria), diatoms, flagellates and green algae were obtained and used as input for the model.

Primary consumers

Grazer biomass is imposed in the ECO model as a forcing condition with the CONSBL module (Blauw et al., 2008; Deltares, 2013b). The user needs to specify the biomass development of filter feeders (zooplankton) over the year. Based on this biomass, the grazing rate on phytoplankton and detritus is simulated by the model. Whenever the food availability is insufficient to sustain the specified biomass development, the zooplankton biomass in the model is corrected (Deltares, 2013b)

Field data of zooplankton densities was used to estimate zooplankton biomass. The transformation from org/m^3 to gC/m^3 was performed using the estimated values found in literature for the zooplankton species. A temporal function for zooplankton biomass development over the year was calculated and built based on temporal densities from a secondary source (Prado et al., 2012). Based on this biomass function, the grazing rate on phytoplankton and detritus is simulated by the model. This imposed calculated zooplankton biomass (Zoo function) is adjusted if algae and detritus are not enough to sustain this imposed biomass (Blauw et al., 2008). Processes considered in the model are: filtration, assimilation, respiration, mortality and excretion (Deltares, 2013b).

The mass fluxes caused by grazing are calculated with the following steps:

- Conversion of the biomass forcing function input to the desired units;
- Adjustment (if necessary) of the imposed grazer biomass according growth and mortality constraints;
- Calculation of the consumption rates for detritus and algae;
- Calculation of the rates of food assimilation and detritus production;
- Correction of the assimilation rates for respiration;
- Adjustment of the grazer biomass;
- Calculation of the detritus production rates according to the food availability constraints;
- Evaluation of the total conversion rates as additional output parameters; and
- Evaluation of the grazer biomass concentrations as additional output parameters.

The consumption rate of the grazers is limited by the filtration rate at low food availability and by the uptake rate at high food availability. The filtration rate and the uptake rate are equal at a certain food concentration. The total food availability is defined as the sum of the concentrations of detritus and phytoplankton groups, adjusted by a preference factor for each food source (Deltares, 2013b).

Detritus (dead organic matter)

The particulate and dissolved organic fractions of carbon, nitrogen and phosphorus were state variables to be modelled. Measured concentrations of particulate organic carbon (POC), total organic nitrogen (TON) and total organic phosphorus (TOP) were available from the sampling campaigns. The ECO model requires the concentrations of these fractions in both dissolved and particulate forms. It is known that the ratio POC:PON is quite constant in river particulate matter (8:1) (Meybeck, 1982; Wetzel, 2001a). Thus, the measured values of these fractions were also compared with this ratio to check if they were in the same order of magnitude.

The dissolved organic fractions of C, N, P were estimated using ratios and proportions from the literature. Initially, different sources were investigated to understand these ratios. Dissolved organic carbon (DOC) was estimated with a ratio DOC:POC (3:1) (Wetzel, 2001e). Dissolved organic nitrogen (DON) was estimated as 75% of the TON, and the rest 25% as PON (which corresponds to a DON:PON ratio (3:1) (Wetzel, 2001a). This calculation of DON was also checked with the average ratio DOC:DON (20:1) (Meybeck, 1982) to confirm that the values were in the same order of magnitude. Dissolved organic phosphorus (DOP) was estimated following the DOC:DOP ratio proposed for the Amazon river (Meybeck, 1982). Finally, the total concentrations of POC, PON, and POP were split in two fractions: the fast decomposing fraction POC1, PON1, POP1), and the medium slow decomposing fraction (POC2, PON2, POP2) before used as input for the model.

Nitrogen loads

The area of influence of the wetland has experienced many land use changes during the last decades. Currently, it is estimated that just 3% of the area is still covered by forest, which has been mainly replaced by short-term crops such us rice, maize and beans (Arias-Hidalgo, 2012). Therefore, it was considered necessary to include the contribution of agriculture activities in the model. An estimation of the nutrient loads generated by the use of fertilizers was developed. A study about nutrients budgets in the Guayas River basin was the basis of these calculations (Borbor-Cordova et al., 2006). Based on a land use map of the area (Arias-Hidalgo, 2012), it was estimated that rice and maize account for 27% and 59% of the total area of the Chojampe subbasin. In the analyzed area of the wetland (built grid), these crops reach 34% and 52% respectively.

The contribution from the Upper Chojampe subbasin is estimated based on the field measurements, nevertheless the load generated inside the wetland is not included since it enters the wetland as a non-point source that must be added separately. This estimation of the nutrient load in the wetland focuses on the nitrogen contributions and is based on the information from Guayas River Basin (Borbor-Cordova et al., 2006). The estimation of the nutrients loads considered the fertilizer application rate, crop yield, moisture content, and nutrient content based on dry weight. The values considered in the analysis are presented in Table 5-3.

Table 5-3 Nutrients loads estimation for the AdM ECO-Model

CROP	FERTILIZER APPLICATION RATE (kg/ha/yr)	YIELD (kg/ha/yr)	MOISTURE (%)	NUTRIENT CONTENT IN DRY CROP (g/kg)	
				NITROGEN	PHOSPHORUS
Rice	68	4200	0,5	5.5	1.3
Maize	46	3636	0,52	16.7	3.3

5.3 Model performance and verification

In order to determine the performance of the ECO model, measured values of both campaigns (2011 and 2012) were compared with the simulation outputs for both years (Figure 5-3 to Figure 5-6). The simulated period presented in this section was weekly averaged until March, since both campaigns were performed during the first trimester of the year. Measured values are displayed in the precise week that were sampled, thus from February 8th till February 11th for 2011, and from March 5th till March 8th for 2012.

5.3.1 Dissolved Oxygen and dissolved inorganic nitrogen

The model reproduces adequately the measurements of dissolved inorganic nitrogen (DIN) and dissolved oxygen, with simulations in the overall range of the measured concentrations. A better correspondence to the DIN measurements was observed for 2012 (Figure 5-3).

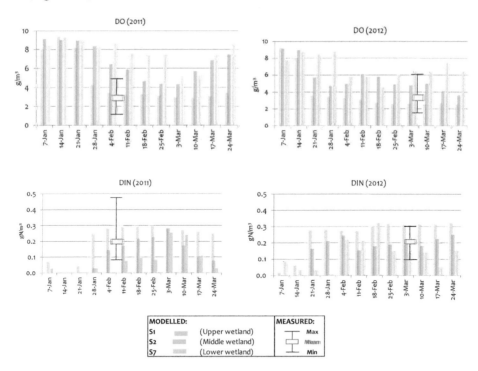

Figure 5-3 Simulated concentrations (weekly averaged) of dissolved oxygen (DO) and dissolved inorganic nitrogen (DIN) in g/m³ (equivalent to mg/l). Simulated period: January-March. Range of measured values correspond to the sampling periods: February 2011 (left) & March 2012 (right). The boxplot represents the mean, minimum and maximum values, calculated from the values measured at all wetland sites.

5.3.2 Nutrients and Chlorophyll-a

Total phosphorus (TotP) is well reproduced by the model, since simulations fall in the range of the measured values (Figure 5-5). For total nitrogen (TotN), the model did not reproduce properly the measured values, overestimating in 2011 and underestimating in 2012 (Figure 5-4). A similar pattern of TotN was observed for Chlorophyll-a, although in March 2012, the simulations were in between the minimum and maximums measured in the whole wetland area (Figure 5-6: upper plot). The spatial representation of chlorophyll-a (2012) is useful to represent the maximum chlorophyll-a values that could not be represented in the temporal plot (Figure 5-6: lower plot)

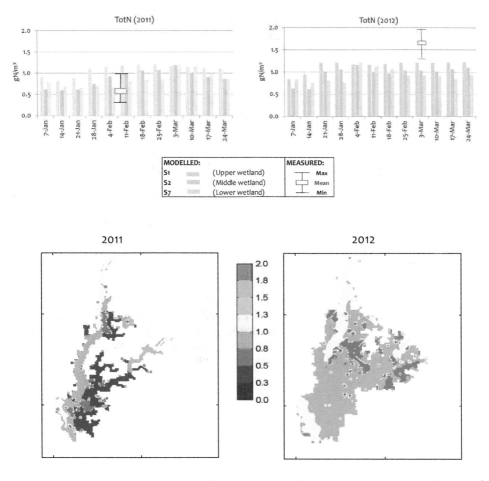

Figure 5-4   Simulated concentrations (weekly averaged) of Total Nitrogen (TotN) in g/m³ (equivalent to mg/l) (upper plots in bars). Simulated period: January-March. Measured values correspond to the sampling periods: February 2011 (left) & March 2012 (right). Lower plots: spatial representation during the sampling days of each year.

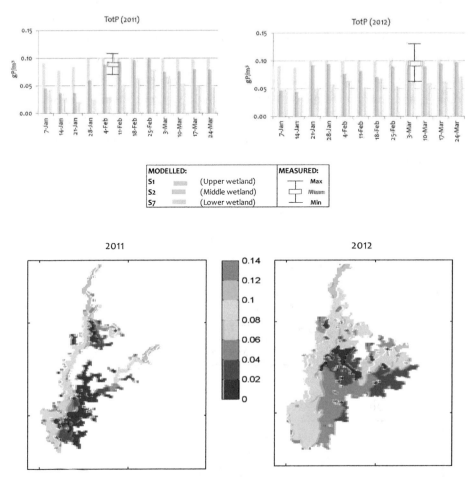

Figure 5-5   Simulated concentrations (weekly averaged) of Total Phosphorus (TotP) in g/m³ (equivalent to mg/l) (upper plots in bars). Simulated period: January-March. Range of measured values correspond to the sampling periods: February 2011 (left) & March 2012 (right). Lower plots: spatial representation during the sampling days of each year.

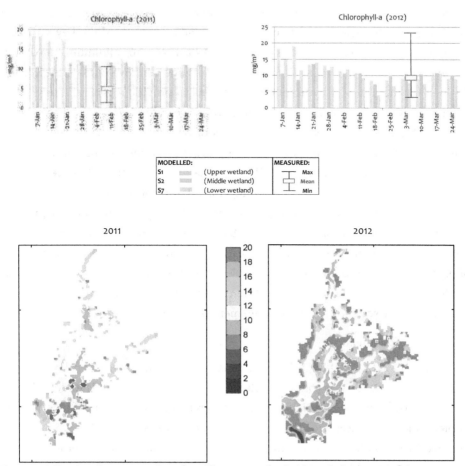

Figure 5-6 Simulated concentrations (weekly averaged) of Chlorophyll-a in mg/m³ (equivalent to μg/l) (upper plots in bars). Simulated period: January-March. Range of measured values correspond to the sampling periods: February 2011 (left) & March 2012 (right). Lower plots: spatial representation during the sampling days of each year.

Overall, the model reproduced quite well inorganic nitrogen, dissolved oxygen and total phosphorus, but has limitations to reproduce total nitrogen and chlorophyll-a at specific observation points, probably due to the complexity that involve the fate of organic components. However, spatial representations are useful to visualize those maximum values that could not be reproduced in the temporal plots, providing an overall range of the variables. Despite these limitations, the model can be used as an indication when applied for different hydrological conditions, since the aim is to evaluate the overall functioning of the wetland.

5.4 Scenarios

The temporal and spatial variability of key environmental variables was evaluated for different hydrological conditions. The assessment for each of the key variables was performed following a number of steps. As a first step, minimum, maximum and average concentrations at different wetland zones, inflows and outflow were obtained for the different hydrological conditions to provide a general overview of the entire range of variability in concentrations. Secondly, a temporal analysis for each variable was performed with time series of one-year simulation. The analysis shows a graphical comparison between the different conditions, using the time series of each hydrological condition. Thirdly, a spatial representation of concentrations was obtained. Finally, nutrient partitioning for total nitrogen and phosphorus was determined, as well as for total organic carbon. Primary production and primary consumers biomass are also illustrated. Since the wetland is a seasonal system, monthly averaged time series were used to illustrate the temporal fluctuations of the different variables. Since the aim was to assess the overall functioning of the wetland for the wet and dry season and for different scenarios, this can be considered an appropriate time scale for general management purposes.

5.4.1 Hydrological conditions

The output of DELFT3D-FLOW simulations for different hydrological years and conditions (Chapter 2) was coupled with DELFT3D-WAQ. The aim is to describe the patterns of the simulated variables under different magnitudes of inflows. In this chapter, additional scenarios have been included (Scen dry 30% smooth and ScenMAXhisto), both smoothed with a polynomial function. This chapter include the results of all the conditions detailed in Table 5-4.

Table 5-4 Different hydrological conditions and scenarios evaluated

HYDROLOGICAL CONDITIONS	DESCRIPTION
1990 (Dry)	A typical dry year selected from the period (1962-2010)
1998 (El Niño)	An extreme wet year from the period (1962-2010)
2006 (Ave)	An average year from the period (1962-2010)
MAXhisto	Scenario built with the MAXIMUM daily discharges (1962-2010)*
MINhisto	Scenario built with the MINIMUM daily discharges (1962-2010)*
Scen dry-30%smooth	Scenario built based on a dry year, then subtracted a 30% of Nuevo river inflow (due to Baba dam)
ScenMAXhisto_smooth	Scenario built based on the MAXhisto, and subsequently smoothed with a polynomial function

(*) The values do not belong to a specific year, but to the whole period (1962-2010)

5.4.2 Temporal and spatial variability of key physico-chemical variables

Temperature and dissolved oxygen

Results from the different simulations showed that there is a narrow range in water temperature variability, which was observed for all conditions simulated. Overall, the three-wetland areas showed minimum water temperatures between 26 °C and 28 °C. Nuevo river inflow showed values not higher than 26.5 °C, while El Recuerdo inflow showed a range similar to the wetland areas. The outflow ranged from 26 to 27 °C (Table 5-5). Inspection of temporal time series results describes the wetland as having a rather constant temperature throughout the year. On the other hand, dissolved oxygen concentrations showed a wide range with minima from 2.2 mg/l to maxima up to 10 mg/l in the wetland areas. El Recuerdo inflow had similar ranges to the wetland areas (from 3 to 8.5 mg/), while for Nuevo river inflow the range was narrower and with higher values (6 to 8.7 mg/l). The outflow showed minimum values similar to the wetland ones, ranging from 3.1 to 8.7 mg/l. Overall, concentrations were similar between the different conditions, however inspection of annual average values suggests lower averages for wettest conditions (1998, MAXhisto) at wetland areas (Table 5-6).

Table 5-5 Simulated water temperature (°C) at observations points in a) wetland areas: upper, middle, and low, and b) inflows and outflow, for the different conditions. Results from one-year simulation.

a) MODTEMP	UPPER S1 MIN	MAX	AVE	MIDDLE S2 MIN	MAX	AVE	ABANICOF MIN	MAX	AVE	LOW S7 MIN	MAX	AVE	S3C MIN	MAX	AVE
1990 (Dry)	26.5	27.9	27.1	26.2	27.7	26.7	26.2	26.9	26.5	26.0	27.2	26.5	26.0	27.2	26.3
1998 (El Niño)	26.5	27.9	27.4	26.1	27.7	27.0	26.1	27.2	26.8	26.0	27.2	26.6	26.0	27.2	26.4
2006 (Ave)	26.5	27.9	27.0	26.2	27.6	26.7	26.2	27.0	26.6	26.0	27.1	26.5	26.0	27.1	26.3
MAXhisto	26.5	27.9	27.3	26.1	27.6	27.0	26.1	27.2	26.7	26.0	27.2	26.5	26.0	27.2	26.2
MINhisto	26.5	27.9	27.1	26.5	27.7	26.8	26.4	27.0	26.5	26.1	27.2	26.6	26.0	27.4	26.5
Scen dry-30% smooth	26.5	27.8	27.1	26.5	27.5	26.8	26.5	26.6	26.5	26.2	27.0	26.6	26.0	26.9	26.5
ScenMAXhisto_smooth	26.4	27.9	27.1	26.3	27.6	26.8	26.4	27.4	26.7	26.0	27.2	26.5	26.0	26.7	26.1

b) MODTEMP	INFLOW (NUEVO RIVER) MIN	MAX	AVE	INFLOW (EL RECUERDO) MIN	MAX	AVE	OUTFLOW MIN	MAX	AVE
1990 (Dry)	26.0	26.1	26.0	26.5	27.9	27.5	26.0	26.5	26.2
1998 (El Niño)	26.0	26.0	26.0	26.9	27.9	27.7	26.0	26.6	26.2
2006 (Ave)	26.0	26.0	26.0	26.7	27.9	27.5	26.0	26.5	26.1
MAXhisto	26.0	26.0	26.0	26.7	27.9	27.6	26.0	26.5	26.1
MINhisto	26.0	26.5	26.2	26.5	27.9	27.5	26.1	26.8	26.3
Scen dry-30%sm	26.0	26.5	26.1	27.1	27.9	27.6	26.1	26.6	26.3
ScenMAXhisto_smooth	26.0	26.0	26.0	26.5	27.9	27.4	26.0	26.3	26.1

Table 5-6 Simulated dissolved oxygen concentrations (mg/l) at observations points in (a) wetland areas: upper, middle, and low, and b) inflows and outflow, for the different conditions. Results from one-year simulation.

a)	UPPER			MIDDLE						LOW					
	S1			S2			ABANICOF			S7			S3C		
DO	MIN	MAX	AVE	MIN	MAX	AVE	MIN	MAX	AVE	MIN	MAX	AVE	MIN	MAX	AVE
1990 (Dry)	2.4	10.0	5.9	3.1	9.6	7.4	4.4	9.2	8.5	3.8	9.6	7.9	3.8	7.8	6.2
1998 (El Niño)	2.3	9.7	4.5	3.0	9.4	6.0	4.9	9.5	7.4	3.8	9.4	7.5	4.2	9.4	6.1
2006 (Ave)	2.4	9.9	6.2	3.0	9.3	7.1	5.4	9.2	8.3	3.7	9.2	7.4	3.9	9.3	6.4
MAXhisto	2.3	9.2	4.8	3.0	9.1	5.9	3.6	9.2	7.4	3.3	9.1	6.3	3.5	8.3	5.9
MINhisto	2.8	10.0	5.9	3.3	9.5	7.2	6.8	9.3	8.6	4.3	9.4	8.2	4.5	9.7	7.2
Scen dry-30%smooth	2.6	10.0	5.6	3.6	9.4	7.0	8.4	9.3	8.7	5.8	9.5	8.3	5.8	9.8	6.8
ScenMAXhistosmooth	2.2	9.8	5.7	3.0	10.9	7.3	4.3	10.9	8.1	3.9	9.3	7.0	5.1	7.7	6.0

b)	INFLOW (NUEVO RIVER)			INFLOW (EL RECUERDO)			OUTFLOW		
DO	MIN	MAX	AVE	MIN	MAX	AVE	MIN	MAX	AVE
1990 (Dry)	6.3	7.3	7.0	3.2	8.3	4.7	3.7	7.4	5.5
1998 (El Niño)	6.1	7.2	6.8	3.2	6.1	3.9	3.5	6.5	5.2
2006 (Ave)	6.1	7.3	7.0	3.2	9.3	4.6	3.7	6.6	5.1
MAXhisto	6.1	7.2	6.5	3.0	7.8	4.4	3.1	6.0	5.0
MINhisto	5.9	8.7	7.5	3.4	8.2	4.7	4.1	8.6	6.4
Scen dry-30%smooth	5.4	8.3	7.1	3.3	5.4	4.2	4.9	8.7	5.8
ScenMAXhisto_smooth	6.1	7.1	6.5	2.9	9.1	5.2	4.6	5.6	5.1

Nitrogen

Simulated total-N (TotN) concentrations varied between 0.6 and 1.5 mg/l at wetland areas, and between 0.7 and 2.3 mg/l at the main inflows. Lower yearly average values occurred in middle wetland areas, probably also influenced by the decrease in the inorganic fraction (DIN) due to higher uptake by phytoplankton in this area. Overall, simulated concentrations did not vary significantly among the different conditions (Table 5-7).

Table 5-7 Minimum, maximum and average simulated concentrations of total nitrogen (TotN), dissolved inorganic nitrogen (DIN) and total organic nitrogen (TON) in gN/m^3 (equivalent to mgN/l) at observations points in a) wetland areas: upper, middle, and low, and b) inflows and outflow, for the different conditions. Results from one-year simulation.

a)	UPPER			MIDDLE						LOW					
	S1			S2			ABANICOF			S7			S3C		
TotN	MIN	MAX	AVE	MIN	MAX	AVE	MIN	MAX	AVE	MIN	MAX	AVE	MIN	MAX	AVE
1990 (Dry)	0.7	1.2	1.0	0.6	1.2	0.8	0.6	1.1	0.7	0.6	1.3	0.8	0.7	1.3	1.0
1998 (El Niño)	0.8	1.2	1.1	0.6	1.2	0.9	0.6	1.0	0.8	0.6	1.1	0.8	0.7	1.1	1.0
2006 (Ave)	0.7	1.2	0.9	0.6	1.1	0.8	0.6	1.0	0.7	0.6	1.3	0.8	0.7	1.3	1.0
MAXhisto	0.7	1.2	1.0	0.6	1.2	0.9	0.6	1.1	0.8	0.7	1.3	0.9	0.7	1.2	1.0
MINhisto	0.7	1.2	1.0	0.7	1.2	0.8	0.6	0.9	0.7	0.6	1.3	0.7	0.6	1.3	0.8
Scen dry-30%smooth	0.8	1.2	1.0	0.7	1.1	0.9	0.6	0.8	0.7	0.6	0.9	0.7	0.6	1.1	0.9
ScenMAXhisto_smooth	0.6	1.2	1.0	0.6	1.4	0.9	0.6	1.5	0.8	0.7	1.3	0.9	0.9	1.1	1.1

DIN

1990 (Dry)	0.0	0.3	0.2	0.0	0.3	0.1	0.0	0.3	0.0	0.0	0.3	0.1	0.1	0.3	0.2
1998 (El Niño)	0.0	0.3	0.2	0.0	0.3	0.1	0.0	0.3	0.1	0.0	0.3	0.1	0.0	0.3	0.2
2006 (Ave)	0.0	0.3	0.1	0.0	0.3	0.1	0.0	0.2	0.0	0.0	0.3	0.1	0.0	0.3	0.2
MAXhisto	0.0	0.3	0.2	0.0	0.3	0.1	0.0	0.3	0.1	0.0	0.4	0.2	0.0	0.3	0.2
MINhisto	0.0	0.3	0.2	0.0	0.3	0.1	0.0	0.1	0.0	0.0	0.3	0.0	0.0	0.3	0.1
Scen dry-30%smooth	0.0	0.3	0.2	0.0	0.2	0.1	0.0	0.1	0.0	0.0	0.2	0.0	0.0	0.2	0.2
ScenMAXhisto_smooth	0.0	0.3	0.2	0.0	0.3	0.1	0.0	0.3	0.0	0.0	0.3	0.1	0.1	0.3	0.3

TON

1990 (Dry)	0.7	0.9	0.8	0.6	0.9	0.7	0.6	0.8	0.7	0.6	1.0	0.7	0.6	1.0	0.8
1998 (El Niño)	0.7	0.9	0.8	0.6	0.9	0.8	0.6	0.8	0.7	0.6	0.8	0.7	0.7	0.8	0.8
2006 (Ave)	0.7	0.9	0.8	0.6	0.9	0.7	0.6	0.8	0.7	0.6	1.0	0.7	0.7	1.0	0.8
MAXhisto	0.7	1.0	0.8	0.6	0.9	0.8	0.6	0.8	0.7	0.7	0.9	0.7	0.7	0.9	0.8
MINhisto	0.7	1.0	0.8	0.6	0.9	0.7	0.6	0.8	0.7	0.6	1.0	0.7	0.6	1.1	0.7
Scen dry-30%smooth	0.7	0.9	0.8	0.7	0.9	0.8	0.6	0.8	0.7	0.6	0.8	0.7	0.6	0.9	0.7
ScenMAXhisto_smooth	0.6	1.0	0.8	0.6	1.1	0.8	0.6	1.3	0.8	0.7	1.0	0.8	0.8	0.8	0.8

b)

	INFLOW (NUEVO RIVER)			INFLOW (EL RECUERDO)			OUTFLOW		
TotN	MIN	MAX	AVE	MIN	MAX	AVE	MIN	MAX	AVE
1990 (Dry)	1.1	1.1	1.1	0.7	1.2	1.1	0.8	1.6	1.0
1998 (El Niño)	1.1	1.1	1.1	0.9	1.2	1.2	0.9	1.3	1.0
2006 (Ave)	1.1	1.1	1.1	0.8	1.2	1.1	0.9	1.6	1.0
MAXhisto	1.1	1.1	1.1	0.8	1.2	1.1	1.0	1.6	1.1
MINhisto	0.8	2.3	1.1	0.7	1.2	1.1	0.6	1.6	1.0
Scen dry-30%smooth	1.1	2.4	1.2	1.0	1.2	1.2	0.6	1.1	1.0
ScenMAXhisto_smooth	1.1	1.1	1.1	0.7	1.2	1.1	1.0	1.2	1.1
DIN									
1990 (Dry)	0.2	0.3	0.2	0.0	0.3	0.2	0.1	0.3	0.2
1998 (El Niño)	0.2	0.3	0.3	0.2	0.3	0.3	0.2	0.4	0.3
2006 (Ave)	0.2	0.3	0.2	0.0	0.3	0.2	0.2	0.3	0.3
MAXhisto	0.2	0.3	0.3	0.1	0.3	0.2	0.2	0.4	0.3
MINhisto	0.0	1.5	0.3	0.0	0.3	0.2	0.0	0.3	0.2
Scen dry-30%smooth	0.2	1.6	0.3	0.2	0.3	0.3	0.0	0.3	0.2
ScenMAXhisto_smooth	0.2	0.3	0.3	0.0	0.3	0.2	0.3	0.3	0.3
TON									
1990 (Dry)	0.8	0.9	0.9	0.7	1.0	0.9	0.6	1.4	0.7
1998 (El Niño)	0.8	0.9	0.8	0.8	1.0	0.9	0.7	0.9	0.8
2006 (Ave)	0.8	0.9	0.9	0.8	1.0	0.9	0.7	1.4	0.8
MAXhisto	0.8	0.9	0.8	0.8	1.0	0.9	0.7	1.3	0.8
MINhisto	0.8	0.9	0.8	0.7	1.0	0.9	0.6	1.4	0.8
Scen dry-30%smooth	0.8	0.9	0.9	0.8	1.0	0.9	0.6	0.8	0.7
ScenMAXhisto_smooth	0.8	0.9	0.8	0.7	1.0	0.9	0.7	0.9	0.8

Nitrogen temporal variability

The temporal variability (monthly averaged) of the simulated total nitrogen concentrations (TotN) at observation points located in the three-wetland areas is presented in Figure 5-7 and for dissolved inorganic nitrogen (DIN) in Figure 5-8. Results from the model showed that despite the different inflow conditions of Nuevo River, the range of concentrations between the different conditions was similar. Other conditions are presented in Appendix E.

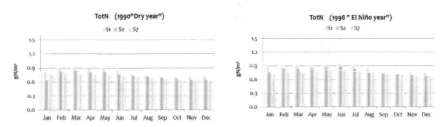

Figure 5-7 Simulated total nitrogen (TotN) concentrations in gN/m³ (equivalent to mgN/l). Observations points located at the Upper (S1), Middle (S2), and Low (S7) wetland areas. For the extreme years: 1990 (dry year) and 1998 (Extreme wet- El Niño year). Results from one-year simulation.

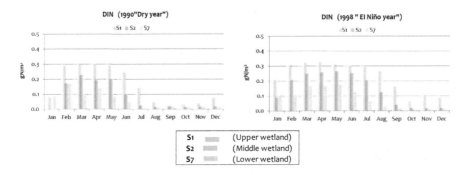

Figure 5-8 Simulated dissolved inorganic nitrogen (DIN) concentrations in gN/m³ (equivalent to mgN/l). Observations points located at the Upper (S1), Middle (S2), and Low (S7) wetland areas. For the extreme years: 1990 (dry year) and 1998 (Extreme wet- El Niño year). Results from one-year simulation.

Nitrogen spatial variability

Spatial variability of total nitrogen (TotN) for the extreme years 1990 (dry) and 1998 (El Niño) is illustrated in Figure 5-9. Higher concentrations usually started during February, which coincide with periods of maximum wetland inundation. During these periods, the highest concentrations (1.3-1.5 mgN/l), were observed at the lower and middle areas of the wetland, indicating an increase in the organic fraction produced in-situ during periods of high floods. As the wet season continues, concentrations (around 1 mgN/l) were observed mainly at the areas located closer to the inflows.

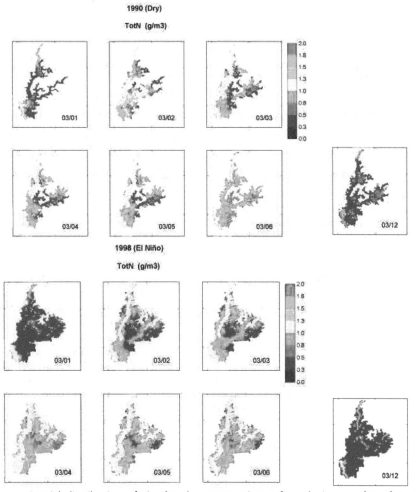

Figure 5-9   Spatial distribution of simulated concentrations of total nitrogen (TotN) in gN/m^3 (equivalent to mgN/l). Output maps extracted the same day every month. From left to right: January-June and December (on the right). For the extreme years: 1990 (dry year) and 1998 (Extreme wet- El Niño year).

The highest concentrations (up to 2 mgN/l) were more frequent at shallow patches and not in the main wetland channels. A decrease in concentrations occurred during the dry season, and was more evident since July, stabilizing in values between 0.5 and 0.8 mgN/l. The last three months of the year showed a regular distribution of low concentrations in the entire wetland area, with higher concentrations occurring only in areas nearby the Nuevo River (Figure 5-9). Other hydrological conditions are presented in Appendix E. The maps showed that during extreme wet conditions like El Niño year-1998 and the Scenario MAXhistoric, concentrations in the main channels remained at higher levels even until June-July (Appendix E). Overall, all the years and conditions simulated showed the lowest concentrations in the last three months of the year.

Phosphorus

Simulated total-P concentrations (TotP) ranged between 0.02 and 0.2 mg/l. The highest values occurred during dry conditions at observation point S3c located in the low wetland area and close to the main inflow, indicating that during driest conditions concentrations at the low areas might increase. Nevertheless, excluding these two maxima, the entire wetland area did not reach concentrations higher than 0.1 mg/l, and overall the model provided simulated results within the range of measured ones (Table 5-8).

Table 5-8 Minimum, maximum and average simulated concentrations of total phosphorus (TotP) in gP/m^3 (equivalent to mgP/l), at observations points in wetland areas: upper, middle, and low, for the different conditions. Results from one-year simulation.

| | UPPER | | | | | | MIDDLE | | | | | | LOW | | | | | |
| | S1 | | | S2 | | | ABANICOF | | | S7 | | | S3C | | | | |
TotP	MIN	MAX	AVE	MIN	MAX	AVE	MIN	MAX	AVE	MIN	MAX	AVE	MIN	MAX	AVE
1990 (Dry)	0.07	0.10	0.09	0.03	0.10	0.08	0.02	0.08	0.05	0.03	0.09	0.06	0.04	0.09	0.07
1998 (El Niño)	0.09	0.11	0.10	0.07	0.10	0.09	0.06	0.10	0.09	0.06	0.10	0.08	0.06	0.10	0.08
2006 (Ave)	0.07	0.10	0.09	0.04	0.10	0.08	0.02	0.10	0.06	0.03	0.09	0.06	0.04	0.09	0.07
MAXhisto	0.07	0.10	0.09	0.05	0.10	0.08	0.04	0.10	0.07	0.04	0.10	0.07	0.05	0.10	0.07
MINhisto	0.08	0.10	0.10	0.05	0.10	0.08	0.02	0.10	0.06	0.03	0.09	0.06	0.04	0.20	0.07
Scendry-30%smooth	0.08	0.10	0.10	0.06	0.10	0.09	0.04	0.10	0.07	0.04	0.09	0.07	0.05	0.23	0.07
ScenMAXhisto_smooth	0.04	0.10	0.09	0.02	0.10	0.07	0.02	0.10	0.06	0.03	0.10	0.06	0.06	0.09	0.07

Phosphorus temporal variability

Simulated total-P concentrations (TotP) (monthly averaged) were up to 0.1 mg/l. The temporal variability of the simulated total phosphorus concentrations (TotP) at observation points located in the three-wetland areas is illustrated in Figure 5-10. Similarly to TotN, results from the model showed that the range of TotP concentrations between different conditions was similar, ranging between 0.05 and 0.1 mgP/l. Other conditions are presented in the Appendix E.

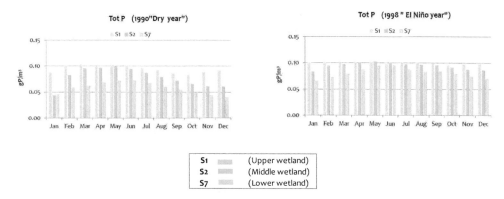

Figure 5-10 Simulated total phosphorus (TotP) concentrations in gP/m³ (equivalent to mgP/l). Observations points located at the Upper (S1), Middle (S2), and Low (S7) wetland areas. For the extreme years: 1990 (dry year) and 1998 (Extreme wet- El Niño year). Results from one-year simulation.

Phosphorus spatial variability

The spatial variability of total phosphorus (TotP) for the extreme years 1990 and 1998 is illustrated in Figure 5-11. Higher concentrations characterize the Upper Chojampe inflow area for both years. For the dry year, in the main wetland channels, concentrations started increasing from February and maintained high till July (> 0.08 mgP/l), when they started decreasing and maintained around 0.05 mgP/l until the end of the year. Therefore, the duration of higher concentrations of TotP in the main channels was longer compared to the duration of higher concentrations of TotN. This pattern might be related to the difference in the fractioning of both nutrients: TotN is mainly composed by the organic fraction, which is higher during the months of high inundation (February-March) due to primary production, while TotP is mainly composed by the inorganic fraction, and probably is being released locally, not depending on flooding (Figure 5-11). The driest conditions (1990 and MINhisto) showed a similar pattern, with maximum concentrations in the main channels until July. On the other hand, during extreme wet conditions like 'El Niño year' and the

Scenario MAXhisto, concentrations in the main channels remained at higher levels longer, reaching September for MAXhisto, and even till December during El Niño year 1998. The spatial maps for the MINhisto and MAXhisto scenarios are presented in Appendix E. Overall, for all conditions simulated, higher concentrations of TotP were maintained for longer periods when compared with high concentrations of TotN, which probably might be related to the differences in partitioning of both nutrients.

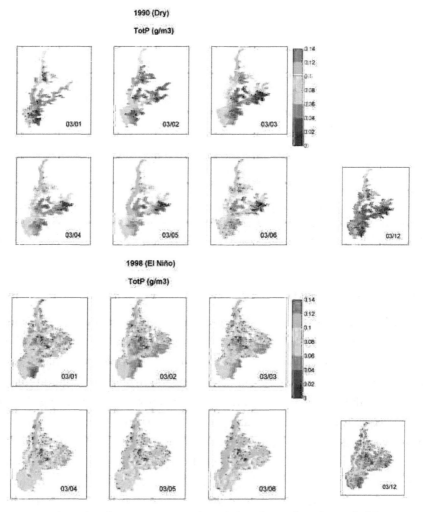

Figure 5-11 Spatial simulated concentrations of total phosphorus (TotP) in gP/m³ (equivalent to mgP/l). Output maps extracted the same day every month. From left to right: January to June, and December (on the right). For the extreme years: 1990 (dry year) and 1998 (Extreme wet- El Niño year).

Silica

Simulated Silica concentrations showed minimum values around 10 mgSi/l (gSi/m³), and average values up to 30 mg Si/l as indicated in Table 5-9.

Table 5-9 Minimum, maximum and average simulated concentrations of silica (Si) in gSi/m³, at observations points (upper, middle, and low wetland areas), for the different conditions. Results from one-year simulation.

	UPPER		MIDDLE				Low			
	S1		S2		ABANICOF		S7		S3C	
Si	MIN	AVE	MIN	AVE	MIN	AVE	MIN	AVE	MIN	AVE
1990 (Dry)	10	23	10	30	12	26	11	27	11	17
1998 (El Niño)	10	17	12	25	14	27	12	26	11	16
2006 (Ave)	10	22	10	30	14	31	12	26	11	16
MAXhisto	10	19	12	19	13	20	12	17	11	14
MINhisto	10	23	10	30	13	28	12	29	11	29
Scen dry-30%smooth	10	20	11	28	14	23	14	25	12	22
ScenMAXhisto_smooth	10	22	11	21	11	21	11	18	12	13

Organic carbon

Simulated total organic carbon concentrations (TOC) varied between 9 and 22 mgC/l, with higher values occurring mainly in the low area and Nuevo River inflow. The upper wetland site (S1) showed similar values to its closer inflow 'El Recuerdo'. The outflow of the wetland showed a combination of values from wetland and main inflow 'Nuevo River' (Table 5-10). At the low wetland area, the average values were lower during dry simulated conditions (MINhisto and Scendry-30%), for the rest no clear differences between the different conditions simulated were observed. Overall the model produced simulated results in the range of measured ones.

Table 5-10 Minimum, maximum and average simulated concentrations of total carbon (TOC) in gC/m³ (mgC/l), at observations points in a) wetland areas: upper, middle, and low, and b) inflows and outflow, for the different conditions. Results from one-year simulation.

	UPPER			MIDDLE						Low					
a)	S1			S2			ABANICOF			S7			S3C		
TOC	MIN	MAX	AVE	MIN	MAX	AVE	MIN	MAX	AVE	MIN	MAX	AVE	MIN	MAX	AVE
1990 (Dry)	11	15	12	10	18	11	9	18	11	10	21	13	12	22	18
1998 (El Niño)	11	15	13	10	18	12	9	19	11	9	21	12	11	22	17
2006 (Ave)	10	15	12	10	19	12	11	17	12	11	21	14	12	22	19
MAXhisto	11	15	13	11	19	13	11	20	13	11	22	15	12	22	19
MINhisto	11	15	12	10	14	11	9	15	10	9	20	11	9	22	14
Scen dry-30%smooth	11	14	13	10	13	11	10	11	11	10	19	11	9	22	15
ScenMAXhisto_smooth	11	15	13	11	16	12	11	14	12	11	21	14	15	22	20

b) TOC	INFLOW (NUEVO RIVER)			INFLOW (EL RECUERDO)			OUTFLOW		
	MIN	MAX	AVE	MIN	MAX	AVE	MIN	MAX	AVE
1990 (Dry)	21	22	22	11	15	14	13	21	19
1998 (El Niño)	22	22	22	12	15	14	17	21	20
2006 (Ave)	22	22	22	11	15	14	13	21	20
MAXhisto	22	22	22	11	15	14	13	21	20
MINhisto	16	22	21	11	15	14	9	21	16
Scen dry-30%smooth	18	22	22	12	15	14	9	21	18
ScenMAXhisto_smooth	22	22	22	11	15	13	19	21	21

Total organic carbon temporal variability

The temporal variability of the simulated total organic carbon concentrations (TOC) (monthly averaged) at observation points located in the three-wetland areas is illustrated in Figure 5-12. Similar to TotN and TotP, results from the model showed that the range of TOC concentrations between different conditions was similar, ranging between 10 and 18 mg C/l. Other conditions are presented in the Appendix E.

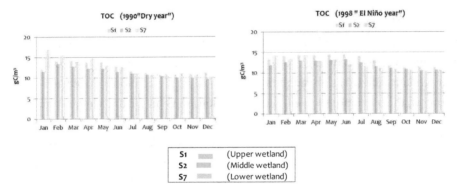

Figure 5-12 Simulated total carbon (TOC) concentrations in gC/m³ (equivalent to mgC/l). Observations points located at the Upper (S1), Middle (S2), and Low (S7) wetland areas. For the extreme years: 1990 (dry year) and 1998 (Extreme wet- El Niño year). Results from one-year simulation.

Total organic carbon spatial variability

The spatial distribution of total organic carbon (TOC) for the dry and extreme wet year is presented in Figure 5-13. Higher concentrations characterized the low wetland areas located close to the main inflow 'Nuevo River', showing extended patches with concentrations up to 20 mgC/l. This pattern was observed along the year for both conditions, excepting for February of the dry year, where the higher concentrations were shown in the main branches. The rest of the wetland area appeared to stabilize in concentrations between 10 and 15 mgC/l. A decrease in the concentrations in the upper wetland area was evident during the last two months of the year. The other hydrological conditions followed a similar pattern, with higher concentrations at low wetland areas, differing only in the surface extension of the patches between simulations. Lower concentrations in the upper areas were also observed during the last months of the year. The exception was the MAXhisto scenario that maintained concentrations above 10 mgC/l till the end of the year, most probably fomented by the high discharges from Nuevo River (Appendix E). Overall, the wetland is characterized by concentrations between 10 and 15 mgC/l, since all simulations showed bigger patches with concentrations in this range.

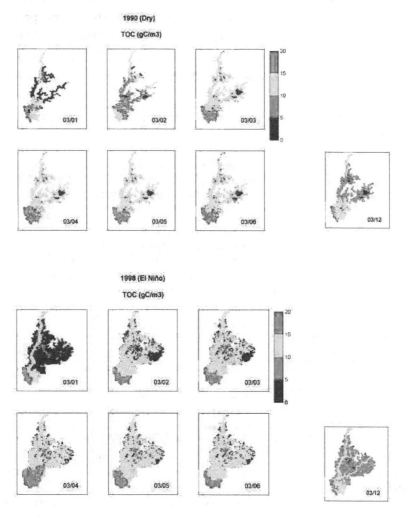

Figure 5-13 Spatial simulated concentrations of total carbon (TOC) in gC/m³ (equivalent to mgC/l). Output maps extracted the same day every month. From left to right: January-June and December. For the extreme years: 1990 (dry year) and 1998 (Extreme wet- El Niño year).

5.4.3 Temporal and spatial variations of primary producers

Primary production was evaluated in terms of Chlorophyll-a and Phytoplankton biomass (Table 5-11 & Table 5-12). Simulated concentrations of Chlorophyll-a were between 3 and 33 mg/m³ at wetland sites, 7 - 19 mg/m³ at the inflows, and 2 - 37 mg/m³ at the outflow (Table 5-11). Despite the wide range at wetland sites, annual average values were similar for all conditions simulated, ranging between 9 and 12 mg/m³ at the three wetland areas. However, based on a closer observation only of the maxima, it can be seen that the maxima of the middle area sites were lower than the ones at the upper area for all simulations but Scen MAXhisto, and probably an indication of higher primary consumption by zooplankton in middle areas. Furthermore, the lowest minimum values were observed at S7 (low wetland area), as a result of the dilution effects due to proximity to the Nuevo River inflow, a high primary consumption, or a combination of both. Simulated concentrations of phytoplankton biomass (Phyt) in gC/m³ followed the same pattern of Chlorophyll-a (Table 5-12).

Table 5-11 Minimum, maximum and average simulated concentrations of Chlorophyll-a in mg/m³ (equivalent to µg/l), at observations points in a) wetland areas: upper, middle, and low, and b) inflows and outflow, for the different conditions. Results from one-year simulation.

a)	UPPER			MIDDLE						LOW					
	S1			S2			ABANICOF			S7			S3C		
Chl-a	MIN	MAX	AVE	MIN	MAX	AVE	MIN	MAX	AVE	MIN	MAX	AVE	MIN	MAX	AVE
1990 (Dry)	10	23	12	8	15	11	8	14	11	5	21	11	7	17	12
1998 (El Niño)	8	21	11	6	17	11	5	15	11	3	14	11	6	18	11
2006 (Ave)	8	19	11	8	18	11	7	14	11	3	20	10	6	21	12
MAXhisto	7	17	10	5	18	10	5	15	11	3	15	9	5	16	9
MINhisto	11	25	12	10	14	11	8	13	11	6	15	11	10	18	12
Scen dry-30%smooth	11	25	12	10	15	11	8	13	11	9	17	11	11	18	12
ScenMAXhisto_smooth	8	24	10	8	24	11	9	33	12	5	14	9	6	13	9

b)	INFLOW (NUEVO RIVER)			INFLOW (EL RECUERDO)			OUTFLOW		
Chl-a	MIN	MAX	AVE	MIN	MAX	AVE	MIN	MAX	AVE
1990 (Dry)	9	19	17	11	25	13	4	35	9
1998 (El Niño)	7	19	14	10	16	13	3	11	7
2006 (Ave)	7	19	16	10	29	13	3	36	8
MAXhisto	7	18	11	9	18	12	2	25	5
MINhisto	11	19	16	11	18	14	5	32	10
Scen dry-30%smooth	11	19	17	12	16	14	5	12	9
ScenMAXhisto_smooth	7	17	11	9	26	12	3	7	5

Table 5-12 Minimum, maximum and average simulated concentrations of phytoplankton biomass (gC/m^3), at observations points in a) wetland areas: upper, middle, and low, and b) inflows and outflow, for the different conditions. Results from one-year simulation.

a)

Phyt	UPPER			MIDDLE						LOW					
	S1			S2			ABANICOF			S7			S3C		
	MIN	MAX	AVE	MIN	MAX	AVE	MIN	MAX	AVE	MIN	MAX	AVE	MIN	MAX	AVE
1990 (Dry)	0.3	0.9	0.3	0.2	0.6	0.3	0.2	0.5	0.3	0.1	0.7	0.3	0.2	0.5	0.3
1998 (El Niño)	0.2	0.8	0.3	0.2	0.5	0.3	0.1	0.5	0.3	0.1	0.5	0.3	0.2	0.5	0.3
2006 (Ave)	0.2	0.8	0.3	0.2	0.7	0.3	0.2	0.5	0.3	0.1	0.6	0.3	0.2	0.8	0.3
MAXhisto	0.2	0.5	0.3	0.1	0.5	0.3	0.1	0.5	0.3	0.1	0.5	0.3	0.2	0.5	0.3
MINhisto	0.3	0.9	0.4	0.3	0.5	0.3	0.3	0.5	0.3	0.2	0.5	0.3	0.3	0.6	0.3
Scen dry-30%smooth	0.3	0.9	0.4	0.3	0.5	0.3	0.2	0.5	0.3	0.3	0.6	0.3	0.3	0.6	0.3
ScenMAXhisto_smooth	0.2	0.8	0.3	0.3	0.8	0.3	0.3	1.0	0.4	0.1	0.4	0.3	0.2	0.4	0.2

b)

Phyt	INFLOW (NUEVO RIVER)			INFLOW (EL RECUERDO)			OUTFLOW		
	MIN	MAX	AVE	MIN	MAX	AVE	MIN	MAX	AVE
1990 (Dry)	0.3	0.5	0.5	0.3	0.7	0.4	0.1	0.9	0.3
1998 (El Niño)	0.2	0.5	0.4	0.3	0.5	0.4	0.1	0.3	0.2
2006 (Ave)	0.2	0.5	0.4	0.3	1.0	0.4	0.1	1.0	0.2
MAXhisto	0.2	0.5	0.3	0.3	0.5	0.4	0.1	0.7	0.1
MINhisto	0.3	0.5	0.4	0.3	0.5	0.4	0.2	0.8	0.3
Scen dry-30%smooth	0.3	0.5	0.5	0.3	0.5	0.4	0.2	0.3	0.3
ScenMAXhisto_smooth	0.2	0.5	0.3	0.3	0.8	0.4	0.1	0.2	0.1

Temporal variability of primary producers

The temporal variability of primary production variables (monthly averaged) is represented in Figure 5-14 and Figure 5-15. Similarly, to nutrients and carbon patterns, simulated results for the three-wetland areas determined that regardless the different conditions, the concentrations between the conditions maintained a similar range. Chlorophyll-a concentrations (monthly averaged) were between 7 and 15 mg/m^3, and phytoplankton biomass from 0.2 to 0.5 gC/m^3. The influence of the main inflows on the chlorophyll-concentrations in the three-wetland areas was also evaluated by correlation analysis (Table 5-13).

At the upper wetland area (S1), correlation analysis showed an inverse relation between Chlorophyll-a concentrations and 'El Recuerdo' discharge. Thus suggesting that a peak in el Recuerdo inflow produced drops in chlorophyll-a not higher than 3 mg/m^3, mainly when the wetland has reaching its maximum inundation capacity. Correlation coefficients were higher when the correlation was calculated only for the wet season (r=-0.5 to -0.7). When the correlation was calculated for the whole year the coefficients showed still an inverse relation, but the coefficients were lower (-0.3 to -0.6) (Table 5-13).

At middle wetland area (S2), correlation analysis also showed an inverse relation between Chlorophyll-a concentrations and El Recuerdo discharge. During a dry year (1990), Nuevo River appeared to have more influence than El Recuerdo inflow in the chlorophyll-a concentrations of this area (Table 5-13). However, coefficients were weak for 1990 and 2012, and intermediate for 1998, suggesting that El Recuerdo had more influence on the middle wetland area during a wettest year (pattern previously explained by the tracer analysis). An inverse relation was also observed with Nuevo River inflow, with higher coefficients during wettest years (1998 & 2012), suggesting that the influence of Nuevo River on the chlorophyll-a concentrations of the middle area was also more pronounced during wettest conditions. Nevertheless, the coefficient of the Nuevo River during a dry year (1990) was higher than the one of El Recuerdo, indicating that the Nuevo River could have more influence on the general pattern of chlorophyll-a concentrations at the middle wetland areas, rather than El Recuerdo during a dry year (Table 5-13). Overall, oscillations of both inflows occurring during the central period of the wet season did not cause drops in chlorophyll-a concentrations higher than 2 mg/m^3. Hence, suggesting that the chlorophyll-a concentrations in the middle wetland area are not dependent only of the inflows influence, but also of in-situ processes that had an important role in defining the concentrations in this area. At the low wetland area (S7), correlation analysis indicated an inverse relation between Chlorophyll-a concentrations and Nuevo River discharge (r=-0.6 to -0.8) for the different conditions tested (Table 5-13). Thus suggesting that peaks in Nuevo River discharge produces a drop in Chlorophyll-a concentrations.

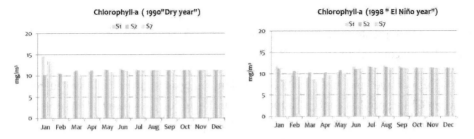

Figure 5-14 Simulated Chlorophyll-a concentrations (mg/m^3). Observations points located at the Upper (S1), Middle (S2), and Low (S7) wetland areas. For the extreme years: 1990 (dry year) and 1998 (Extreme wet- El Niño year). Results from one-year simulation.

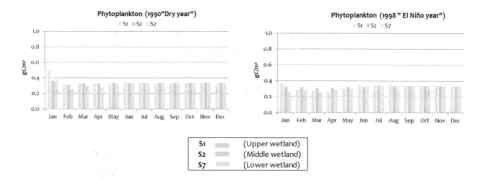

Figure 5-15 Simulated phytoplankton biomass (Phyt) (gC/m³). Observations points located at the Upper (S1), Middle (S2), and Low (S7) wetland areas. For the extreme years: 1990 (dry year) and 1998 (Extreme wet- El Niño year). Results from one-year simulation.

Table 5-13 Correlation coefficients: Chlorophyll-a vs. main inflows discharges, at upper, middle and low wetland observation points and different conditions. (Results from one year and only wet season)

	EL RECUERDO		NUEVO RIVER	
WETLAND AREAS/CONDITIONS	WHOLE	WET SEASON	WHOLE YEAR	WET SEASON
S1 (Upper wetland)				
1990 (dry year)	-0.3	-0.5		
1998 (El Niño)	-0.6	-0.7		
2012	-0.4	-0.7		
S2 (Middle wetland)				
1990 (dry year)	-0.2	0.2	-0.3	0.1
1998 (El Niño)	-0.4	-0.1	-0.5	-0.2
2012	-0.3	-0.1	-0.6	-0.5
S7 (Low wetland)				
1990 (dry year)			-0.6	-0.7
1998 (El Niño)			-0.8	-0.7
2012			-0.8	-0.8

Spatial variability of primary production

The spatial distribution of chlorophyll-a and phytoplankton biomass for the dry and extreme wet year is presented in Figure 5-16 & Figure 5-17. Lower concentrations in the main branches were observed during February for the dry year and April for 'El Niño year. During these periods, the wetland experienced dilution in the concentrations due to the higher inflows entering the system. Afterwards, main channels of the wetland stabilized in concentrations around 12 mg/m³ for chlorophyll-a and 0.4 gC/m³ for phytoplankton biomass (concentrations observed until the end of the year). Higher concentrations were observed in shallow marginal areas. The other simulated conditions MINhisto and MAXhisto followed a similar pattern, with lower concentrations produced during peak inflows (Appendix E). The difference between conditions was only in the surface extension of the patches concentrations, due to the difference in inundation between simulations. During wettest simulations (1998 & MAXhisto), lower concentrations were quite evident in the low wetland area during peak inflows of Nuevo River. This dilution effect reached up to the middle wetland area, as it is observed for the MAXhisto scenario. Driest conditions (1990 & MINhisto) showed a homogeneous pattern in patches concentrations around 12 mg/m³ (Appendix E). Overall, the wetland is characterized by concentrations around 12 mg/m³, since all simulations showed higher extension of patches areas with this concentration.

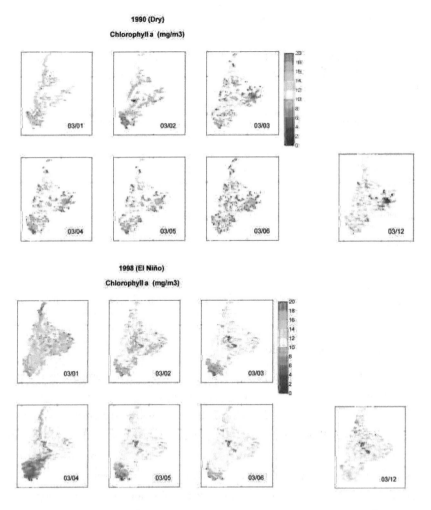

Figure 5-16 Spatial simulated concentrations of Chlorophyll-concentrations (mg/m³) (equivalent to µg/l). Output maps extracted the same day every month. From left to right: January-June and December. For the extreme years: 1990 (dry year) and 1998 (Extreme wet- El Niño year).

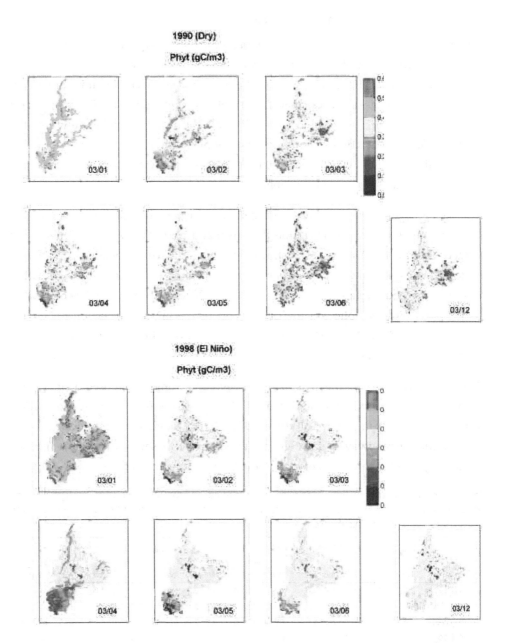

Figure 5-17 Spatial simulated concentrations of phytoplankton biomass (Phyt) (gC/m³). Output maps extracted the same day every month. From left to right: January-June and December. For the extreme years: 1990 (dry year) and 1998 (Extreme wet- El Niño year).

5.4.4 Temporal and spatial variations primary consumers

Primary consumers were modelled as zooplankton biomass (gC/m^3) with an imposed biomass function. Simulated biomass concentrations deviated from the imposed concentrations when algae and detritus where not in a sufficient amount to maintain the imposed biomass. The calculated concentrations varied from 0.5 gC/m^3 to 3 gC/m^3 (Table 5-14). These values coincide with the low and upper limits of the imposed forcing function. The range of zooplankton concentrations occurring between these values correspond to the adjustments that the biomass experienced due to the fluctuations in food availability.

Annual average biomass values were similar between the conditions simulated, ranging from 1.4 to 1.9 gC/m^3 at the three wetland areas (Table 5-14). The site S3c located close to the Nuevo River showed the highest average values (1.9 gC/m^3). Concentrations at Nuevo River inflow showed maxima of 3 gC/m^3, indicating that there was no food limitation at this point. Thus, at Nuevo River inflow, the calculated zooplankton biomass followed the same trend as the imposed function for all the conditions tested but MINhisto, also confirmed by inspection of the corresponding inflow time series (figure not presented).

During the dry season of the MINhisto, the deviation from the imposed biomass was from 3 to 2.4 gC/m^3. On the other hand, 'El Recuerdo' inflow showed a deviation from the imposed function for all the conditions. This deviation was reflected in the maxima that were always below 3gC/m^3 (maximum imposed in the function), thus, suggesting possibly limitation of food at the Upper Chojampe inflows.

Table 5-14 Minimum, maximum and average simulated zooplankton biomass (gC/m^3), at observations points in a) wetland areas: upper, middle, and low, and b) inflows and outflow, for the different conditions. Results from one-year simulation.

a)	UPPER			MIDDLE						Low					
	S1			S2			ABANICOF			S7			S3c		
CZooplank	MIN	MAX	AVE	MIN	MAX	AVE	MIN	MAX	AVE	MIN	MAX	AVE	MIN	MAX	AVE
1990 (Dry)	0.5	2.5	1.6	0.5	2.1	1.5	0.5	2.2	1.5	0.5	2.5	1.6	0.5	3.0	1.9
1998 (El Niño)	0.5	2.1	1.5	0.5	2.0	1.5	0.5	2.0	1.5	0.5	2.5	1.6	0.5	3.0	1.9
2006 (Ave)	0.5	2.8	1.6	0.5	2.1	1.6	0.5	2.4	1.6	0.5	3.0	1.7	0.5	3.0	1.9
MAXhisto	0.5	2.1	1.6	0.5	2.1	1.5	0.5	2.0	1.6	0.5	3.0	1.8	0.5	3.0	1.9
MINhisto	0.5	2.6	1.5	0.5	2.1	1.5	0.5	2.1	1.5	0.5	2.2	1.6	0.5	3.0	1.6
Scendry-30%smooth	0.5	2.6	1.5	0.5	2.1	1.5	0.5	2.1	1.4	0.5	2.1	1.6	0.5	3.0	1.9
ScenMAXhisto_smooth	0.5	2.8	1.6	0.5	2.1	1.6	0.5	2.5	1.7	0.5	2.0	1.5	0.5	3.0	1.9

B)	INFLOW (NUEVO RIVER)			INFLOW (EL RECUERDO)			OUTFLOW		
CZooplank	MIN	MAX	AVE	MIN	MAX	AVE	MIN	MAX	AVE
1990 (Dry)	0.5	3.0	1.9	0.5	2.1	1.5	0.5	3.0	1.9
1998 (El Niño)	0.5	3.0	1.9	0.5	2.4	1.5	0.5	3.0	1.9
2006 (Ave)	0.5	3.0	1.9	0.5	2.3	1.5	0.5	3.0	1.9
MAXhisto	0.5	3.0	1.9	0.5	2.3	1.5	0.5	3.0	1.9
MINhisto	0.5	3.0	1.8	0.5	2.1	1.5	0.5	3.0	1.4
Scen dry-30%smooth	0.5	3.0	1.9	0.5	2.2	1.5	0.5	3.0	1.8
ScenMAXhisto_smooth	0.5	3.0	1.9	0.5	2.3	1.7	0.5	3.0	1.9

Temporal variability of primary consumers

The temporal variability of primary consumers (Zooplankton biomass) at observation points located in the three-wetland areas is illustrated in Figure 5-18. Zooplankton biomass appeared to be driven by food availability rather than to changes in inflows. Overall, the pattern at the three-wetland areas was similar between the conditions. During the wet season, biomass experienced drops from 3 gC/m^3 (maximum concentration imposed) to values between 2.0 and 2.7 gC/m^3. While during the dry season, biomass stabilized in concentrations around 2 gC/m^3. Figure 5-19 illustrates the deviation of the concentrations at the three observation points from the imposed function.

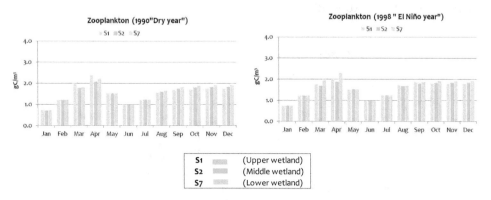

Figure 5-18 Simulated zooplankton biomass (Zoo) (gC/m^3). Observations points located at the Upper (S1), Middle (S2), and Low (S7) wetland areas. For the extreme years: 1900 (dry year) and 1998 (Extreme wet- El Niño year). Results from one-year simulation.

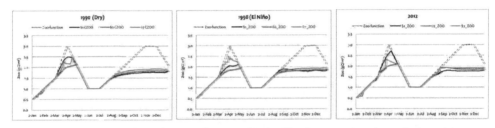

Figure 5-19 Comparison of zooplankton function (dashed line) vs. simulated zooplankton biomass (gC/m³) at observations points located at the Upper (S1), Middle (S2), and Low (S7) wetland areas. For: dry year (1990), extreme wet year (1998), and sampling year 2012. Results from one-year simulation.

Spatial variability of primary consumers

The spatial distribution of zooplankton biomass for the dry and extreme wet year is presented in Figure 5-20. Temporal patterns showed higher concentrations during April (up to 3 gC/m³), and lower ones (0.5 gC/m³) at the beginning of the wet season and during June-July (Figure 5-20). Spatial patterns indicated that areas located close to the main inflow 'Nuevo River' maintained the imposed high concentrations during some months (green patches). This pattern in temporal and spatial distribution was similar to the rest of the conditions simulated (Appendix E), and appeared to be independent of the different peak inflows proper of each condition. Nevertheless, the difference among all conditions was related to the surface extension of these patches concentrations, due to the difference in inundation areas between simulations (Appendix E).

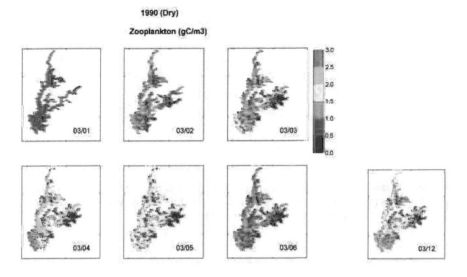

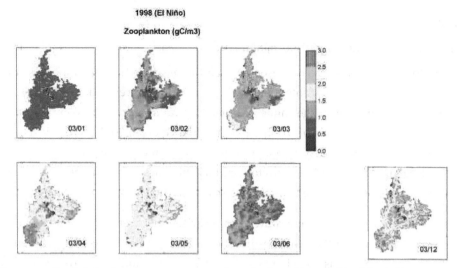

1998 (El Niño)

Zooplankton (gC/m3)

Figure 5-20 Spatial simulated concentrations of zooplankton biomass (Zoo) (gC/m³). Output maps extracted the same day every month. From left to right: January-June and December. For the extreme years: 1990 (dry year) and 1998 (Extreme wet- El Niño year).

Evaluation of temporal and spatial representations

Overall, the ecological model presented in this chapter represented well the measured concentrations in this wetland system. Abras de Mantequilla wetland exhibits concentrations of nutrients and primary production in the range of other tropical systems. Temporal analysis of nutrients (N, P) determined that higher concentrations were mainly observed during the wet season driven by the high inflows occurring during this period, as observed in other tropical systems. The model reproduced better inorganic nitrogen, dissolved oxygen and total phosphorus, but has limitations to reproduce total nitrogen and chlorophyll-a at specific observation points, probably due to the complexity that involve the fate of organic components. However, spatial representations were useful to visualize those maximum values that could not be reproduced in the temporal plots, providing an overall range of the variables. Despite these limitations, the model can be used as an 'indication' when applied for different hydrological conditions, since the aim is to evaluate the overall functioning of the wetland. Concentrations of N, P and C were similar between the different hydrological conditions simulated, as a result, the system stabilized and reached similar concentrations despite the different magnitude in the boundary conditions (namely inflows) imposed. A mass balance analysis (yearly based) is implemented in Chapter 7 as a tool for further evaluation of differences between hydrological conditions.

5.5 Discussion

5.5.1 Temporal and spatial variability of nutrients

Overall, the ecological model represented well the measured concentrations in this wetland system. Abras de Mantequilla wetland exhibits concentrations of nutrients and primary production in the range of other tropical systems. The comparison of AdM results with standards of trophic classification (Wetzel, 2001b; Wetzel, 2001d), suggests that AdM wetland-system falls between two categories: mesotrophic due to total nitrogen, chlorophyll-a, phytoplankton biomass, and annual primary production, and eutrophic due to total phosphorus concentrations.

A narrow range of water temperatures characterizes this wetland system with no marked variations along the year. Results showed similarities between the different conditions simulated. This corroborates the pattern found in humid tropics located close to sea levels, where daily and annual water temperature fluctuations are minor (Lewis, 2008). Thus, in tropical systems, nutrients availability may gain more importance over temperature in driving primary production (Lewis, 1987).

Results indicated that the high dynamics of this wetland system allow oxygenated conditions along the year and sites, with higher annual average values at some sites. Thus, the main inflow to the system 'Nuevo River' showed always high DO concentrations, thus bringing high oxygenated waters to the wetland system. Nonetheless, the wetland-inundated area presented a wider range of concentrations driven not only by hydrodynamics but also by local processes. Upper inflows and the wetland outflow showed intermediate concentrations between Nuevo River and wetland inundated-area, which was in agreement with measured values.

The computational results confirm the observations from Chapter 3 in that overall, nitrogen and phosphorus concentrations were typical for tropical systems and that total nitrogen and fractions and total dissolved fraction of phosphorus were slightly higher than for undisturbed rivers in the tropics (Lewis, 2008), thus suggesting a certain degree of disturbance in the AdM wetland.

On the other hand, nitrogen has been found as the limiting nutrient in other tropical South-American lentic environments (Lewis, 1986). Thus, in tropical waters, the ratio dissolved inorganic nitrogen (DIN) to soluble reactive phosphorus (SRP) is usually lower than in temperate waters (Lewis, 1996).

Temporal analysis of nutrients (N, P) determined that higher concentrations were mainly observed during the wet season driven by the high inflows occurring during this period, as observed in other tropical systems where higher nutrients concentrations are linked to the higher peaks of the hydrograph (Lewis, 2008), and to storm events delivering pulses of N and P (Saunders et al., 2006). However, phosphorus maintained higher concentrations until the middle of dry season for some simulations. The end of the year showed the lowest concentrations of nutrients, due to the minimal inflows entering the wetland during this period. Nevertheless, processes as algae uptake and grazing, also play a role in nutrients cycle during this dry phase.

Spatial analysis determined that nutrient concentrations at the different wetland areas were influenced by the nearest inflows. Thus, upper and middle wetland areas were more related to the discharges of 'El Recuerdo', and lower wetland areas by the Nuevo River. This relation has been also observed in other tropical systems, where nitrogen delivery is mainly controlled hydrographically, and where peaks of nitrogen concentrations take place during the rising limb of the hydrograph (Lewis, 2008).

Furthermore, during dry conditions Nuevo River had a more influence in middle areas, confirming the results of the tracer analysis developed in Chapter 2. The influence of the Nuevo River in low wetland areas was also observed for total organic carbon (TOC). The wetland is characterized with concentrations between 10 and 20 mgC/l, which are considered as characteristic of eutrophic lakes (Wetzel, 2001d). Since the majority of the variables classified AdM as a mesotrophic system, the values of TOC should be considered with care, especially because its main component is DOC, which was derived from measured values of POC, and not directly measured.

Overall, the concentrations of N, P and C were similar between the different hydrological conditions simulated, as a result, the system stabilized and reached similar concentrations despite the different magnitude in the boundary conditions (namely inflows) imposed. A mass balance analysis (yearly based) was implemented as a tool for further evaluation of differences between hydrological conditions.

5.5.2 *Temporal and spatial variability of primary producers and consumers*

Primary production was evaluated with chlorophyll-a concentrations and phyto biomass, both variables exhibiting a similar trend. Yearly average chlorophyll-a concentrations (9-12 mg/m^3), described the system as mesotrophic (Wetzel, 2001d). Overall, yearly average values were similar in the three-wetland areas, however, maximum concentrations were low at middle areas, which might be an indication of higher consumption by zooplankton probably due to the lower velocities and higher residence time that characterized this area. This is also in agreement with the higher densities of zooplankton collected in the middle area. However, the system can reach maximum concentrations up to 37 mg/m^3, which are typical of eutrophic environments.

The relation of primary production variables with the inflows followed a different pattern than the one of nutrients. At the upper wetland area, the pattern of chlorophyll-a can be divided in two periods, one at the beginning of the wet season when the wetland starts its filling, when higher concentrations were observed most likely linked to the initial high inflows peaks. Afterwards, at the middle of the wet season, a despite higher inflows, the chlorophyll-a concentrations remain more or less stable, suggesting an stabilization of this area. The middle area is influenced by both inflows, with the Nuevo River gaining importance during dry years and El Recuerdo during wettest year. The low wetland area is strongly influenced by Nuevo River, with drops in concentration occurring when higher inflows enter the wetland, suggesting a fast dilution and flushing in this area. Spatial analysis determined that lower areas were characterized by lower concentrations during the peak months of the wet season, due to dilution, and probably also related to the low retention times that characterizes this area (up to 5 days). This relation with the inflows was observed with the spatial dynamic simulations.

The concentrations at the outflow of the wetland showed for some simulations the higher maxima compared to the rest of locations evaluated, most likely due to the sudden flushing of inoculums. One of the advantages of the outflow characteristics of AdM is that this outflow branch converts almost immediately into a fast flowing river with the associated short retention time. In rivers with short retention time, planktonic algae will be added to total turbidity and may not increase noticeably during the transport further downstream (Hilton et al., 2006). Thus, no eutrophication is expected to occur in the downstream river reaches of the wetland.

To sum up, the variations that chlorophyll-a concentrations experience in relation to the inflows describe this system as highly linked to the seasonal hydrology, as well as has been found in other tropical systems, where seasonal hydrology is a main driver shaping food webs (Douglas et al., 2005). Overall, the wetland stabilizes in concentrations around 12 mg/m^3.

Inspection of algae limiting factors, determined that algae was not limited by nutrient availability in this model set up. Constant input of nutrients via the inflows, and in-situ mineralization of the organic matter (dissolved and particulate organic fractions of carbon, nitrogen, and phosphorus that were also input variables for this model) were key factors providing nutrients to this system. The latter is supported by studies that confirm that recycling of nutrients can be more efficient at lower latitudes (Lewis, 1987). The only limiting factor was growth, as a result of the grazing function imposed.

Model outcome suggests that there was no limitation in primary production resources available for next trophic level (primary consumers/grazers). As a result, primary consumers represented by zooplankton maintained the imposed biomass concentration almost all the time during the simulated period.

5.5.3 Nutrients partitioning

Total nitrogen was mainly composed by the organic fraction in all wetland areas (>80%). However, middle and low areas showed higher contributions than the upper area, suggesting a higher productivity in middle and low areas. The low contribution of the inorganic fraction in these areas would suggest an active uptake by algae. On the other hand, upper wetland areas showed higher contributions of the inorganic fraction, suggesting a lower uptake of algae in this area. The inorganic fraction DIN was mainly composed by nitrates (85-95%), slightly higher but overall following the trend of what has been reported in undisturbed watersheds (Lewis et al., 1999; Meybeck, 1982).

Total organic nitrogen was mainly composed by the dissolved fraction DON (>75%) with the particulate fraction between 19 and 25%. This is consistent with the river continuum concept that propose a higher nutrient processing at the lower sections of a river (Vannote et al., 1980), and that the particulate fraction is usually more representative at high stream orders. Furthermore, this fractioning is also comparable with the average ratio PN/TN (0.3) found in these undisturbed watersheds ratio that increases as the watershed area increases (Lewis et al., 1999). Particulate organic nitrogen (PON) showed a similar partitioning between the living fraction (N in algae or AlgN) and the non-living one (N in detritus or PONnoa).

Total phosphorus was mainly composed by the inorganic fraction (phosphates > 72%), with higher contributions at the upper wetland, suggesting a lower uptake by algae in this area. The organic fraction had particulate organic phosphorus (POP) as main component (> 60%), which in turn was primarily composed by the living fraction (P in algae or AlgP > 67%).

The fact that at upper wetland areas, the inorganic fraction of both nutrients had a higher contributions than in the other two areas, it is an indication of a lower uptake by algae in this area, which is in agreement with the lower phytoplankton densities at upper wetland sites compared to the middle and low areas (See Chapter 4). It appears that the water is transported from upper to middle areas, thus, the inorganic nutrient uptake is fomented at middle areas.

Total organic carbon (TOC) was mainly composed by the dissolved fraction (DOC > 78%). The fractioning of particulate organic carbon (POC) indicated a higher contribution of the non-living fraction (POCnoa) over the living-one (Phyto biomass), especially during wet season with contributions up to 89%, while the living fraction up to 48%.

Overall, the system did not show extreme differences between wetland areas and simulations, but although changes in percentage of the nutrient fractioning contributions were not so different between wetland areas, a tendency can be seen. This occurs also for some processes in the mass balances of algae.

"Success is a science; if you have the conditions, you get the result"

Oscar Wilde

6
EVALUATION OF HABITAT SUITABILITY CONDITIONS FOR FISH

6.1 Background and scope

Tropical rivers and associated floodplain areas provide dynamic habitats for fish (Winemiller and Jepsen, 1998) and contribute to maintaining the biodiversity of the whole river ecosystem, provided that connectivity is maintained. Connectivity is a key issue for floodplains, since richness and diversity of species decrease with decreasing hydrological connectivity (Aarts et al., 2004). Tropical floodplains faced gradual drying due to anthropogenic activities such as dam construction and irrigation, causing impacts on fish communities. A reduction in the inundated areas of floodplains decreases the habitat availability for fish communities. Reduction in habitat areas in turn produces an increase in fish densities (per unit surface area), intensification in species interaction and competition for resources (Winemiller and Jepsen, 1998). In South-America, designation of aquatic protected areas has recently started, and fish studies have been focusing more frequently at local rather than at river basin scale (Barletta et al., 2010). Furthermore, in tropical systems few studies have identified habitat preferences and developed habitat suitability criteria for fish availability (Costa et al., 2013; Teresa and Casatti, 2013). These studies developed habitat suitability criteria in the form of 'preference curves' for several fish species based on hydraulic features (water-depth, velocity) and substrate.

A number of habitat suitability index models were developed in the early eighties to provide habitat information of several wildlife species in the United States (Schamberger et al., 1982). Initial habitat studies started in the 50's to determine suitable areas for salmon spawning (Jowett, 1997). Habitat suitability is conventionally indicated by a numerical index on a normalised 0.0 to 1.0 scale, with the assumption that there is a positive relationship between the index and the habitat carrying capacity of the selected species (Schamberger et al., 1982). Habitat suitability criteria can be expressed in different categories and formats (Bovee, 1982). A first category is based on expert opinions (professionals, stakeholders) instead of data. A second category is based on data of the target species collected. These are known as 'utilization or habitat-use functions' because they represent the conditions that the target species faced at the time of observation or sampling. However, these criteria can be biased by the environmental conditions available at the observation time, since organisms could be forced to use suboptimal conditions when optimal ones are not available. In order to correct this functional bias and be less site specific, a third category named 'preference functions' was created (Bovee, 1986; Bovee et al., 1998).

The development of a habitat suitability analysis has to be related to the fact that species are distributed according to their preferences for feeding and reproduction

(Teresa and Casatti, 2013). Linking target species to their physical, chemical and biotic conditions is the basis of habitat assessment. Physical aspects include hydrology and geomorphology, and hydrological indicators can explain physical, chemical and biological processes in wetlands (Funk et al., 2013). Thus, hydrological conditions are key drivers for the wetland's structure and function. Hydrology influences several abiotic factors that determine which biota will develop in the wetland. *'Hydrology is probably the single most important determinant of the establishment and maintenance of specific types of wetlands and wetland processes'* (Mitsch and Gosselink, 2007). Specific mesohabitat characteristics such as water-depth and velocity have been found to play a key role in explaining fish communities structure (Arrington and Winemiller, 2006), with several studies considering them as the main variables for fish habitat analysis (Freeman et al., 2001; Schneider et al., 2012; Teresa and Casatti, 2013; Van de Wolfshaar et al., 2010). In this chapter, the following key research questions are investigated:

- What hydrological conditions are important for the AdM habitats?
- Is there an optimal period during the year that provides a higher extension of suitable areas?
- Are there specific regions in the wetland that are more suitable than others?

This study proposes a methodology to quantify the extension of suitable habitat areas for the fish communities of AdM wetland based on hydrodynamic features. A measure related to the percentage of suitable habitat areas (PSA) is proposed as a tool to explore the temporal and spatial variability of the habitat through the year for different hydrological conditions.

6.2 Study area

6.2.1 A Ramsar site

The AdM wetland is located at the centre of the Guayas River Basin in the Coastal Region of Ecuador. The wetland was declared a Ramsar site in 2000 due to the important role in conservation of bird fauna biodiversity, and especially because it supports three migratory species of birds (Ramsar, 2014). It is also an IBA site (Important Bird and Biodiversity Area) with 127 bird species reported. The wetland was selected as the South-America case study for the WETwin Project, a project funded by the European Commission (FP7) to enhance the role of the wetlands in integrated water resource management. The wetland, part of the Chojampe sub-basin, consists of branching water courses surrounded by elevations between 5–10 m (Quevedo, 2008). Agriculture in the surrounding wetland area mainly consists of short-term crops (rice, maize). Littoral areas of the wetland are covered by banks of macrophytes (Figure 6-1a) populated by small fish (Characidae family) (Figure 6-1d). Collected species from this family included common ones like *Astyanax festae*, and endemics like *Landonia latidens* and *Hyphessobrycon ecuadoriensis* (Figure 6-2). The wetland also provides a habitat for several fish species of commercial interest for local communities (Barnhill and Lopez, 1974; Florencio, 1993; Quevedo, 2008; Revelo, 2010).

Figure 6-1 Study area: Sampling sites located in the upper (a) and middle (b) wetland areas. Fish sampling (c, d)

Figure 6-2 Some species of small littoral fish species of Characidae family collected in Abras de Mantequilla wetland (upper panel) during sampling campaigns 2011-2012. Species of commercial interest for local communities reported in the wetland (low panel). Photos source (Aguirre, 2014).

Due to the proximity to the Equator, there are only two climatic periods: the wet season (mid-December up to mid-May), and the dry season (July-November). The seasonal variability and inter-annual variability of hydrological conditions was presented in Chapter 2. The highest annual precipitation was observed in 1997 and during the 1998 'El Niño event', with annual values of 4736 and 4790 mm. Figure 6-3 (same as Fig. 2-7) is repeated here to highlight the four hydrological years evaluated in this chapter (1990, 1998, 2011, and 2012) and relate them to the historical rainfall variability.

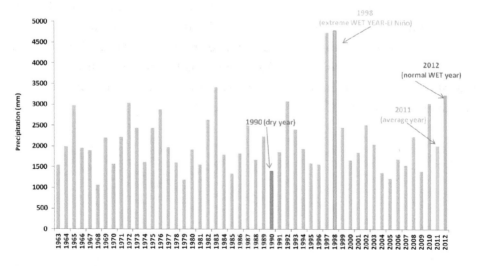

Figure 6-3 Annual precipitation in Quevedo-Vinces Basin. Pichilingue station (1963-2012)

6.2.2 *The hydrodynamics of AdM wetland*

The 2D hydrodynamic model of the AdM wetland was built with the Delft3D-FLOW software, based on a 1:10000 topography. The model considered the wetland extension and the location of the discharges. According to this topography, the wetland area recorded levels between 6 and 34 m.a.s.l. The main inflow to the wetland is the Nuevo River that flows through the Estero Boquerón and contributes with 85% of the total wetland inflow (Chapter 2). During a strong rainy year like El Niño 1998, the inflow discharges from Nuevo River to the wetland can reach maximum values up to 650 m³/s, while during a dry year, maximum discharges are up to 260 m³/s. The wetland also receives rainfall-run off from the Chojampe subbasin with a contribution of around 15% (Figure 2-29 in Chapter 2). These contributions slightly fluctuate according to the type of year (dry or wet). During the dry season, the water level in the wetland decreases considerably, and water remains only in the deep central channels, reducing the inundated area to around 10% compared with the wet season. The total modelled wetland area was 4029 hectares (40.3 km²) (Figure 2-17 in Chapter 2). Results from the hydrodynamic model show that the wetland is flooded up to 27 km² (Figure 6-4).

The natural variability of the wetland inundation area was presented in Chapter 2. Monthly averages of inundated areas indicate that historically the wetland experiences flooding from 5 to 23 km². A high variability between the different simulations during the wet season is evident. On the other hand, during the dry season, the inundation areas do not differ among the simulations, all reaching a value of about 5 km. Nevertheless, the exception is the maximum historical level since this time series includes the complete set of extreme wet conditions for a long historical period (1962-2010).

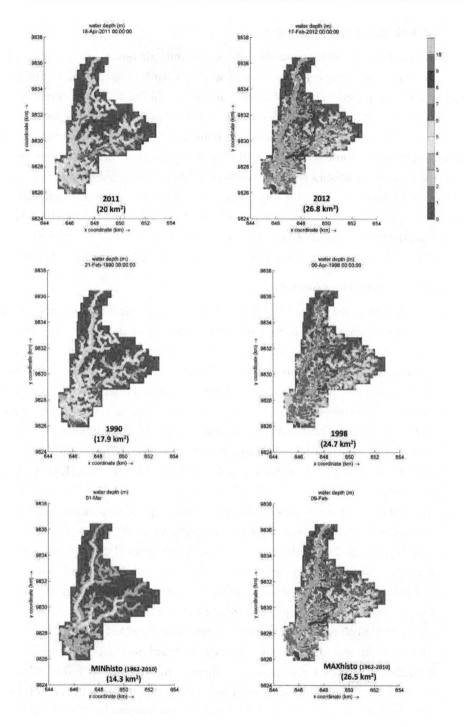

Figure 6-4 Maximum wetland inundated areas (km²) from Delft3D-FLOW simulations for 2011& 2012 (sampling years), 1990 (dry year), 1998 (wet year), MINIMUM and MAXIMUM historical (period 1962-2010). Scale bar indicates the water depth range (m).

6.3 The habitat suitability index

The aim to develop a habitat suitability index is to indicate how suitable a particular area is for a determined species or group of species. Clearly, this implies a number of assumptions, for example: it is not clear if an index will indicate with certainty the presence or absence of neither these species nor the quantity of the species. On the other hand, to be able to determine a species habitat, it is important to know in which period a species distributes. Thus, by exploring the space-time variability of an index, the presence of these species could be estimated. The hydrological functioning of the wetland explained in the previous section provides the base for our habitat suitability approach, given that the hydrology shapes the habitat. The following section describes the multiple tests performed to determine the relation between water-depth, velocity and the habitat for the fish community in this tropical wetland.

6.3.1 Steps for habitat index construction

1. Calculation of the habitat index was based on a general rule obtained from literature of the natural behaviour of fish in the study area. Some key aspects explored include seasonal behaviour, spawning, food availability and inundation area fluctuations. Field information and expert knowledge was used to validate this information. In situ measurements of water depth and velocity were compared with literature values to assess the distribution and habitat preferences of the overall fish community. As a result, response curves (knowledge rules) for these two variables were derived (Figure 6-5).

2. A dynamic HABITAT modelling tool was built with the MATLAB toolbox (Figure 6-6).

3. The habitat suitability was evaluated by relating in situ measurements of water depth and flow velocity from sampling years 2011 and 2012 with the results of the 2D hydrodynamic model (Delft3D-FLOW). Furthermore, extreme conditions (dry and wet years), and minimum and maximum historical conditions were also modelled to account for natural variability.

4. Output maps of water depth and velocity from the 2D hydrodynamic model (Delft3D-FLOW) were used as input for the dynamic HABITAT model.

5. The wetland was divided into five areas considering the influence of the boundary conditions and residence times. This division criterion allowed evaluation of the response of each area according to the influence of each boundary on the two hydrodynamic variables (water depth and velocity) (Figure 6-7).

6. The overall habitat analysis was performed for the total wetland area and for each area independently.

6.3.2 The habitat index formulation

The habitat suitability index formulation was developed with the following steps:

a) A response curve for water depth and for velocity was developed (Figure 6-5).

b) A 'Habitat Index' (HI) was calculated independently for water depth (HI-WD) and for velocity (HI-vel) for each cell of the grid (Figure 6-6). Delft3D-FLOW output maps of water depth and velocity were combined with their corresponding response curves (Figure 6-5). Cells with an index > 0.7 were given a value of 1 and considered for further calculation of the HSI:

Selection of the cells with HI_WD > 0.7:

$$HI\_WD > 0.7 = 1$$
$$HI\_WD < 0.7 = 0$$

Selection of the cells with HI_Vel > 0.7:

$$HI\_vel > 0.7 = 1$$
$$HI\_vel < 0.7 = 0$$

c) A 'Combined Habitat Index' (HSI) was calculated for each cell of the grid (Figure 6-5). In this step, the HABITAT model selected the minimum of both HI (HI-WD and HI-Vel) (Figure 6-6). Thus, the total habitat suitability is the minimum of the results of both rules (Equation 6-1). The results of the HSI were expressed in terms of percentages of suitable areas (PSA) with HSI > 0.7. PSA was calculated for each time step (Equation 6-2).

$$HSI = Min (HI\_WD, HI\_vel) \qquad \text{Eq (6-1)}$$

$$PSA_t = \frac{\sum_{k=1}^{n} HSI \geq 0.7}{N} \times 100 \qquad \text{Eq (6-2)}$$

Where:

HSI= Habitat Suitability Index

n= each cell

N= total number of cells

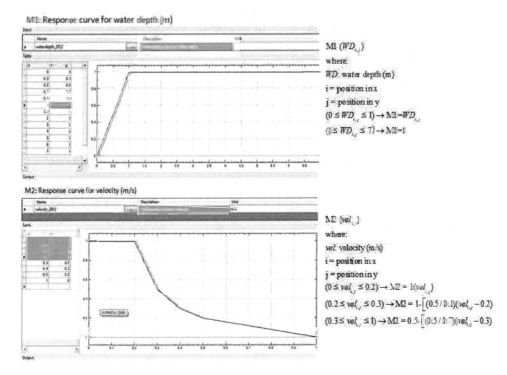

Figure 6-5 Response curves for water depth (M1) and velocity (M2). The x axis presents the variable values: water depth (m), velocity (m/s); the y axis presents the habitat index score

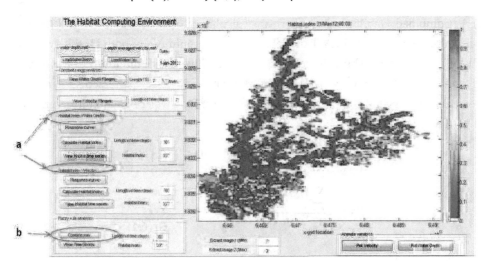

Figure 6-6 The dynamic Habitat Computing tool. a) Constant range analysis, b) Variable habitat index (VHI), c) Combined habitat index (Comb-HI). Colour bar indicates the Habitat index scale (0: no suitable 1: most suitable).

Furthermore, in order to have a large scale and overall HSI for each of the five areas, a second approach was applied. In this second approach, the cell values of water depth and velocity of each area were averaged before the calculation of the HI_WD and HI_Vel. Subsequently, the HSI was calculated in the same way as the previous approach by selecting the minimum of both. Results of this second approach were expressed in HSI (with a scale from zero to one), both temporally and spatially. All the calculations for approach 1 and 2 were performed for the total wetland area, and also for each of the five areas independently.

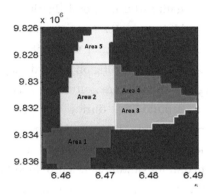

Figure 6-7 Wetland areas delimitation for habitat analysis

6.4 Results

The two different approaches introduced above were evaluated to explore the habitat suitability conditions of this tropical wetland. Results of the first approach are expressed in terms of percentage of suitable areas (PSA) with a HSI above 0.7 (sections 6.4.1 to 6.4.3). The second approach evaluated the wetland in terms of HSI scores (sections 6.4.4 to 6.4.6).

6.4.1 Natural variability of suitable areas

The analysis started by evaluating the temporal distribution of the percentage of suitable areas (PSA) with a HSI above 0.7 (first approach). Different hydrological years were simulated to describe the natural variability. Results described a high variation in terms of suitable areas depending on the hydrological conditions. During a dry year, the percentage of suitable areas for fish was up to 40% of the total wetland area, increasing to around 70% during wet years and for the historical maximum. Sampling years 2011 and 2012 were between both extreme conditions, with 2012 presenting higher percentages of suitable areas (up to 60%) compared to 2011 (up to 50%). Minimum and maximum values provided the limits to describe the historical thresholds that the wetland had experienced. The simulation of the minimum time series shows that historically the wetland had always provided at least a 25% of suitable area even in this extreme condition. For all conditions, higher percentages of suitable areas occurred during the wet season (January-May) (Figure 6-8).

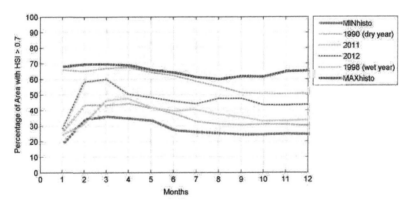

Figure 6-8 Temporal distribution of the Percentage of suitable wetland area (PSA) with Habitat suitability index (HSI) > 0.7. For: sampling years (2011 & 2012), compared with extreme dry and wet years (1990&1998), and minimum and maximum historical conditions (period 1962-2010).

6.4.2 Contribution of each wetland to the total wetland suitable area

Figure 6-9 illustrates the contribution of each of the five areas to the total wetland area with a HSI > 0.7. From the results, it can be seen that the areas 1 and 2 are the ones with a higher contribution, while the rest of the areas contribute less. The proportion of this contribution is maintained throughout the years analysed. The timing, at which the maximum of suitable areas occurred during the wet season, differed between years. For example, in 2011 it occurred during March and April, and in 2012 in February and March.

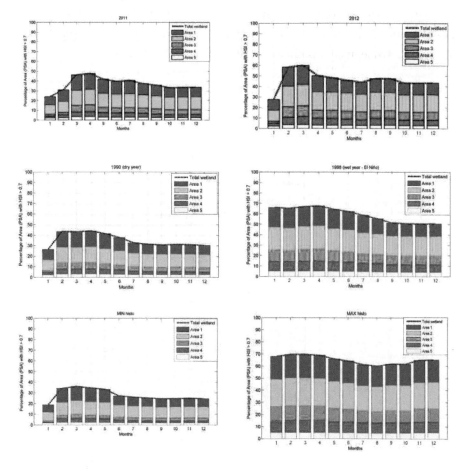

Figure 6-9 Contribution of each wetland area to the total wetland area with HSI > 0.7. For sampling years (2011&2012) see upper panel, dry and wet (1990&1998) see middle panel, and for Historical minimum and maximum conditions (1962-2010) see lower panel. The sum of the five areas equals the total wetland area (dashed black line). Months: January (1) to December (12).

6.4.3 Independent analysis of the PSA per area

Each of the five wetland areas was also analysed independently in terms of percentage of suitable areas (PSA). For this analysis, each area was compared to its own total area. Figure 6-10 illustrates the temporal pattern of each area for sampling years 2011 and 2012. From the analysis, it is shown that areas 1 and 2 are the ones with higher percentage of suitable areas (HSI > 0.7), with percentages up to 70% and 50% in 2011, and 80% and 65% in 2012, respectively. Wetland area 3 showed an intermediate pattern during both years, with percentages around 45% in 2011 and 60% in 2012. Lower percentages were observed for areas 4 and 5. Area 4 showed values around 30% in 2011 and up to 40% in 2012, showing a clear separation from area 5 in 2012, while in 2011 both areas followed a similar pattern.

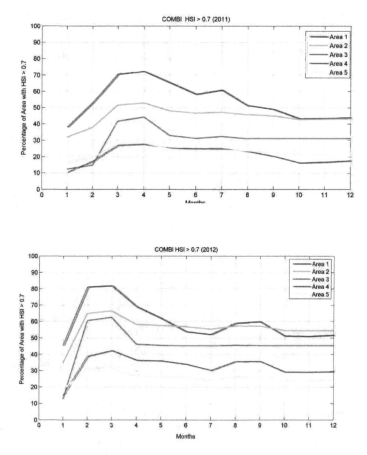

Figure 6-10 Percentage of suitable area (PSA) with a Habitat suitability index (HSI) > 0.7 for each wetland area. Sampling years 2011 & 2012. Months: January (1) to December (12).

6.4.4 Natural variability of HSI

The second approach of this chapter analysed the wetland in terms of HSI scores. Like in the first approach, different hydrological years were simulated to evaluate the natural variability of the index. The dry year maintained an HSI score of 1 from February till May, decreasing to 0.5 from July on. On the other hand, a wet year (El Niño) maintained a HSI of 1 for a longer period, decreasing slightly to 0.8 from September on till December. Sampling years 2011 and 2012 reached the maximum score in the months of March-April and February-March, respectively. The year 2011 followed a similar trend of a dry year, with values around 0.5 during the dry season. Extreme scenarios indicated that during maximum conditions the wetland can maintain a HSI of 1 during the whole year. The time series for Minimum Historical illustrates that in the most unfavourable conditions, the HSI score in the wetland is around 0.4 (Figure 6-11).

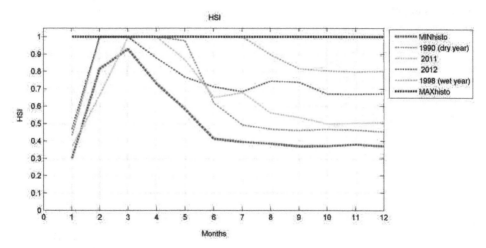

Figure 6-11 Temporal distribution of the Habitat suitability index (HSI). For: sampling years (2011 & 2012), dry year (1990), wet year (1998), and minimum and maximum historical conditions (period 1962-2010). Months: January (1) to December (12).

6.4.5 Independent analysis of the HSI per area

The five wetland areas were also analysed independently in terms of their individual HSI scores. Figure 6-12 illustrates the temporal pattern of HSI in each area for different hydrological conditions. From this analysis, areas 1 and 2 were again seen to be the ones with higher HSI scores; area 3 exhibited intermediate scores, while areas 4 and 5 had the lowest ones, for all simulated conditions. During a dry year, a clear separation between the areas was observed over the whole simulation period, while during a wet year this separation was only evident during the dry season period. The maximum historical condition displayed a constant highest score of 1 during the whole simulation, with the exception of areas 4 and 5 that slightly decreased to 0.9 during August. Interesting to see was that higher scores for areas 1 and 2 were also reached during the minimum historical simulation. Overall and from a temporal perspective, all wetland areas reached higher scores of HSI during the wet season period.

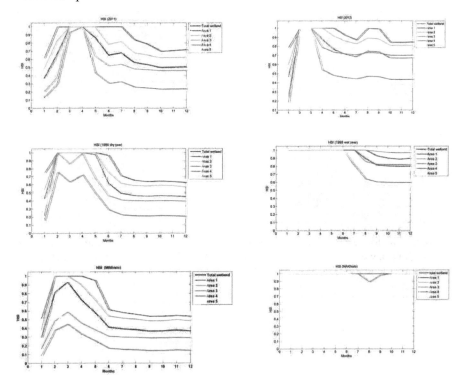

Figure 6-12 Temporal distribution of Habitat suitability index (HSI) for each wetland area (colour lines), and total wetland area (dashed black line). For: sampling years (2011 & 2012), dry year (1990), wet year (1998), and minimum and maximum historical conditions (period 1962-2010). Months: January (1) to December (12).

6.4.6 Spatial and temporal variation of HSI

Figure 6-13 displays the spatial and temporal variation of the habitat suitability index (HSI) for the different hydrological conditions. During the first six months of a WET year (El Niño year - 1998), all areas reached a HSI of 1, and were > 0.6 even during the dry season. The Maximum Historical showed a constant HSI above 0.9 during the whole year for all the areas. Overall, spatial HSI results for simulations 1990, 2011, 2012 and Minimum Historical described areas 1 and 2 as the ones with higher HSI scores (even during dry season (> 0.6), while Areas 4 and 5 were the ones with the lowest HSI scores (HSI values < 0.5 during the dry season). Area 3 presented intermediate HSI values. These results are suggesting that wetland areas 1 and 2 are the ones that provide better conditions for fish. Since wetland areas 4 and 5 showed lower HSI scores, these areas may require special attention in terms of management. Temporal pattern of HSI (for all simulations) defined the wet season period (February to April), as the key period providing suitable habitat conditions for the entire fish community of Abras de Mantequilla wetland.

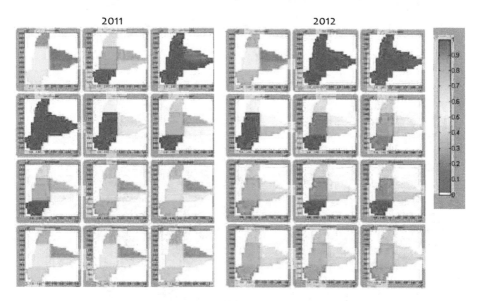

Dry year_1990   El Niño year_1998

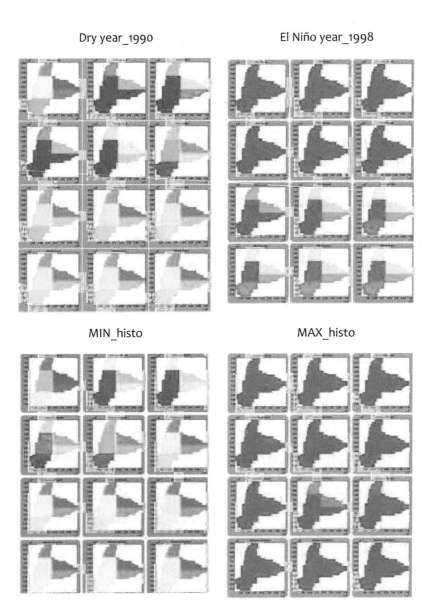

MIN_histo   MAX_histo

Figure 6-13 Spatial and temporal distribution of Habitat suitability index (HSI). For: sampling years (2011 & 2012), dry year (1990), wet year (1998), and minimum and maximum historical conditions (period 1962-2010). Months: top left (January), low right (December). Colour bar indicates the Habitat index scale from 0 (no suitable *in red*) to 1 (most suitable *in dark green*).

6.5 Discussion

6.5.1 The habitat index approach

This study described a methodology to evaluate the temporal and spatial distribution of habitat suitable areas for the overall fish community of the Abras de Mantequilla wetland. The response curves in the present study were developed with the aim to include the overall fish assemblage (for littoral and pelagic / limnetic). The criteria for the development of these rules were based on field sampling and literature for the littoral fish community, while for the pelagic community literature was the main source. Both communities utilize shallow littoral areas, the first ones as habitat during their entire life period, and the second one mainly to protect their eggs after spawning. Thus, a general criterion requiring 1 meter of water depth was assumed to be optimal in combination with velocities not higher than 0.2 m/s. Small littoral fish from the Characidae family were collected during both sampling campaigns in shallow littoral areas up to 1.5 m, combined with velocities not higher than 0.2 m/s. These hydrodynamic values are in agreement with the findings of other studies of Neotropical Characids about their habitat, distribution and feeding ecology (Casatti et al., 2003; Ferreira et al., 2012; Maldonado-Ocampo et al., 2012; Teresa and Casatti, 2013) and suggest both variables to be good predictors of community structure and species abundance (Teresa and Casatti, 2013). Regarding the influence of both hydrodynamic variables, results of the HI for water depth (HI-WD) and velocity (HI-Vel), showed that HI-WD was the main variable driving the habitat suitability index (HSI).

6.5.2 Percentage of suitable areas

The findings revealed a high natural variability of the percentage of suitable areas (PSA) depending on the different hydrological conditions simulated. Thus, from a historical perspective, the results showed that the wetland can provide a range between 25–70 % of suitable area (given the response curves implemented for this study). These limits can be used as minimum and maximum thresholds for management purposes. Regarding temporal patterns, results obtained for the wet season period (January-May) are the ones with a higher percentage of suitable areas (PSA) for all simulations. Nevertheless, during extreme wet conditions, a higher PSA was also observed during what can be considered normal for the months of the dry season period. The spatial analysis described the areas 1 and 2 as the ones providing more suitable habitat conditions and contributing with higher percentages to the total wetland habitat suitability. These areas are the ones that best fulfil the conditions described by the response curves. Local physical characteristics of these areas, like topography and proximity of the main inflow (Nuevo River) appeared to be the main drivers of the results. This is in agreement with the higher catch per effort for fishing activities reported in San Juan de Abajo (Florencio, 1993). This location belongs to Area 1 of our study (low wetland area). On the other hand, the areas 4 and 5 related to the Upper Chojampe inflows were the ones with a lower percentage of suitable areas. In these areas, the main source of water is related to run-off, and not to river inflow. Thus, these areas will require specific management measures in the future, in order to maintain their inflow contribution.

6.5.3 HSI scores

When the wetland was evaluated in terms of HSI scores, a similar pattern was observed for both spatial and temporal results. Higher HSI scores were obtained for areas 1 and 2 despite their hydrological conditions, and in general the months corresponding to the wet season period were the ones exhibiting higher scores for all simulations. From a historical perspective, and considering the whole wetland area, HSI scores were not lower than 0.4 even in the most unfavourable conditions (Minimum Historical).

6.5.4 Temporal availability of suitable areas

The temporal availability of suitable areas plays an important ecological role in the basin, since the majority of fish in the Vinces River and associated floodplains exhibit one reproductive cycle per year. At the end of the dry season, several fish species have a mature state ready for spawning (Barnhill and Lopez, 1974). These species usually have a high fecundity (high number of eggs to assure adequate repopulation). However, there are also species like *Aequidens rivulatus* (Vieja Azul) and *Cichlasoma festae* (Vieja roja), that spawn during the transition periods between wet and dry season, and others like the Brycon dentex that has been reported in mature stages also during the dry season (Barnhill and Lopez, 1974). Thus, both seasons are important but for different species, and therefore it is important to maintain the natural timing of inflows as a management measure for the AdM river-wetland system.

6.5.5 Littoral areas and vegetation

Several studies acknowledged the importance of littoral areas as habitat for fish communities (Arrington and Winemiller, 2006; Teixeira-de Mello et al., 2009), and the association of fish to the macrophytes (Agostinho et al., 2007; Meerhoff et al., 2007a; Meschiatti et al., 2000). Shallow areas of the AdM wetland are predominantly populated by small sized fish from the Characidae family (Alvarez-Mieles et al.,

2013). Characids are an important source of food for higher trophic levels (top fish predators that have a value for local communities) and important seed dispersers in Neotropical floodplains. Previous studies in the wetland and associated 'Guayas River basin' reported the presence of this family (Florencio, 1993; INP, 2012; Laaz et al., 2009; Prado, 2009; Prado et al., 2012). Some species of this family are common in all western basins of Ecuador (Gery, 1977; Glodek, 1978; Laaz et al., 2009; Loh et al., 2014), but others are endemic of the 'Guayas Basin' (Laaz and Torres, 2014; Roberts, 1973). However, information about the ecology or the evolutionary history of most fish species in the region is very limited and even lacking (Aguirre et al., 2013).

Littoral fish assemblages in AdM included both common and endemic species. At the middle and lower wetland areas endemic species like *Phenacobrycon henni, Landonia latidens, Iotabrycon praecox, Hyphessobrycon ecuadoriensis* were collected, indicating the importance of actually assessing the habitat conditions in this tropical wetland. Furthermore, the wetland provides a habitat for fish of commercial interest for local communities: *Aequidens rivulatus, Cichlasoma festae, Curimatorbis boulengeri, Brycon dentex, Ichthyoelephas humeralis*. These species have been collected in the main channels of the wetland, and move freely in the pelagic areas, but also utilize littoral vegetated areas, to protect their eggs after spawning (Barnhill and Lopez, 1974; Florencio, 1993; Quevedo, 2008; Revelo, 2010).

During fish sampling, another important characteristic that was observed in the littoral areas of the AdM wetland was the presence of associations of aquatic macrophytes. Floating macrophytes from the species *Eichornia crassipes* (Pontederiaceae), commonly known as 'water hyacinth' represented around 80% of the total macrophytes biomass in the wetland. *Salvinia auriculata, Pistia stratiotes, Ludwigia peploides, Lemna aequinoctialis, Paspalum repens,* and *Panicum frondescens* represented the other 20%. Thus, sampling results confirm the findings of other authors (Agostinho et al., 2007; Meerhoff et al., 2007a; Meschiatti et al., 2000), that recognized the association of small size species from the Characidae family to macrophytes banks that colonize littoral shallow areas and their essential role as shelter and food provider. Since juvenile and adults stages of small size species and eventually also juveniles of larger species are typical in macrophytes banks present in lentic shallow habitats (Meschiatti et al., 2000), their shelter role to protect small fish from higher predators plays an important role.

Shallow areas are also important for the pelagic community. Pelagic species such as *Aequidens rivulatus* (Vieja Azul) and *Cichlasoma festae* (Vieja roja), both typical of this wetland area, utilize the littoral areas mainly during and after spawning. These species present a high parental care after spawning because they produce a low number of eggs (Barnhill and Lopez, 1974). This is contained in the general criterium of defining the water depth optimal from 1 meter onwards.

6.5.6 Fish studies in the AdM wetland and associated basin

A study about biological aspects of the fish community in the basin revealed that 70% of the specimen sampled during the months of January to March reported an advanced stage of sexual maturity (stages III to V) (Revelo, 2010), while the rest of the specimen were already in stages of post spawning (I and II). When the analysis becomes more specific per species, the timing of mature stages differed slightly between the months of the wet season. For instance, Brycon dentex (Dama) reported a higher number of specimen with an advanced mature stage in January (III, IV and V), while in February and March immature stages were more frequent, indicating that the spawning probably occurred between January and February. *Ichthyoelephas humeralis* (bocachico) was reported in advanced mature stages (III and IV) during January and February. Other species of less commercial interest like *Hoplias microlepis* (Guanchiche), were reported in mature stages during the first 3 months of the year (Jan-March), and immature during April. *Aequidens rivulatus* (Vieja azul) was reported in advanced maturity stage (III, IV ad V) during January and February, and had immature stages in April and May. In addition, a smaller percentage of immature specimens were reported during October and November, possibly explaining that this species has two reproductive cycles per year. *Curimatorbis boulengeri* (Dica), reported advanced mature stages during February, and immature stages during March and April (Revelo, 2010). The last one is confirmed by the sampling performed during the present research where small immature specimens of *Curimatorbis boulengeri* were collected during March 2012.

All these findings provided the evidence that the wet season and associated high flows, represents an important period for the development and increase of the fish population in the study area, which is consistent with findings of other tropical systems that acknowledge the importance of high flows and floods in supporting the gonadal maturation of fish (McClain et al., 2014).

Fishing activities in the wetland occurs during 10 moths over the year, but starts usually at the end of the wet season (Florencio, 1993). Local farmers from El Recuerdo village have also reported the catch of bigger size fish during the dry season (T. Estrella, pers comm, 2016). They mentioned that they wait until the sizes are big enough to catch them in order to allow the growing of the fish population. In this regard, there is also a regulation in Los Rios Province that establish a ban ('veda') for fishing activities from January 10 till March 10 (Revelo, 2010).

The information in Chapter 4 about biotic communities structures determined that although the different areas of the wetland share similar fish species from the Characidae family, there were species that seem to typify middle areas where higher residence times take place. Other species typified the lower area, which is an area influenced by the dynamics of the river inflow, while other species typified the river itself. These findings at species level can provide the basis for future research on the construction of new rules and habitat assessment at a lower taxonomic level for the fish community in this tropical wetland.

6.5.7 Overall findings

The present study is the first attempt in providing an assessment of the temporal and spatial variability of suitable habitat areas for the fish community in Abras de Mantequilla wetland. Results indicate how hydrodynamic variables can facilitate the definition of suitable habitat areas in this wetland, both in terms of PSA and HSI scores. The study describes areas with HSI > 0.7 as optimal habitat for the entire fish community. In this study, wetland areas with HSI >0.7 were described as an optimal habitat for this fish community. However, areas with a HSI <0.7 can not necessarily be considered as uninhabitable. One of the limitations of the approach followed here is that the rules were developed for the whole fish community, rather than for specific species. For this, more extensive sampling in the area is required to measure the habitat preference of different species. Still, the present methodology can provide an initial basis for future habitat assessments of specific fish communities in the area.

The combination of hydrodynamic variables was useful for an initial habitat assessment of the fish communities in this wetland. However, should be acknowledged that other physical, chemical and biotic variables play an important role in defining the habitat preferences and therefore should be gradually included for an integrated ecological habitat assessment. The habitat tool developed in this study is quite flexible for adding more variables and their correspondent rules.

The high flow phase of the wet season was recognized as the period with a higher percentage of suitable areas and HSI scores for all the simulated conditions. Spatial zonation defined the areas close to the main inflow as the ones providing better habitat conditions, and areas related to Chojampe subbasin as the ones that will require special attention in terms of management measures.

Based on the results of this study, it is recommended to maintain the timing and magnitude of the natural flows especially during the periods with higher percentage of suitable areas (high flows of the wet season), since this period is crucial to foment the spawning and development of the fish community in the AdM wetland.

"Great things are done by a series of small things brought together"

Vincent Van Gogh

7

DISCUSSION AND SYNTHESIS

7.1 Sustainability of the AdM wetland hydrodynamics

Based on the results of previous chapters containing data and model results, an assessment can be made for Abras de Mantequilla wetland in terms of sustainability. The Abras de Mantequilla (AdM) wetland is subject to two major environmental disturbances i.e. (i) an infrastructure project located in the upper catchment of the Quevedo-Vinces River (construction of the Baba dam), and (ii) short term enhanced agriculture inside the wetland area. Initially DAUVIN, a water transfer project (ACOTECNIC, 2010), was also expected to have an impact on the wetland. However, because its design follows the old river path of the Nuevo River that does not interfere with the current flows entering and leaving the wetland, this project has neither positive nor negative effect on the system (Arias-Hidalgo, 2012).

In order to achieve sustainable solutions for AdM wetland, management options should focus on maintaining natural variation in hydrodynamic conditions throughout the entire catchment, as well as implement good practices in agriculture and reforestation using native species. Local and national authorities should support continuous monitoring programmes, taking account of seasonal variation and of future impacts from flow reduction and nutrient enrichment. (Alvarez-Mieles et al., 2013).

The wetland functioning was seen to mainly depend on the Nuevo River which determines the timing and magnitude of the inundation patterns. Since the Nuevo River is diverted from the Vinces River, it is expected that the effect of the Baba dam, located in the upper Vinces River catchment, may reduce the Nuevo River flow. Previous research in the study area determined that the dam operation might reduce the flow of the Vinces River immediately down the dam by about 70%. Nevertheless, due to the contribution of the two rivers Lulu and San Pablo, the Vinces River is expected to recover in the downstream section and the final annual reduction of flow may be about 30% (Arias-Hidalgo, 2012; Arias-Hidalgo et al., 2013). Although this reduction may affect the overall hydrodynamic pattern, the analysis of historical minimum flows indicated that the wetland has experienced even lower values than the ones that could be attributed to the dam operation that started during the period 2012-2013.

AdM wetland is a system that experiences a marked and strong seasonal variations in flushing rates, and as a result, in inundation areas. During the wet season (January-April), the system exhibits a great variability in inundation area due to the rainfall peak events during a given year (intra-annual), and inter-annual rainfall variability. During the dry season, the wetland decreases dramatically and water remains only in the main channels (observations in situ), and with a similar pattern from year to year. Thus, AdM wetland can be classified as a temporary wetland with a predictable pattern (Brock et al., 2003).

Usually, longer dry periods can affect aquatic biota. The impact of drought on the biota is influenced by factors such as previous hydrological history (system used to dry recurrently), the timing of this drying period, and availability of refuges for the biota during the dry period (Boulton, 2003). Longer-lived pools and even the main river/wetland channel can also serve as refugia. Aquatic communities of temporary wetlands have several mechanisms to survive the dry periods. For instance, zooplankton and aquatic plants are known to rely on their egg and seeds as an instrument to survive the dry periods, and being able to recover after these dry periods by patterns of dormancy, hatching, germination, establishment and reproduction (Brock et al., 2003). Macroinvertebrates survive dry periods by sheltering below cobbles, debris, or among macrophytes, while other penetrate the hyporheic zone (Boulton, 2003). The availability of refugia is crucial for the survival of some aquatic biota, providing the capacity to recover from the droughts once they pass (Lake, 2003). For fish, droughts lead to (i) a reduction in the surface area/volume, and (ii) an increase in physicochemical extremes. These abiotic disturbances increases biotic interactions such as competition and predation, which in turn also leads to an increase in mortality, migration and a decline in birth rates (Magoulick and Kobza, 2003).

There is also a difference between seasonal droughts, e.g. predictable dry periods in tropical areas), and the supra-seasonal droughts that are 'unpredictable'. This distinction is crucial because a 'seasonal drought' is simply an extreme of the hydrological pattern of a specific water system, with perhaps littler impact on the occurrence of biota, even if having a major effect on the relative abundance of species (Lake, 2003). Natural droughts also cause a temporal variety of habitats and diversity in aquatic environments (Boulton et al., 2000). Nonetheless, many ecological responses depend on the timing of hydrological transitions across certain thresholds (Boulton, 2003). Furthermore, droughts have not only 'direct' impacts (reduction of flow and habitat), but also indirect impacts such as predation, competition among species (Lake, 2003).

From the previously exposed about how species can adapt to 'predicted seasonal droughts', it is possible to presume that the aquatic species at AdM wetland are already adapted to the seasonal dry period occurring from May to December. Thus, the sustainability of the AdM wetland system will depend on the maintenance of the wet-dry cycles under the range that the system has experienced historically, as explained in Chapter 2.

The application in this thesis of the hydrodynamics model DELFT3D-FLOW was provides with an indication of how this tropical river-wetland system works and how the magnitude of the inflows that the system has faced historically, can be expected under future conditions, depending on effects of dam operation and climate change. This way of numerical modelling allows exploring water system functioning under extreme limiting conditions and evaluating the response to changing conditions. Such results provide with a basis for assessing this tropical river-wetland system in terms of altering hydrodynamic conditions that in turn influence the nutrients and plankton. The temporal and spatial patterns of nutrients and chlorophyll-a in AdM (Chapter 5), identified nitrogen as the limiting nutrient in the system, and intermediate in terms of trophic status. Furthermore, hydrodynamics outputs provided also with the basis for setting a habitat model for fish (Chapter 6), suggesting which could be the important periods during the year that flows should be maintained, and also the identification of which areas require special attention for future management.

An integrated mass balance analysis for nutrients components and primary production was applied for this wetland in order to evaluate the system on a 'yearly' basis. A yearly basis could provide with a useful approach to see the system in terms of overall productivity. This mass balance analysis was possible by integrating sampling results and the outputs from the eco-model application (Chapter 5). Relevant processes and associated fluxes for nutrients and primary production are also quantified. Section 7.2 shows overall yearly balances of nitrogen and phosphorus, and evaluate the possible contribution of internal loads versus external loads in this system. Section 7.3 evaluates a mass balance for primary producers. Comparison with other water systems and further analysis is presented in sections 7.4 and 7.5.

7.2 Mass balances of nutrients

Annual mass balances of nutrients in AdM wetland system were evaluated for the different hydrological conditions described in chapters 2 and 5. Output files (.bal) from the ECO-model (Chapter 5) containing annual mass balances information for each variable were used to calculate the mass balance for each hydrological condition Mass balances for Total Nitrogen and Total Phosphorus were evaluated within this thesis in order to quantify the magnitude of external loads, internal loads, and relevant processes and fluxes associated to both nutrients (Figure 7-1 & Figure 7-5).

7.2.1 Total Nitrogen (TN)

Results of yearly nitrogen mass balances determined that external loads (Total inflow + diffusive waste NO_3) entering the wetland system has been in the order of 51 $gN/m^2/year$ for a dry year, and up to 144 $gN/m^2/year$ for El Niño year, thereby increasing about three fold from dry to wet conditions. When also extreme scenarios (MINhisto and MAXhisto) were included this range was wider, increasing up to ten times (24-266 $gN/m^2/$ year). Nitrate concentrations in the wetland are low (Chapter 3), thus these mass balances results are mainly driven by the organic fractions of nitrogen.

The main external source of N was the Nuevo River for all conditions. As explained in data (Ch3) and modelling (Ch5) chapters, the organic nitrogen fraction was the main component of total N, especially during periods of high inundation (sampling of 2012); while nitrates concentrations were usually low and in the range of the ones suggested in the paper by Meybeck (1982) for unpolluted tropical waters (Meybeck, 1982) .

Internal loads were not higher than 11 gN/m^2 and, therefore, seemed to not represent a major source for the system (Figure 7-1 & Figure 7-8). Denitrification in water was almost negligible and total denitrification (water+sediment) was low, most likely due to the oxygenated conditions of the system (Figure 7-1). As a result, the modelled denitrification flux (water+sediment) was not higher than 9% of the external loads. Despite these low percentages, a particular decreasing pattern from driest to wettest conditions is suggested. Thus, driest conditions showed higher percentages, suggesting that, although denitrification in water was almost negligible, total denitrification processes (water+sediment) appeared to be more relevant during years of low inflows (Figure 7-2).

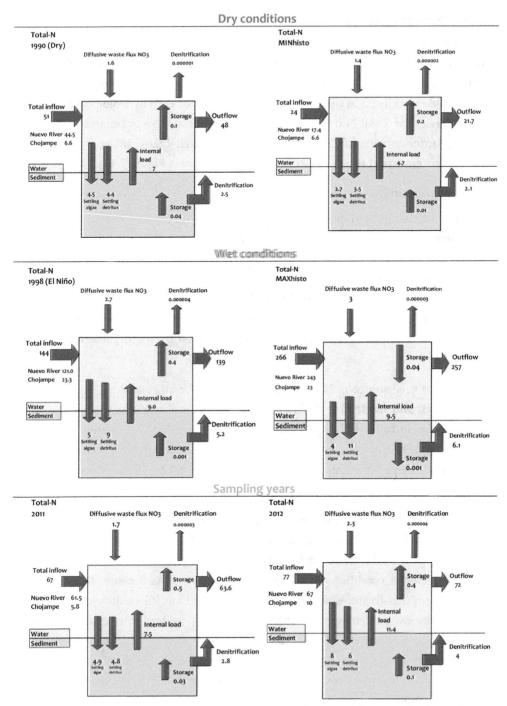

Figure 7-1 Total Nitrogen mass balance for one year simulation and entire wetland area (Units: gN/m^2/year). Inflows, outflows and main processes depicted for the different hydrological conditions and sampling years 2011 & 2012.

Storage (accumulation) in the water and sediment compartments was not important with values close to zero (Figure 7-1), indicating that the system shows features of dynamic steady state system (Smits and van Beek, 2013). Outflow of nitrogen was always over 90% of the inflows for all the conditions, indicating a very dynamic system with high flushing patterns (Figure 7-1). Total N settling (N in algae +N in detritus) appeared to be higher during wettest conditions (up to 16 gN/m²/year) (Figure 7-3). The relative importance of each settling component was calculated (Figure 7-4). Results suggest that the proportion of N settling as detritus was higher in five of the conditions simulated; especially during wettest conditions. Furthermore, two of the wettest simulations (1998 and MAXhisto) showed a lower proportion of nitrogen settling as algae (Figure 7-4), perhaps suggesting that algae may be subject to more consumption by grazers during wettest periods and therefore the nitrogen settling as algae is lower.

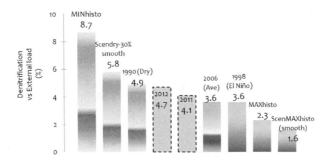

Figure 7-2 Total denitrification (water+sediment) with respect to external loads (%). Brown bars (driest conditions); blue bars (wettest conditions); light orange with dashed lines (sampling years 2011& 2012).

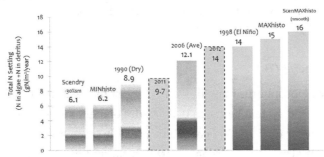

Figure 7-3 Total Nitrogen settling (N in algae+N in detritus) in gN/m²/year. Brown bars (driest conditions); blue bars (wettest conditions); light orange with dashed lines (sampling years 2011& 2012).

a)

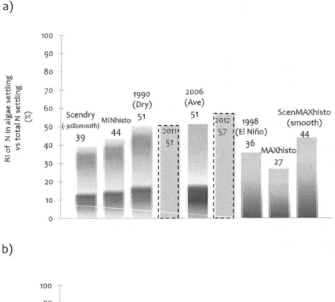

b)

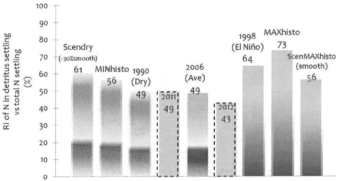

Figure 7-4 Relative importance (%) of: a) N in algae settling and b) N in detritus settling; both versus Total N settling. Brown bars (driest conditions); blue bars (wettest conditions); light orange with dashed lines (sampling years 2011& 2012).

7.2.2 Total Phosphorus (TP)

Results of yearly phosphorus mass balances suggest that external loads (total inflow) entering the wetland system has been in the order of 3 gP/ m²/year for a dry year and up to 10 gP/ m²/year for El Niño year. Thus, possibly three times higher from dry to wet years. When also extreme scenarios (MINhisto and MAXhisto) were evaluated this range appeared to be wider (2 to 18 gP/ m²/ year). For all conditions, the main external source of P was the Nuevo River. Internal loads were not higher than 0.9 gP/ m² (Figure 7-5 & Figure 7-8). Similar to the nitrogen balance, storage in water and sediment appeared to be not important, and the P outflows were over 93% of the P inflows; possibly indicating a highly dynamic system (Figure 7-5). Total P settling (P in algae + P in detritus) was also higher during wettest conditions (up to 1.4 gP/m²/year) (Figure 7-6). The proportion of P settling as detritus was also higher in five of the conditions simulated; and the proportion of phosphorus settling as algae was also lower during the wet simulations (1998 and MAXhisto) (Figure 7-7).

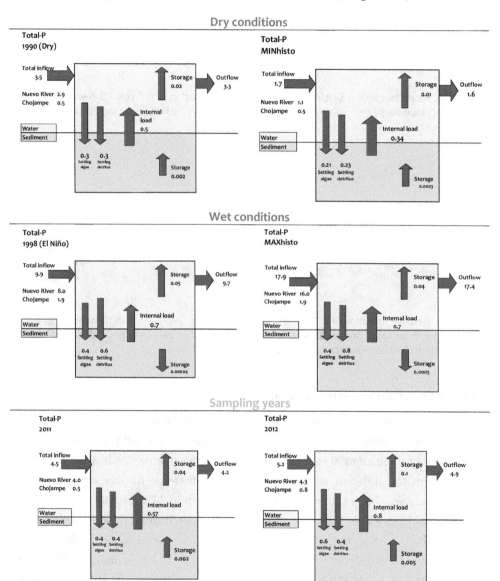

Figure 7-5 Total Phosphorus mass balance for one year simulation and entire wetland area (Units: in gP/m²/year). Inflows, outflows and main processes depicted for the different hydrological conditions and sampling years 2011 & 2012.

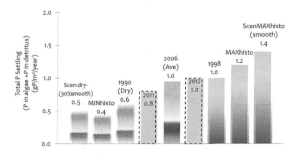

Figure 7-6 Total Phosphorus settling (P in algae + P in detritus) in gP/m²/year. Brown bars (driest conditions); blue bars (wettest conditions); light orange with dashed lines (sampling years 2011& 2012).

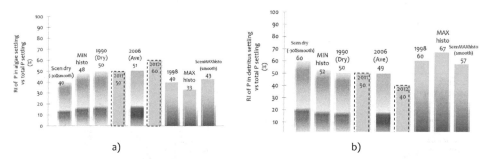

Figure 7-7 Relative importance (%) of: a) P in algae settling and b) P in detritus settling; both versus Total P settling. Brown bars (driest conditions); blue bars (wettest conditions); light orange with dashed lines (sampling years 2011& 2012).

7.2.3 *Relative importance of internal loads*

Nutrient return fluxes from the sediments to the water column were considered in this analysis as 'internal loads' for the water column compartment. The relative importance of these internal loads (%) compared to the external loads was evaluated. Results from the different simulations suggested that the relative importance of internal loads was higher during driest conditions up to 20% for both nitrogen and phosphorus (Figure 7-8). Thus processes such as 'mineralization fluxes in sediments' that contribute with nutrients to the water column appeared to be more relevant during years of low inflows. On the other hand, during wettest conditions (El Niño and both MAXhisto scenarios) the relative importance of these sediment-water fluxes decreased and was not higher than 6% for both nutrients, suggesting that during wettest conditions the wetland system might be less dependent on internal processes (at least regarding to sediment-water column exchanges).

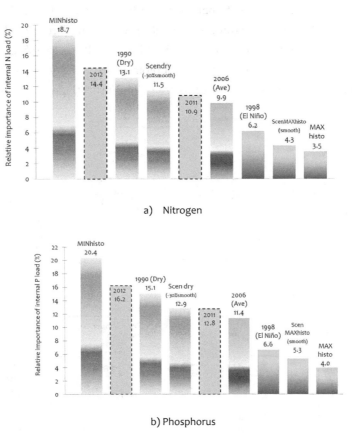

a) Nitrogen

b) Phosphorus

Figure 7-8 Relative importance (%) of internal loads of nutrients (N, P) with respect to external loads. Brown bars (driest conditions); blue bars (wettest conditions); light orange with dashed lines (sampling years 2011& 2012).

7.3 Mass balances of primary producers

Annual mass balances of primary producers (phytoplankton) in AdM wetland system were evaluated for the different hydrological conditions described in chapters 2 and 5. Output files (.bal) from the eco-model (Chapter 5) containing the annual mass balances information for 'total phytoplankton' were used to calculate the mass balance for each hydrological condition. Results of analysing yearly phytoplankton mass balances indicated that external loads (total inflow) of phytoplankton may increase three times from dry to wet years: from 3.5 gC/m²/year (dry year-1990) to 10 gC/ m²/year (El Niño-1998). A broader range seemed to occurred by including also extreme scenarios (1.7 to 18 gC/ m²/year).

Nuevo River was the main external source to the system; although it was minimal when compared with the magnitude of in-situ production (Figure 7-9 & Figure 7-11). Internal algae production was between 89 (dry year-1990) to 165 gC/ m²/year (El Niño-1998), thus suggesting a doubling in biomass from a dry to a wet year; with a wider range for extreme scenarios (61 to 221 gC/ m²/year) (Figure 7-9 & Figure 7-10).

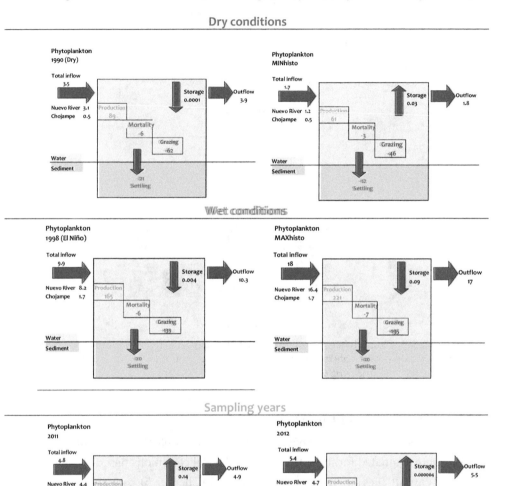

Figure 7-9 Primary producers mass balance for one year simulation and entire wetland area (Units: in gC/m²/year). Inflows, outflows and main processes depicted for the different hydrological conditions and sampling years 2011 & 2012.

Algae consumption by grazers appeared to follow a similar pattern, doubling from a dry year (62gC/m²/year) to El Niño year (139 gC/m²/year) and increasing up to 195 gC/ m²/year in a extreme wet scenario (MAXhisto) (Figure 7-9 & Figure 7-10). Sedimentation and mortality showed similar values for a dry and El Niño year; 21 gC/ m²/year and 6 gC/ m²/year, respectively, so no clear trend was observed when absolute values for these years were analyzed. However, when extreme scenarios were also considered, both processes showed lower values during the driest scenarios (Figure 7-9). Storage was unimportant with modelled values close to zero. Outflows were in the same magnitude of inflows, and even slightly higher than the inflows for some conditions. The latter may suggest that the phyto-biomass that is not consumed in the wetland is probably flushed downstream (Figure 7-9).

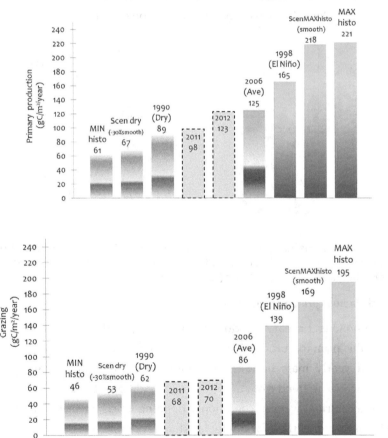

Figure 7-10   In-situ Primary production (upper panel) and Primary consumption (lower panel) (gC/m²/year). Brown bars (driest conditions); blue bars (wettest conditions); light orange with dashed lines (sampling years 2011& 2012).

7.3.1 Allochthonous vs autochthonous primary production

The relative importance of primary production entering the system via the inflows compared with *in situ* primary production was calculated from the mass balance outputs of the primary production (section 7.3) for each hydrological condition (Figure 7-11). Results suggested that during wettest conditions, allochthonous sources of phytoplankton were more relevant in the overall balance of phytoplankton production than during driest conditions. Nevertheless, for all the conditions this external contribution was not higher than 8.2%, indicating that autochthonous primary production may be more important than allochthonous in this wetland system. These findings contribute by illustrating the importance of this wetland system as a key source of autochthonous primary production relevant for the next trophic levels. Sampling years 2011 and 2012 showed intermediate results with percentages closer to an average year simulation (Figure 7-11).

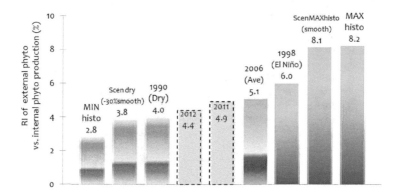

Figure 7-11 Relative importance (%) of external phytoplankton biomass (inflows), with respect to internal phytoplankton biomass production. Brown bars (driest conditions); blue bars (wettest conditions); light orange with dashed lines (sampling years 2011& 2012).

7.3.2 Autochthonous primary production

The autochthonous primary production was calculated from the mass balance outputs of the primary production (section 7.3) for each hydrological condition. Results from yearly mass balances determined that internal primary production showed an increasing trend from dry to wettest conditions (Figure 7-12). This finding indicates that during wettest conditions, flooding areas increased and therefore provided more water surface for phytoplankton development.

Furthermore, during wettest conditions, the system is expected to be more dynamic in terms of water exchanges that promote frequent input of nutrients, and increases the algae turnover. The range of algae production can increase around four times from the driest (2462 tons/year) to almost 9000 tons/ year for the wettest condition (Figure 7-12).

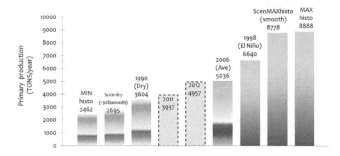

Figure 7-12 Internal primary production biomass (Tons/per year). Calculated from mass balances outputs of the different conditions, and sampling years 2011& 2012. Brown bars (driest conditions); blue bars (wettest conditions); light orange bars with dashed lines (sampling years 2011& 2012).

7.3.3 Primary producers and associated processes

Algae biomass in the water column is subject to several processes: gross primary production, respiration, excretion, mortality, grazing, resuspension and settling (Deltares, 2013c). The phytoplankton mass balances presented in this chapter illustrate the relevance of the processes related to control of algae biomass produced in the wetland. Results from the current eco-model set up (Chapter 5) suggested that grazing by zooplankton, was the main process controlling the algae biomass in this system and given the conditions established during the model set up. Grazers consumed 57 to 88% of the total phytoplankton biomass produced in situ, depending on the simulated condition.

Possible explanations for these results are detailed in section 7.5. Wettest simulations showed the higher percentages, suggesting that grazing might be maximized during conditions with higher inflows. On the other hand, algae sedimentation appears to be more relevant during driest conditions (16-23%), compared with 9-15% during wettest conditions. Overall, algae mortality was not higher than 11% of the total algae biomass produced in the wetland (Figure 7-13).

Algae mortality in the DELFT3D BLOOM module is caused by temperature dependent natural mortality, salinity stress mortality and grazing by consumers

(Deltares, 2013b). In this thesis the zooplankton module (CONSBL) was also implemented (section 5.2.6 - Chapter 5). This module modelled the 'grazing' separately from 'mortality' (Deltares, 2013b). Thus, the mass balances outputs from the eco-model (Chapter 5) separated: the algae biomass decrease due to mortality, the algae decrease due to sedimentation, and the algae decrease due to grazing as depicted graphically in Figure 7-13.

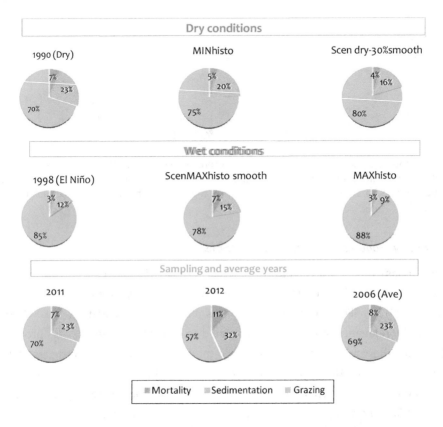

Figure 7-13 Main processes controlling algae biomass (%) calculated from mass balances outputs of the different conditions and sampling years 2011& 2012.

7.4 Nutrient balances and their variations between different hydrological conditions

External loads of nutrients (N, P) appeared to increase three-fold from a dry to a wet year and up to ten times if maximum historical scenarios were considered. Thus, suggesting that the wetland may have a high variation in terms of loads entering the system yearly, which could be mainly influenced by the magnitude of flows, and therefore highly dependent on the hydrodynamics. Evaluating the system only with measured data of nutrients (Chapter 3) and temporal simulations (Chapter 5), were not sufficient to clarify the overall functioning of the system in terms of nutrient loads entering the system yearly. Specially, because simulated nutrient concentrations were similar among the different hydrological conditions evaluated. The application of mass balance analysis suggested that the main external source of nutrients entering the wetland is the Nuevo River, thus proposing that this wetland system may be highly depending on this inflow in terms of nutrients. The relative importance of internal sources (specifically sediment return fluxes) compared to external sources (nutrients entering via the inflows) appeared to be low, indicating that this system could be mainly driven by external rather than internal sources of nutrients. This finding contrasts with results from a tropical reservoir where internal loads almost double external loads due to long residence times (up to 340 days) (Smits, 2007), while in the AdM wetland it does not exceed 30 days. In shallow lakes, a higher release of phosphorus from the sediments is usually expected (Jensen and Andersen, 1992; Jeppesen et al., 2014). In contrast, the flow dynamic characteristics in AdM and resulting lower residence times differ from the ones of lake systems, and perhaps these factors are limiting the P release. Internal loads in AdM seemed to gain some importance only during dry years and extreme minimum historical scenarios, probably due to less flow dynamics, resulting in less water flushing, but even then, they just represented up to a maximum of 20% of the external loads. This finding suggests more dependency on internal processes during driest conditions.

High denitrification is expected to occur in tropical lakes, since higher temperatures foment the speed and duration of redox mechanisms as well as denitrifiers metabolism rate (Lewis, 1996). However, in AdM the importance of total denitrification (water+sediment) is suggested to be relatively low, most probably attributed to the high water dynamics that supports constant oxygenation and limited time with stagnant conditions. A study of several catchments in United States determined that watersheds with greater precipitation and discharge have low denitrification due to fast flushing through the wetlands and streams (Howarth et al., 2006). Although high denitrification is expected to occur in wetlands, the amount of nitrogen that can be removed depends on the residence time. Higher residence times promote an increase in the settling of particulates and the amount of nitrate (per unit volume of water) that diffuse into sediments, resulting in higher nitrogen removal (Saunders and Kalff, 2001; Seitzinger et al., 2002; Van Breemen et al., 2002). Thus, possibly the residence times in AdM, that range from 5 to a maximum 30 days (depending on the location) may not be high enough to foment this process.

When comparing with external sources, denitrification was not higher than 9% of the external loads, and this occurred only during extreme dry scenarios. This finding suggests also more dependency on internal processes during driest conditions. Nutrient storage was almost negligible suggesting that the system is in balance (steady state system), and thus nutrient loads entering in the system are uptake by primary producers which are in turn consumed by grazers, and therefore the storage of nutrients is close to zero.

Sedimentation of nutrients was low compared with the external loads, most likely influence by the high water dynamics and grazing, since grazers also consumed detritus (PON-POP), which are also components of the total nitrogen and total phosphorus balances. In wetlands, sedimentation of nitrogen has been usually found lower than denitrification (Saunders and Kalff, 2001). Results from the yearly nitrogen balance in AdM wetland differed with this pattern, showing that yearly settling rate of nitrogen (N algae + N in detritus) was on average three times higher than yearly denitrification rate (water+sediment). Two factors could be responsible for this pattern: (i) the fact that some wetland areas have higher residence times would promote settling, but on the other hand (ii) oxygenated conditions due to constant floods would prevent denitrification.

The higher settling of total nitrogen during wettest simulated conditions is perhaps the result of great nutrient availability (that is not uptake by algae) during high inflow conditions. The proportion of N settling as detritus was higher than the N in algae.

Outflows fluxes were almost in the same magnitude as the inflows, suggesting that the system has high flushing patterns of nutrients. Thus, the areas located downstream the wetland could receive similar concentrations to the ones measured in the wetland. All these characteristics may describe the system as a highly dynamic wetland.

7.5 Wetland productivity and related processes

Phytoplankton mass balances established that external loads (total inflow) of phytoplankton might increase three times from dry (3.5 gC/m²/year) to wet years (10 gC/m²/year). However, the magnitude of external loads could be low if compared to the in-situ production, since internal algae production in the wetland may reach up to 165 gC/m³/year during El Niño year. Thus, given the high algae biomass produced in-situ, the wetland could be considered as a highly valuable environment in providing resources for the upper trophic levels.

The autochthonous algae production in AdM wetland may double from a dry year (89 gC/m²/year=3604 tons/year) to El Niño year (165 gC/m²/year=6640 tons/year) and can increase up to almost 9000 tons/ year for the wettest condition scenario (MAXhisto). These rates are in between the ranges reported for oligotrophic (up to 300 mgC/m²/d=110 gC/m²/y) and mesotrophic lakes (up to 1000 mgC/m²/d=365 gC/m²/year) (Wetzel, 2001d), and below the maximum reported for relatively unpolluted tropical rivers (up to 1000 mgC/m²/d =365 gC/m²/year) (Dudgeon, 2011). Thus, it could be assumed that AdM system is not polluted. Rates of polluted rivers can go beyond 6000 mgC/m²/d (2190gC/m²/year) (Dudgeon, 2011).

If compared with tropical lakes productivity, AdM rates appeared low compared to the highly productive Lake Victoria and Lanao (both around 600 gC/m²/y); but share similar rates as temperate lakes Clear and Erken (up to 160 g C/m²/y), considered as productive lakes for temperate standards (Lewis, 1974) and classified as mesotrophic (Wetzel, 2001d). If AdM productivity rates are positioning between the aforementioned rivers and lakes rates, it could be proposed that AdM may show rates similar to lotic tropical systems and lentic temperate ones; thus, intermediate position between river and lake, and this can be expected due to the high water dynamics of the system.

Sedimentation of phytoplankton was found also low, and probably driven by the high water dynamics and high grazing rates that may prevent algae to settle. Inspection of algae limiting factors, suggested that algae was not limited by nutrient availability.

Several studies indicated that the biomass of phytoplankton is usually higher than the one of zooplankton (Auer et al., 2004; Havens and Beaver, 2011; Havens et al., 2007; Havens et al., 2009). Results from the present study differed from this pattern. Thus, the zooplankton function implemented for the model may have been overestimated. Nevertheless, this function maintained chlorophyll-a concentrations and nutrients in the range of the measured values, and appeared more important than calibration with bottom-up variables. Thus, the phytoplankton mass balance analysis proposes grazing as a key processes controlling algae biomass, suggesting that the system may be top-down controlled.

Different studies on bottom-up and top-down theories determined different outcomes for different systems, and related also the trophic status of the system to the prevalence of either one of these theories (McQueen et al., 1989). The top-down theory suggests that changes in piscivore abundance will cascade down through the food web and will cause chlorophyll-a concentration to deviate from nutrient-predicted concentrations. Thus, an increase in piscivore abundance will decrease planktivorous fish that feed on zooplankton, decreasing subsequently the grazing pressure of zooplankton over phytoplankton, resulting in an increase in chlorophyll-a concentrations.

There is another theory that combines both regulators (the predicted influences of both predators (top-down TD) and resource availability (bottom-up BU). This BU-TD theory predicts that top-down forces should be strong at the top of the food web and weaken towards the bottom, while BU forces should be strong at the bottom ad weaken towards the top. This leads to the prediction that in higher trophic water systems (not the case of AdM wetland), the top down control apparently exerts limited influence over the chlorophyll-a biomass. In these systems, the changes in piscivore biomass may damp out as they cascade down through the food web so that there will be little or no influence on chlorophyll-biomass (McQueen et al., 1989). On the other hand, since AdM is a system that could be classified as mesotrophic (due to their chlorophyll-a concentrations), top down control is perhaps occurring.

However, in wetlands much of the primary production is not eaten by herbivores and ultimately becomes litter (Van der Valk, 2012). Thus, there are two different paths for the transferring of energy from one trophic level to another: autotrophic and heterotrophic and the relative importance of both paths has been argued (Newman, 1991). The autotrophic pathway involves the consumption of living algae or living macrophytes by herbivores (thus grazers). The heterotrophic involves the decomposition involves the consumption of macrophytes, algae, and animal litter by microorganisms, and the microorganisms by invertebrates (detritivores) (Van der Valk, 2012). Historically, the heterotrophic pathway has been recognized as dominant in wetlands, however, if the magnitude of algae production became better documented, the autotrophic pathway is increasingly considered equally or even more important (Van der Valk, 2012).

Furthermore, majority of authors recognize the higher effectiveness in grazing pressure of large zooplankton species (cladocerans as *Daphnia*) due to their apparently 'more effective' filtration rates compared with smaller sized cladocerans, suggesting that only high abundances of *Daphnia* can have a grazing impact (Auer et al., 2004). However, in a study including 55 lakes in Germany, no difference between the grazing effect of large cladocerans and other smaller cladocerans over other plankton components was detected (Auer et al., 2004).

The zooplankton community of AdM wetland is mainly composed by small sized cladocerans, thus, these smaller size cladocerans probably have also an important grazing function in the wetland. The impact of the grazing function over phytoplankton biomass may be limited when high densities of cyanobacteria (blue-green algae) are present in a system, because of their inedible characteristics, thus reducing the grazing impact of zooplankton over the phytoplankton (Auer et al., 2004). In contrast, in AdM wetland, the densities of blue-green algae were extremely low compared with the other three algae groups (with more edible characteristics), which may lead to greater grazing consumption.

Some possible limitations from the grazing function are described as follows. First, to obtain the zooplankton biomass required by the 'grazing' module of DELFT-3D, the total density of zooplankton was converted from org/m³ to biomass gC/m³. Secondly, the zooplankton forcing function in the eco-model does not split the zooplankton biomass in small sized or bigger sized zooplankton; thus it does not work as for the algae component of the model that is split into four algae types. Thus, the integration of all the components (species) of the zooplankton in one unique biomass (as is required by the model function) may have overestimated the zooplankton biomass and thus the grazing capabilities over the algae biomass. Thirdly, the temporal pattern along a year was calculated from the zooplankton densities of the two campaigns for the months of February and March, since these were the months of the monitoring campaigns 2011 and 2012. Thus, for the other months of year it was used a secondary source of zooplankton densities data (Prado et al., 2012), that was also converted to biomass. Lastly, some other factors that may influence and need to be considered for future studies are the adjustment of the consumption rates of zooplankton, a deeper evaluation towards possible adjustment of the bottom-up processes that control the nutrients and algae biomass in the model.

Based on these previous theories and limitations, there is definitely room for improvement in the present model set up. Further calibration of bottom-up processes and rates need special attention for future research. Fishes and invertebrates, although not modelled in the present research, have also a strong influence in controlling the standing stocks of algae, organic matter and nutrients in tropical systems (Dudgeon, 2011; Ortiz-Zayas et al., 2005).

Detrás de cada línea de llegada, hay una de partida.
Detrás de cada logro, hay otro desafío

Madre Teresa

8

CONCLUSIONS AND RECOMMENDATIONS

8.1 Research approach

In this research, different approaches were explored to develop a knowledge base for the Abras de Mantequilla (AdM) wetland, a tropical wetland system that belongs to the most important coastal river basin of Ecuador. The wetland is of international importance, hosting many migratory birds and being a nursery ground for a number of fish species. It was declared a Ramsar site in the year 2000. It was the South American case of the EU-FP7 WETwin project, which provided the starting point of this thesis. The research involved a combination of primary data collection (two fieldwork campaigns), secondary data acquisition (notably from literature), multivariate analyses, and numerical modelling to explore the characteristics of this wetland system in terms of hydrological conditions, hydrodynamic patterns, biotic communities, chemical and ecological processes, and fish-habitat suitability. The findings on each of these aspects are presented here. Also, some management considerations are presented, as well as a reflection on the role of mathematical modelling tools in eco-hydraulics research.

8.2 Spatio-temporal variability of the AdM wetland hydrodynamics

The AdM wetland is subject to hydrological forcing that exhibits a clear seasonal variability. The annual precipitation may vary from relatively dry conditions during the dry season period to extremely wet events during El Niño years. There are clear connections between the AdM wetland and the river system with its tributaries. As a consequence, the water depth and inundated area in the wetland exhibit extreme changes during the year: from low depths and almost stagnant conditions during the dry season to a very dynamic system with high water levels and large inundation areas during the wet season. During the wet season (January-April) the main inflow source of the wetland is the Nuevo River, which is a tributary of the Vinces River.

Explorations with a 2D numerical hydrodynamic model revealed that the Nuevo River contributes around 86% of the total wetland inflow, while the four tributaries of the Upper Chojampe subbasin contribute around 11%. Also, the timing of peak discharges was seen to vary from year to year, but occurred usually during the months of February and March.

Spatial pattern analysis using numerical tracer simulations revealed that the wetland can be divided into three main areas (upper, middle and low) based on the influence of the aforementioned inflows. The upper wetland area receives its dominant source of water from the Chojampe subbasin.

The middle wetland is a transition area with inflows from both the Chojampe subbasin as well as the Nuevo River. It exhibits a longer retention time (up to 30 days) compared with the other areas. The lower part is predominantly influenced by the Nuevo River.

The maximum discharge entering from the Nuevo River into the wetland system showed considerable variations depending on the hydrological conditions. Thus, the volumes and inundation areas were seen to vary by more than a factor of three between a dry and a wet year, with discharge values from the river into the wetland reaching up to 650 m³/s in an extreme wet year. As a result, the wetland exhibits large variations in inundation area (from 5 to 27 km²), water depths (from 0.4 to 9 m) and flow velocities (up to 0.9 m/s). Overall, it can be concluded that the wetland is a highly dynamic system in terms of its hydrological forcing and hydrodynamic response. Based on the above, together with considerations on data availability, computation time and preferred accuracy, the length and time scales for the numerical hydrodynamic modelling studies were chosen to be (75m x 75m) grid size and a time step of one day. Since the aim was to have a general view of the longer-term ecological processes, these scales were considered adequate for that purpose.

8.3 Dominant and key species in the AdM wetland system

A first step in assessing the AdM wetland biodiversity was to evaluate the densities and distribution of the taxa collected for the different biotic assemblages during the two measuring campaigns in 2011 and 2012. Clear differences in densities were observed between sites located in the wetland area itself (lentic sites) and in the inflow areas (lotic sites) (Alvarez-Mieles et al., 2013). In the wetland area, the highest densities were recorded in the middle, which can be attributed to the higher retention time at this location. Regarding temporal variation and extreme changes in inundation area from wet to dry season, a clear reduction in habitat availability is observed. As a result, migration of fish species downstream can be expected when the wetland starts decreasing its inundation area by the end of May. Higher densities of zooplankton, macroinvertebrates and fish were observed in the middle area where higher retention times occurred. On the other hand, higher nutrient concentrations were observed at the inflow areas (Nuevo River and Chojampe) where phytoplankton was observed in higher densities. Phyto- and zooplankton communities showed an inverse pattern: at the inflows, phytoplankton was high and zooplankton low, while in the middle area zooplankton densities as well as the next levels in the trophic chain (macro invertebrates and fish) were found to be higher. Dominant species for the different biotic groups were: *Cryptomonas*, *Fragilaria longissima, Nitzchia* (for phytoplankton). The dominance of diatoms in the phytoplankton had been reported previously for the study area and downstream rivers (Guayas, Daule, Babahoyo). High densities of *F.longissima* are associated with the high concentrations of dissolved organic matter (INP, 2012). For zooplankton, rotifers dominated during the first campaign, while *Chydorus sphaericus* and *Mesocyclops venezolanus* were prominently present during the second campaign. The high densities of *Ch.sphaericus* could be indicating some degree of eutrophication in the wetland since this species has been reported as a tolerant species which is often present in eutrophic lakes (de Eyto et al., 2002).

Fish was dominated by the family *Characidae* during both campaigns, which is in agreement with previous studies in the wetland (Florencio, 1993). Species of this family are largely widespread in the neotropics, mainly omnivorous and of small size (Moraes et al., 2013). Some species of this family as *Astyanax* move along considerable distances and occupy a variety of habitats (Vilella et al., 2002). The dominance of omnivorous fish species of the family Characidae as *Astyanax festae* is important because they are source of food for carnivorous fishes (*Brycon dentex, Cichlasoma festae*). Macroinvertebrates were identified to family level, and dominant families were Batidae and Chironomidae (Alvarez-Mieles et al., 2013). In general, the system is dominated by few species, a pattern that was observed for both phytoplankton and zooplankton assemblages. A range of 4 to 8 species usually contributed more than 70% of the total community density, while a high number of species are present in percentages lower than 3%. This dominance pattern has been observed in other tropical areas as well.

The zooplankton community was dominated by small-bodied species. This likely reflects a high grazing pressure from planktivorous fish, which tend to be size-selective (Auer et al., 2004; Havens et al., 2007). Although there were no independent measures of fish predation rates, it is reasonable to assume an importance of the zooplankton in supporting the next trophic level (fish), which in turn are also important food sources for endemic and migratory birds. The importance of the AdM wetland as a bird sanctuary supporting the bird fauna was a central motivation to declare this area a Ramsar site in 2000.

8.4 Spatial patterns in the distribution of the environmental variables and biotic communities

From Principal Component Analyses (PCA), key variables for the water column were found to be temperature, total suspended solids, DO, turbidity, alkalinity, nitrogen and phosphorus (organic and inorganic), and the ratio (N/P) between the nutrients. Furthermore, during high inundation conditions, silicates and flow velocity were also found to become important descriptors of the environment. For the sediment (both sand and silt) nitrogen and phosphorus content (inorganic and organic), organic matter and organic carbon were found to have a dominant effect. Thus, the system shows a clear environmental gradient between the river sites with higher concentrations of DO, TSS, organic phosphorus, higher N/P ratios and flow velocities and the wetland sites with higher concentrations of organic nitrogen, alkalinity, chlorophyll-a, turbidity.

From clustering and ordination, the distribution patterns of the biotic communities show that river and wetland sites typically cluster separately. However, the similarity level at which they cluster varied according to the biotic community. Similarity levels that produce these two main clusters (river/wetland) were generally around 20% for all communities during both conditions. Nevertheless, a more detailed inspection revealed that the similarities at which initial splits occurred for planktonic pelagic communities were always lower than the ones of littoral communities. This could indicate that littoral communities due to their more specific zonation could be more similar than planktonic ones that are driven by the flow and, therefore, experience more water mixing.

From the SIMPER analysis, different species from different biotic communities were found to be key discriminators among wetland areas. As a key outcome it was found that average dissimilarities among wetland areas were lower during high inundation conditions than during low inundation conditions for all biotic groups except fish. This reflects a more homogeneous system in terms of species distribution when the wetland is at its maximum inundation capacity.

The multivariate analysis of biotic and abiotic variables resulted in achieving a better understanding of the most important environmental factors influencing the biotic communities distribution and the overall functioning of the river and wetland ecosystems. Flow velocity and sediment type (river or wetland) influence the taxa distribution, their abundance, richness and diversity. The riverine sites with sandy substrates and high velocities had lower species richness and abundance than the wetland sites with fine particle substrate (silt, clay) and low velocities. Even though both ecosystems share some species, mostly because of river and wetland connectivity, the highest densities and number of taxa were found in the wetland sites.

8.5 Main physico-chemical and ecological processes

The AdM wetland has concentrations of nutrients and primary production in the range of other tropical systems and can be classified as a mesotrophic system. Temporal results from the eco-model indicated that generally the wet season is characterized for higher concentrations of nutrients, primary producers and consumers. Spatial analyses indicated that nutrient concentrations at the three wetland areas are influenced by the nearest inflows. Thus, upper and middle wetland areas are more affected by the discharges of El Recuerdo, and lower wetland areas by the Nuevo River (Chapter 5).

The mass balance analysis performed with the output of the eco-model (Chapter 7) was a key tool in identifying the main processes important for wetland characteristics. Processes such as denitrification were found to be not important when comparing with the external loads, perhaps due to the constant oxygenated conditions, gaining slightly in importance only during driest scenarios. Sedimentation processes for nutrients and primary producers were found to be low, most probably influenced by the dynamics of the system in combination with the high grazing rates. Processes associated with primary production suggested that grazing was the key processes controlling algae biomass in the water column. However, this finding has some possible limitations: First, to obtain the zooplankton biomass required by the 'grazing' module of DELFT-3D, the total density of zooplankton was converted from org/m^3 to biomass gC/m^3. Secondly, the zooplankton forcing function in the eco-model does not split the zooplankton biomass in small sized or bigger sized zooplankton, thus, the integration of all the components (species) of the zooplankton in one unique biomass may have overestimated the zooplankton biomass and thus the grazing capabilities over the algae biomass. Thirdly, the temporal pattern along a year was calculated from the zooplankton densities of the two campaigns for the months of February and March, since these were the months of the monitoring campaigns 2011 and 2012. Thus, for the other months of year it was used a secondary source of zooplankton densities data (Prado et al., 2012), that was also converted to biomass. Lastly, some other factors that are relevant to be considered in future studies are: (i) the adjustment of the consumption rates of zooplankton, and (ii) a deeper evaluation towards possible adjustment of the bottom-up processes that control the nutrients and algae biomass in the model. Algae sedimentation and mortality also played a role but to a minor extent. Results of the simulations suggested that nutrient availability did not appear as a limiting factor for algae growth. Thus, algae limitation was more linked to growth limitation due to grazing pressure, rather than nutrient availability.

8.6 Spatio-temporal variability of Fish-Habitat suitability

A habitat suitability analysis was performed for the overall fish community in AdM wetland. Major environmental variables defining the presence of fish communities in water systems are the hydrodynamic variables: water depth and flow velocity. Thus, response curves based on these variables were built based on field sampling and literature. The potential extension of suitable areas was calculated for different hydrological conditions and scenarios.

The high flow phase of the wet season was recognized as a period with a higher percentage of suitable areas reflected in high Habitat Suitability Index (HIS) scores for all simulated conditions. Spatial zonation defined the areas close to the main inflow as the ones providing better habitat conditions, and areas related to Chojampe subbasin as the ones that will required special attention in terms of management. Based on the results of the present study, it is recommended to secure the timing and magnitude of natural flows especially during periods with higher percentage of suitable areas (high flows during the wet season), since this period is crucial for the spawning and development of fish community.

The combination of hydrodynamic variables was useful for an initial habitat assessment of the fish communities in this wetland. However, other physical, chemical and biotic variables do play an important role in defining the habitat preferences and therefore should be gradually included in an integrated ecological habitat assessment. In this regard, the habitat tool developed for this study is quite flexible for adding more variables and their correspondent rules.

8.7 Management measures for the AdM wetland

One of the issues discovered during a recent visit to the wetland (February 2016) is that the implementation of management measures proposed by WETwin project has not started yet, or at least in so far as the local inhabitants know. Thus, apparently local authorities are still not fully involved in the management of this area. According to local farmers the authorities do not visit the area. Furthermore, some local farmers have even developed unfriendly measures against some 'birds spots' known as 'El Garzal', which are a type of floating islands where aquatic birds build their nests. Apparently a couple of these spots were destroyed by a local farmer with the use of chemicals, and even more of these kind of harmful activities have not been penalized by any authority. On the other hand, there is another group of local farmers that is aware of the ecological importance of the wetland and perform fisheries activities that are sustainable with the environment. Thus, they use nets with special mesh sizes in order not to capture smaller fishes. Furthermore, they have also two main periods for fishing which they know are productive but do not affect the overall production of the wetland. Ecotourism is still a main activity for a few farmers in the main locality named 'El Recuerdo'. The people of this area are the ambassadors of the wetland and know almost every detail about what is going on there. Considering flows and habitat conditions for fish communities, an initial measure could be to maintain the timing and magnitude of the natural flow variability especially during the periods with higher suitable habitat areas (February and March). This period is crucial for the spawning and development of the fish species.

The perception of local farmers about Baba dam is that it has not affected the area as expected. They have not seen big changes in terms of water and fish availability, probably because there has not been a dry year since the dam started operating in 2012. However, they are quite concerned about a project in the Upper Chojampe basin named 'Pacalori', which seems to be a project with mini dams. The management of the Abras de Mantequilla wetland requires that not only local but also national authorities are involved in the management of this valuable spot in the short term. Even though studies like the present research can be used as a tool to develop more awareness about the environmental services the wetland provide, it can be of minimal help if authorities are not aware themselves of the value, step in for the management and penalized not environmental activities in the area. The importance of this wetland as a fauna sanctuary should be continuously highlighted by environmental authorities to increase the awareness from all stakeholders to work towards a sustainable management of this valuable Ramsar site.

8.8 Numerical modelling as a tool to describe wetland dynamics

Since the measurement data obtained from the two field campaigns (just one week in Feb 2011 and in Feb 2012) were rather sparse, numerical models were evoked to try to describe the annual hydrodynamic cycle and natural variability of the AdM wetland system. For the abiotic processes of inundation and transport of solutes, numerical hydrodynamic simulations were able to provide adequate indications of the relative importance of the different river inflow conditions due to time varying hydrological conditions. For aspects related to water quality and water chemistry, numerical models were quite capable of capturing the dynamic features of the wetland, showing that comparing concentrations of water chemistry variables was not enough to identify changes due to different inflow conditions. Assessing the system in terms of yearly mass balances provided a clearer perspective for understanding how different water inflow conditions can affect the mass of the different variables within the water body. Also, modelling showed that the variability in water quality conditions over time and space is clearly related with the hydrodynamics.

By using numerical models to assess the wetland functioning over a much longer time span than the brief (one week) measurement campaigns, a better understanding of the main ecological processes governing primary production (chlorophyll-a and phytoplankton biomass) was obtained for the system. Results suggested that this river-wetland system is dominated by top down zooplankton grazing, rather than bottom-up nutrient availability. Sedimentation and mortality of algae are secondary processes influencing the algae standing biomass. In both cases, numerical modelling facilitated a better understanding of the ecological processes considered in this thesis.

From a management perspective, models can show a possible trend, pattern that could become key in the design and implementation of management options for such valuable ecosystem. Primary data collected in the field only provided snapshots of the wetland functioning while numerical modelling was an important tool to describe changes in the system functioning over the longer time scale of the full annual cycle. The combination of field measurements with numerical models was extremely useful and relevant during this research and confirmed that they complement each other to obtain a better understanding of the dynamics of freshwater river–wetland systems.

8.9 Recommendations for further research

Based on the experience gained in this research, several recommendations can be made to further improve the understanding of the wetland and develop appropriate management scenarios by collecting additional field measurements and enhancing the mathematical modelling.

In order to verify the hydrological regime and hydrodynamic flows in the river-wetland system, installing a discharge station at the entrance of the Nuevo River is strongly recommended. In this study, the Nuevo River inflow was derived from a correlation analysis using the Quevedo gauging station, and not with a direct measurement at the entrance of the Nuevo River inflow. In the wetland, the limnometric station stopped functioning in 2007, so the modelled water level fluctuations from 2007 till the present could not be verified with measured water levels. Thus, a new water level station is needed for continuous measuring of water level fluctuations in the wetland. Furthermore, since the hydrological regimes were seen to exhibit very different characteristics in magnitude, timing, frequency, duration and rate of change, future research should include proper characteristics in the analysis.

For both the biotic and abiotic variables, extending the sampling along the entire year is strongly recommended to account for seasonal variations. During dry season, the wetland reduces dramatically so two intermediate points located in the middle wetland area and the two main inflows would suffice. Synchronized sampling for both biotic and abiotic variables is recommended in order to complement the findings of the present study. Based on the results of the modelling chapters provided in this research new sampling points for future campaigns can be identified. Since changes in the AdM wetland may occur between the present research and future monitoring campaigns, it is always recommended to consult local community knowledge before instrumenting new sampling sites.

From the ecological point of view, it is recommended to identify specific valuable areas in the wetland, like the ones that have remnants of forest patches that host high terrestrial fauna. New sampling sites should include such spots. Moreover, the eco-model seems to suggest the system is governed by top down grazing. Thus, a more detailed analysis for bottom up processes is required to confirm or refute this hypothesis. For that purpose, a deeper check of the coefficients that regulate the bottom up processes is recommended.

"The whole of science is nothing more than a refinement of everyday thinking"

Albert Einstein

REFERENCES

Aarts, B. G. W., F. W. B. Van Den Brink & P. H. Nienhuis, (2004). Habitat loss as the main cause of the slow recovery of fish faunas of regulated large rivers in Europe: the transversal floodplain gradient. *River Research and Applications*, 20(1):3-23. doi:10.1002/rra.720.

Ackerman, C. T., (2009). HEC-GEO-RAS: GIS Tools for support of HEC-RAS using ArcGis, User's Manual, Davis, California, USA.

ACOTECNIC, (2010). Estudios de Factibilidad del proyecto DAUVIN: Informe de Hidráulica fluvial (in Spanish). SENAGUA, Guayaquil, Ecuador.

Agostinho, A., S. Thomaz, L. Gomes & S. S. M. A. Baltar, (2007). Influence of the macrophyte Eichhornia azurea on fish assemblage of the Upper Paraná River floodplain (Brazil). *Aquatic Ecology*, 41(4):611-619. doi:10.1007/s10452-007-9122-2.

Aguirre, W., (2014). The Freshwater Fishes of Western Ecuador. In: http://condor.depaul.edu/waguirre/fishwestec/intro.html.

Aguirre, W. E., V. R. Shervette, R. Navarrete, P. Calle & S. Agorastos, (2013). Morphological and Genetic Divergence of Hoplias microlepis (Characiformes: Erythrinidae) in Rivers and Artificial Impoundments of Western Ecuador. *Copeia*, 2013(2):312-323.

Alonso, M., (1996). Crustacea, Branchiopoda In: M.A. Ramos, J. A., X. Bellés, J. Gonsálbes, A. Guerra, E. Macpherson, F. Martin & J. Serrano Templado (ed), *Fauna Ibérica Vol 7(In Spanish)*. vol 7. Museo Nacional de Ciencias Naturales-CSIC, Madrid, Spain, 486.

Alvarez-Mieles, G., G. Corzo & A. E. Mynett, (2019a). 6 - Spatial and Temporal Variations' of Habitat Suitability for Fish: A Case Study in Abras de Mantequilla Wetland, Ecuador. In: Corzo, G. & E. A. Varouchakis (eds), *Spatiotemporal Analysis of Extreme Hydrological Events*. Elsevier, 113-141. doi:https://doi.org/10.1016/B978-0-12-811689-0.00006-9.

Alvarez-Mieles, G., E. Galecio & A. E. Mynett, (2019b). Natural hydrodynamic variability of a tropical river-wetland system: a case study for the Abras de Mantequilla wetland, Ecuador (submitted). *Wetlands*.

Alvarez-Mieles, G., K. Irvine, A. V. Griensven, M. Arias-Hidalgo, A. Torres & A. E. Mynett, (2013). Relationships between aquatic biotic communities and water quality in a tropical river–wetland system (Ecuador). *Environmental Science & Policy*, 34(0):115-127. doi:http://dx.doi.org/10.1016/j.envsci.2013.01.011.

Alvarez-Mieles, G., K. Irvine & A. E. Mynett, (2019c). Community structure of biotic assemblages in a tropical wetland under different inundation conditions (in preparation). *Limnologica - Ecology and Management of Inland Waters.*

Alvarez-Mieles, G., C. Spiteri & A. E. Mynett, (2019d). Water quality and primary production dynamics in a tropical river-wetland system (submitted). *Journal of Ecohydraulics.*

Alvarez, G., (2007). Biological indicators, a tool for assessment of the present state of a river. A pre-impoundment study in the Quevedo River, Ecuador. Master, MSc, Environmental Sciences, UNESCO-IHE, Institute for Water Education, Delft. The Netherlands, 181.

Arias-Hidalgo, M., (2012). A Decision Framework for integrated wetland-river basin management in a tropical and data scarce environment. PhD, UNESCO-IHE, Institute for Water Education. PhD thesis, Delft, The Netherlands, 180.

Arias-Hidalgo, M., G. Villa-Cox, A. V. Griensven, G. Solórzano, R. Villa-Cox, A. E. Mynett & P. Debels, (2013). A decision framework for wetland management in a river basin context: The "Abras de Mantequilla" case study in the Guayas River Basin, Ecuador. *Environmental Science & Policy*, 34(0):103-114. doi:http://dx.doi.org/10.1016/j.envsci.2012.10.009.

Arrington, D. A. & K. O. Winemiller, (2006). Habitat affinity, the seasonal flood pulse, and community assembly in the littoral zone of a Neotropical floodplain river. *Journal of the North American Benthological Society*, 25(1):126-141. doi:10.1899/0887-3593(2006)25[126:hatsfp]2.0.co;2.

Auer, B., U. Elzer & H. Arndt, (2004). Comparison of pelagic food webs in lakes along a trophic gradient and with seasonal aspects: influence of resource and predation. *Journal of Plankton Research*, 26(6):697-709. doi:10.1093/plankt/fbh058.

Barletta, M., A. J. Jaureguizar, C. Baigun, N. F. Fontoura, A. A. Agostinho, V. M. F. Almeida-Val, A. L. Val, R. A. Torres, L. F. Jimenes-Segura, T. Giarrizzo, N. N. Fabré, V. S. Batista, C. Lasso, D. C. Taphorn, M. F. Costa, P. T. Chaves, J. P. Vieira & M. F. M. Corrêa, (2010). Fish and aquatic habitat conservation in South America: a continental overview with emphasis on neotropical systems. *Journal of Fish Biology*, 76(9):2118-2176. doi:10.1111/j.1095-8649.2010.02684.x.

Barnhill, B. & E. Lopez, (1974). Estudio sobre la biologia de los peces del Rio Vinces. Instituto Nacional de Pesca Boletin Cientifico y Tecnico 3(1):1-40.

Barriga, R., (1994). Peces del Noroeste del Ecuador. Politecnica 19(2):43-154, Escuela Politecnica Nacional, Quito, Ecuador.

Bauman, A. G., D. A. Feary, S. F. Heron, M. S. Pratchett & J. A. Burt, (2013). Multiple environmental factors influence the spatial distribution and structure of reef communities in the northeastern Arabian Peninsula. *Marine Pollution Bulletin*, 72(2):302-312. doi:http://dx.doi.org/10.1016/j.marpolbul.2012.10.013.

Bellinger, E. G. & D. C. Sigee, (2010). Freshwater algae: identification and use as bioindicators. Wiley-Blackwell, UK.

Blauw, A. N., H. F. J. Los, M. Bokhorst & P. L. A. Erftemeijer, (2008). GEM: a generic ecological model for estuaries and coastal waters. *Hydrobiologia*, 618(1):175. doi:10.1007/s10750-008-9575-x.

Boltovskoy, D. (ed) 1981. Atlas del Zooplancton del Atlántico Sudoccidental y métodos de trabajo con el zooplancton marino (In Spanish). Publicación Especial del Instituto Nacional de Investigación y Desarrollo Pesquero (INIDEP), Mar del Plata, Argentina, 1938.

Borbor-Cordova, M. J., E. W. Boyer, W. H. McDowell & C. A. Hall, (2006). Nitrogen and phosphorus budgets for a tropical watershed impacted by agricultural land use: Guayas, Ecuador. *Biogeochemistry*, 79(1):135-161. doi:https://doi.org/10.1007/s10533-006-9009-7.

Boulton, A., F. Sheldon, M. Thoms & E. Stanley, (2000). Problems and constraints in managing rivers with variable flow regimes. In: P.J. Boon, B. R. Davies & G. E. Petts (eds), *Global Perspectives on River Conservation: Science, Policy and Practice*. Wiley, Chinchester, 441-426.

Boulton, A. J., (2003). Parallels and contrasts in the effects of drought on stream macroinvertebrate assemblages. *Freshwater Biology*, 48(7):1173-1185. doi:10.1046/j.1365-2427.2003.01084.x.

Bourrelly, P., (1966). Les Algues D'eau Douce. Initiation a la Systématique. Tome I: Les algues Vertes vol 1, N. Boubée & Cie. edn, Paris, 572.

Bourrelly, P., (1968). Les Algues D'eau Douce. Initiation a la Systématique. Tome II:Les algues jaunes et brunes Chrysophycées, Phéophycées, Xanthophycées et Diatomées, vol 2, N. Boubée & Cie. edn, Paris, 438.

Bourrelly, P., (1970). Les Algues D'eau Douce. Initiation a la Systématique. Tome III:Les algues bleues et rouges Les Eugléniens, Peridiniens et Cryptomonadines., vol 3, N. Boubée & Cie edn, Paris, 1-512.

Bovee, K. D., (1982). A guide to stream habitat analysis using the Instream Flow Incremental Methodology. IFIP No. 12, 248.

Bovee, K. D., (1986). Development and evaluation of habitat suitability criteria for use in the instream flow incremental methodology, Washington, D.C., 235.

Bovee, K. D., B. L. Lamb, J. M. Bartholow, C. B. Stalnaker & J. Taylor, (1998). Stream habitat analysis using the instream flow incremental methodology. U.S. Geological Survery,Biological Resources Division Information and Technology Report USGS/BRD-1998-0004, 143.

Boyero, L., A. Ramírez, D. Dudgeon & R. G. Pearson, (2009). Are tropical streams really different? *Journal of the North American Benthological Society*, 28(2):397-403. doi:10.1899/08-146.1.

Brock, M. A., D. L. Nielsen, R. J. Shiel, J. D. Green & J. D. Langley, (2003). Drought and aquatic community resilience: the role of eggs and seeds in sediments of temporary wetlands. *Freshwater Biology*, 48(7):1207-1218. doi:10.1046/j.1365-2427.2003.01083.x.

Brues, C., A. L. Melander & F. M. Carpenter, (1954). Classification of Insects. Keys to the living and Extinct Families of insects, and to the Living Families of Other Terrestrial Arthropods. Bulletin of the Museum of Comparative Zoology. Harvard College. vol 108, Cambridge, Mass., U.S.A., 916.

Brunner, G. W., (2010). HEC-RAS River Analysis System, User's Manual Version 4.1. U.S. Army of Engineers (Hydrologic Engineering Center -HEC), Davis, CA.

Casatti, L., H. F. Mendes & K. M. Ferreira, (2003). Aquatic macrophytes as feeding site for small fishes in the Rosana Reservoir, Paranapanema River, Southeastern Brazil. *Brazilian Journal of Biology*, 63:213-222. doi:http://dx.doi.org/10.1590/S1519-69842003000200006

Chen, Q., W. Wu, K. Blanckaert, J. Ma & G. Huang, (2012). Optimization of water quality monitoring network in a large river by combining measurements, a numerical model and matter-element analyses. *Journal of Environmental Management*, 110:116-124. doi:http://dx.doi.org/10.1016/j.jenvman.2012.05.024.

Chow, V. T., (1959). Open channel hydraulics. McGraw-Hill, New York, NY, USA.

Clarke, K. & R. Gorley, (2005). PRIMER: Getting started with v6.

Clarke, K. R., P. J. Somerfield & R. N. Gorley, (2008). Testing of null hypotheses in exploratory community analyses: similarity profiles and biota-environment linkage. *Journal of Experimental Marine Biology and Ecology*, 366(1–2):56-69. doi:http://dx.doi.org/10.1016/j.jembe.2008.07.009.

Clarke, K. R. & R. M. Warwick, (2001). Change in marine communities: an approach to statistical analysis and interpretation, 2nd edn. PRIMER-E, Plymouth., UK, 176.

Clint Dawson & C. M. Mirabito, (2008). "The Shallow Water Equations" http://users.ices.utexas.edu/~arbogast/cam397/dawson_v2.pdf.

Davidson, N. C., (2014). How much wetland has the world lost? Long-term and recent trends in global wetland area. *Marine and Freshwater Research*, 65(10):934-941. doi:https://doi.org/10.1071/MF14173.

Davies, B. R. & K. F. Walker (eds), 2013. The ecology of river systems. Springer Netherlands.

de Eyto, E. & K. Irvine, (2001). The response of three chydorid species to temperature, pH and food. *Hydrobiologia*, 459(1-3):165-172. doi:10.1023/a:1012585217667.

de Eyto, E., K. Irvine & G. Free, (2002). The Use of Members of the Family Chydoridae (Anomopoda, Branchiopoda) as an Indicator of Lake Ecological Quality in Ireland. *Biology and Environment: Proceedings of the Royal Irish Academy*, 102B(2):81-91.

De Eyto, E., K. Irvine, F. García-Criado, M. Gyllström, E. Jeppensen, R. Kornijow, M. R. Miracle, M. Nykänen, C. Bareiss, S. Cerbin, J. Salujõe, R. Franken, D. Stephens & B. Moss, (2003). The distribution of chydorids (Branchiopoda, Anomopoda) in European shallow lakes and its application to ecological quality monitoring. *Archiv für Hydrobiologie*, 156(2):181-202. doi:10.1127/0003-9136/2003/0156-0181.

De Pauw, N. & G. Vanhooren, (1983). Method for biological quality assessment of watercourses in Belgium. *Hydrobiologia*, 100(1):153-168.

De Saint-Venant, A. J. C. B. d., (1871). "Théorie du mouvement non permanent des eaux, avec application aux crues des rivières et a l'introduction de marées dans leurs lits", Comptes rendus de l'Académie des Sciences, 73: 147-154 and 237-240.

Deltares, (2013a). Delft3D-FLOW User Manual. Simulation of multi-dimensional hydrodynamic flows and transport phenomena, including sediments., Delft, The Netherlands, 702.

Deltares, (2013b). D-Water Quality. Processes Library Description. Technical Reference Manual, Delft, The Netherlands, 471.

Deltares, (2013c). D-Water Quality. Processes Library Tables Description, Delft, The Netherlands, 234.

Deltares, (2013d). D-Water Quality. Documentation of the input file- User manual Delft, The Netherlands, 414.

Deltares, (2014). D-Water Quality. Versatile water quality modelling in 1D, 2D or 3D systems including physical, (bio)chemical and biological processes. User Manual, Delft, The Netherlands.

Deng, Z.-Q., V. P. Singh & L. Bengtsson, (2004). Numerical solution of fractional advection-dispersion equation. *Journal of Hydraulic Engineering,* 130(5):422-431. doi:10.1061/(ASCE)0733-9429(2004)130:5(422).

Desikachary, T. V. (ed) 1959. Cyanophyta. Botany departament-University of Madras. Indian Council of Agricultural Research New Delhi, 686.

Dodson, C. H. & A. H. Gentry, (1991). Biological Extinction in Western Ecuador. *Annals of the Missouri Botanical Garden,* 78(2):273-295. doi:10.2307/2399563.

Dominguez-Granda, L., K. Lock & P. L. M. Goethals, (2011). Application of classification trees to determine biological and chemical indicators for river assessment: case study in the Chaguana watershed (Ecuador). *Journal of Hydroinformatics,* 13(3):489-499. doi:https://doi.org/10.2166/hydro.2010.082.

Douglas, M. M., S. E. Bunn & P. M. Davies, (2005). River and wetland food webs in Australia's wet–dry tropics: general principles and implications for management. *Marine and Freshwater Research,* 56(3):329-342. doi:http://dx.doi.org/10.1071/MF04084.

Downing, J. A. & E. McCauley, (1992). The nitrogen : phosphorus relationship in lakes. *Limnology and Oceanography,* 37(5):936-945. doi:10.4319/lo.1992.37.5.0936.

Dudgeon, D. (ed) 2011. Tropical stream ecology. Elsevier, 370.

Eaton, A. D., L. S. Clesceri, A. E. Greenberg & M. A. H. Franson, (2005). Standard methods for the examination of water and wastewater. *American public health association,* 21:1600.

Edmondson, W. T., (1966). Fresh water biology, 2nd edn. John Wiley and Sons Inc., New York & London, 1248.

Efficacitas, (2006). Estudio de Impacto Ambiental Definitivo: Proyecto Multiproposito BABA (in Spanish). Consorcio Hidroenergetico del Litoral, Guayaquil.

Eigenmann, C. H., (1922). The fishes of western South America, Part I. The Fresh-water fishes of northwestern South America, including Colombia, Panama, and the Pacific slopes of Ecuador and Peru, together with an appendix upon the fishes of the Rio Meta in Colombia. *Memoirs of the Carnegie Museum*, 9(1):1-347.

Eigenmann, C. H. & G. S. Myers, (1927). The American Characidae. Memories of the Museum of Comparative Zoology at Harvard College, vol 43 (5). Cambridge, Printed for the Museum, 1917-1929.

Ferreira, A., F. R. de Paula, S. F. de Barros Ferraz, P. Gerhard, E. A. L. Kashiwaqui, J. E. P. Cyrino & L. A. Martinelli, (2012). Riparian coverage affects diets of characids in neotropical streams. *Ecology of Freshwater Fish*, 21(1):12-22. doi:10.1111/j.1600-0633.2011.00518.x.

Field, J., K. Clarke & R. Warwick, (1982). A practical strategy for analysing multispecies distribution patterns. *Marine ecology progress series*, 8(1):37-52. doi:10.3354/meps008037.

Finlayson, C. M. & A. G. van der Valk, (1995). Wetland classification and inventory: A summary. *Vegetatio*, 118(1):185-192. doi:10.1007/bf00045199.

Florencio, A., (1993). Estudio bioecológico de la laguna Abras de Mantequilla. Vinces-Ecuador. *Revista de Ciencias del Mar y Limnologia*. vol 3 (1). Instituto Nacional de Pesca (INP), Guayaquil-Ecuador, 171-192.

Freeman, M. C., Z. H. Bowen, K. D. Bovee & E. R. Irwin, (2001). Flow and Habitat Effects on Juvenile Fish Abundance in Natural and Altered Flow Regimes. *Ecological Applications*, 11(1):179-190. doi:10.2307/3061065.

French, R. H., (1986). Open-channel hydraulics. McGraw-Hill New York, United States, 704.

Funk, A., P. Winkler, T. Hein, M. Diallo, B. Kone, G. Alvarez-Mieles, B. Pataki, S. Namaalwa, R. Kaggwa, I. Zsuffa, T. D'Haeyer & J. Cools, (2013). Balancing ecology with human needs in wetlands. Available in: http://www.iwmi.cgiar.org/news/campaigns/world-wetlands-day/wetwin-fact-sheets/.

Galecio, E., (2013). Hydrodynamic and Ecohydrological modelling in a tropical wetland: The Abras de Mantequilla wetland (Ecuador). MSc thesis, Water Sciences and Engineering Department, UNESCO-IHE, Delft, The Netherlands, 139.

Gervais, F., (1998). Ecology of cryptophytes coexisting near a freshwater chemocline. *Freshwater Biology*, 39(1):61-78. doi:10.1046/j.1365-2427.1998.00260.x.

Gery, J., (1977). Characoids of the world. Neptune City: T.F.H. Publications, 672.

Glodek, G. S., (1978). The freshwater fishes of western Ecuador (MSc thesis). MSc, Department of Biological Sciences, Department of Biological Sciences. Northern Illinois University, Dekalb, Illinois, 407.

Gopal, B., (2009). Biodiversity in Wetlands. In: E. Maltby & T. Barker (eds), *The Wetlands Handbook*. Wiley-Blackwell, Oxford, UK, 65-95. doi:doi:10.1002/9781444315813.ch3.

Gyllström, M., L. A. Hansson, E. Jeppesen, F. G. Criado, E. Gross, K. Irvine, T. Kairesalo, R. Kornijow, M. R. Miracle, M. Nykänen, T. Nõges, S. Romo, D. Stephen, E. V. Donk & B. Moss, (2005). The role of climate in shaping zooplankton communities of shallow lakes. *Limnology and Oceanography*, 50(6):2008-2021. doi:10.4319/lo.2005.50.6.2008.

Halls, A. J. (ed) 1997. Wetlands, Biodiversity and the Ramsar Convention: The Role of the Convention on Wetlands in the Conservation and Wise Use of Biodiversity, Ramsar Convention Bureau. Gland, Switzerland.

Hansson, L.-A., M. Gyllström, A. Ståhl-Delbanco & M. Svensson, (2004). Responses to fish predation and nutrients by plankton at different levels of taxonomic resolution. *Freshwater Biology*, 49(12):1538-1550. doi:10.1111/j.1365-2427.2004.01291.x.

Havens, K. E., (1991). Summer zooplankton dynamics in the limnetic and littoral zones of a humic acid lake. *Hydrobiologia*, 215(1):21-29. doi:10.1007/bf00005897.

Havens, K. E. & J. R. Beaver, (2011). Composition, size, and biomass of zooplankton in large productive Florida lakes. *Hydrobiologia*, 668(1):49-60. doi:10.1007/s10750-010-0386-5.

Havens, K. E., J. R. Beaver & T. L. East, (2007). Plankton biomass partitioning in a eutrophic subtropical lake: comparison with results from temperate lake ecosystems. *Journal of Plankton Research*, 29(12):1087-1097. doi:10.1093/plankt/fbm083.

Havens, K. E., A. C. Elia, M. I. Taticchi & R. S. Fulton, (2009). Zooplankton–phytoplankton relationships in shallow subtropical versus temperate lakes Apopka (Florida, USA) and Trasimeno (Umbria, Italy). *Hydrobiologia*, 628(1):165-175. doi:10.1007/s10750-009-9754-4.

Hilton, J., M. O'Hare, M. J. Bowes & J. I. Jones, (2006). How green is my river? A new paradigm of eutrophication in rivers. *Science of The Total Environment*, 365(1–3):66-83. doi:http://dx.doi.org/10.1016/j.scitotenv.2006.02.055.

Howarth, R. W., D. P. Swaney, E. W. Boyer, R. Marino, N. Jaworski & C. Goodale, (2006). The influence of climate on average nitrogen export from large watersheds in the Northeastern United States. *Biogeochemistry*, 79(1):163-186. doi:10.1007/s10533-006-9010-1.

Huisman, J., F. J. Weissing & L. D. Associate Editor: Donald, (2001). Fundamental Unpredictability in Multispecies Competition. *The American Naturalist*, 157(5):488-494. doi:10.1086/319929.

Humphries, P., A. King & J. Koehn, (1999). Fish, Flows and Flood Plains: Links between Freshwater Fishes and their Environment in the Murray-Darling River System, Australia. *Environmental Biology of Fishes*, 56(1-2):129-151. doi:10.1023/a:1007536009916.

Huszar, V. L. M., L. H. S. Silva, M. Marinho, P. Domingos & C. L. Sant'Anna, (2000). Cyanoprokaryote assemblages in eight productive tropical Brazilian waters. In: Reynolds, C. S., M. Dokulil & J. Padisák (eds), *The Trophic Spectrum Revisited: The Influence of Trophic State on the Assembly of Phytoplankton Communities Proceedings of the 11th Workshop of the International Associationof Phytoplankton Taxonomy and Ecology (IAP), held at Shrewsbury, U.K., 15–23 August 1998*. Springer Netherlands, Dordrecht, 67-77. doi:10.1007/978-94-017-3488-2_6.

Huszar, V. M., N. Caraco, F. Roland & J. Cole, (2006). Nutrient–chlorophyll relationships in tropical–subtropical lakes: do temperate models fit? *Biogeochemistry*, 79(1-2):239-250. doi:10.1007/s10533-006-9007-9.

Ilmavirta, V., (1988). Phytoflagellates and their ecology in Finnish brown-water lakes. *Hydrobiologia*, 161(1):255-270. doi:10.1007/BF00044116.

INP, (1998). Comportamiento temporal y espacial de las características físicas, químicas y biológicas del Golfo de Guayaquil y sus afluentes Daule y Babahoyo (In Spanish). Instituto Nacional de Pesca (INP), Guayaquil, Ecuador, 418.

INP, (2012). Monitoreo de organismos bioacuáticos y recursos pesqueros en el Rio Baba. Preparado para HIDROLITORAL Instituto Nacional de Pesca, Guayaquil, Ecuador, 484.

Irvine, K., B. Moss & H. Balls, (1989). The loss of submerged plants with eutrophication II. Relationships between fish and zooplankton in a set of experimental ponds, and conclusions. *Freshwater Biology*, 22(1):89-107. doi:10.1111/j.1365-2427.1989.tb01086.x.

Jacobsen, D., (2008). Tropical High-Altitude Streams. In: Dudgeon, D. (ed), *Tropical stream ecology*. Elsevier, London, 216-256.

Jennings, E., P. Mills, P. Jordan, J.-P. Jensen, M. Søndergaard, A. Barr, G. Glasgow & K. Irvine, (2003). Eutrophication from agricultural sources: seasonal patterns and effects of phosphorus. Environmental Protection Agency, Johnstown Castle, Co. Wexford, Ireland.

Jensen, H. S. & F. O. Andersen, (1992). Importance of temperature, nitrate, and pH for phosphate release from aerobic sediments of four shallow, eutrophic lakes. *Limnology and Oceanography*, 37(3):577-589. doi:10.4319/lo.1992.37.3.0577.

Jeppesen, E., M. Meerhoff, T. A. Davidson, D. Trolle, M. Søndergaard, T. L. Lauridsen, M. Beklioglu, S. Brucet, P. Volta, I. González-Bergonzoni & A. Nielsen, (2014). Climate change impacts on lakes: an integrated ecological perspective based on a multi-faceted approach, with special focus on shallow lakes. *2014*, doi:10.4081/jlimnol.2014.844.

Jeppesen, E., M. Meerhoff, K. Holmgren, I. González-Bergonzoni, F. Teixeira-de Mello, S. A. J. Declerck, L. De Meester, M. Søndergaard, T. L. Lauridsen, R. Bjerring, J. M. Conde-Porcuna, N. Mazzeo, C. Iglesias, M. Reizenstein, H. J. Malmquist, Z. Liu, D. Balayla & X. Lazzaro, (2010). Impacts of climate warming on lake fish community structure and potential effects on ecosystem function. *Hydrobiologia*, 646(1):73-90. doi:10.1007/s10750-010-0171-5.

Jeppesen, E., M. Meerhoff, B. A. Jacobsen, R. S. Hansen, M. Søndergaard, J. P. Jensen, T. L. Lauridsen, N. Mazzeo & C. W. C. Branco, (2007). Restoration of shallow lakes by nutrient control and biomanipulation—the successful strategy varies with lake size and climate. *Hydrobiologia*, 581(1):269-285. doi:10.1007/s10750-006-0507-3.

Jeppesen, E., M. SØNdergaard, J. P. Jensen, K. E. Havens, O. Anneville, L. Carvalho, M. F. Coveney, R. Deneke, M. T. Dokulil, B. O. B. Foy, D. Gerdeaux, S. E. Hampton, S. Hilt, K. Kangur, J. A. N. KÖHler, E. H. H. R. Lammens, T. L. Lauridsen, M. Manca, M. R. Miracle, B. Moss, P. NÕGes, G. Persson, G. Phillips, R. O. B. Portielje, S. Romo, C. L. Schelske, D. Straile, I. Tatrai, E. V. A. WillÉN & M. Winder, (2005). Lake responses to reduced nutrient loading – an analysis of contemporary long-term data from 35 case studies. *Freshwater Biology*, 50(10):1747-1771. doi:10.1111/j.1365-2427.2005.01415.x.

Johnston, R., J. Cools, S. Liersch, S. Morardet, C. Murgue, M. Mahieu, I. Zsuffa & G. P. Uyttendaele, (2013). WETwin: A structured approach to evaluating wetland management options in data-poor contexts. *Environmental Science & Policy,* 34(0):3-17. doi:http://dx.doi.org/10.1016/j.envsci.2012.12.006.

Jones, R. I. & V. Ilmavirta, (1988). Flagellates in freshwater ecosystems — Concluding remarks. In: Jones, R. I. & V. Ilmavirta (eds), *Flagellates in Freshwater Ecosystems.* vol 45. Springer Netherlands, Dordrecht, 271-274. doi:10.1007/978-94-009-3097-1_22.

Jorgensen, S. E., S.N. Nielsen & L. A. Jorgensen, (1991). Handbook of ecological parameters and ecotoxicology. Elsevier, Amsterdam, 1263.

Jowett, I. G., (1997). Instream flow methods: a comparison of approaches. *Regulated Rivers: Research & Management,* 13(2):115-127. doi:10.1002/(SICI)1099-1646(199703)13:2<115::AID-RRR440>3.0.CO;2-6.

Junk, W. J., (2002). Long-term environmental trends and the future of tropical wetlands. *Environmental Conservation,* 29(4):414-435. doi:10.1017/S0376892902000310.

Junk, W. J., P. B. Bayley & R. E. Sparks, (Year). The flood pulse concept in river-floodplain systems. In: Dodge, D. P. (ed) Proceedings of the International Large River Symposium (LARS), 1989. vol 106. Canadian special publication of fisheries and aquatic sciences, p 110-127.

Kasangaki, A., L. J. Chapman & J. Balirwa, (2008). Land use and the ecology of benthic macroinvertebrate assemblages of high-altitude rainforest streams in Uganda. *Freshwater Biology,* 53(4):681-697. doi:10.1111/j.1365-2427.2007.01925.x.

Kibichii, S., W. A. Shivoga, M. Muchiri & S. N. Miller, (2007). Macroinvertebrate assemblages along a land-use gradient in the upper River Njoro watershed of Lake Nakuru drainage basin, Kenya. *Lakes & Reservoirs: Research & Management,* 12(2):107-117. doi:10.1111/j.1440-1770.2007.00323.x.

Kilham, P., (1971). A hypothesis concerning silica and the freshwater planktonic diatoms. *Limnology and Oceanography,* 16(1):10-18. doi:10.4319/lo.1971.16.1.0010.

King, A. J., P. Humphries & P. S. Lake, (2003). Fish recruitment on floodplains: the roles of patterns of flooding and life history characteristics. *Canadian Journal of Fisheries and Aquatic Sciences,* 60(7):773-786. doi:10.1139/f03-057.

Kingsford, R. T., (2000). Ecological impacts of dams, water diversions and river management on floodplain wetlands in Australia. *Austral Ecology,* 25(2):109-127. doi:10.1046/j.1442-9993.2000.01036.x.

Komárek, J. & K. Anagnostidis, (2005). Cyanoprokaryota 2. Oscillatoriales, *Süßwasserflora von Mitteleuropa 19/2 (In German).* Büdel, B., Gärtner, G., Krienitz, L. & Schlagerl, M. edn. Elsevier GmbH, München, Germany, 760.

Laaz, E., V. Salazar & A. Torres, (2009). Guía Ilustrada para la identificación de peces continentales de la Cuenca del Río Guayas (In Spanish). Facultad de Ciencias Naturales-Universidad de Guayaquil, Ecuador.

Laaz, E. & A. Torres, (2014). Lista de Peces continentales de la Cuenca del Río Guayas. In: The freshwater fishes of western Ecuador, http://condor.depaul.edu/waguirre/fishwestec/intro.html.

Lake, P. S., (2003). Ecological effects of perturbation by drought in flowing waters. *Freshwater Biology,* 48(7):1161-1172. doi:10.1046/j.1365-2427.2003.01086.x.

Leira, M., G. Chen, C. Dalton, K. Irvine & D. Taylor, (2009). Patterns in freshwater diatom taxonomic distinctness along an eutrophication gradient. *Freshwater Biology,* 54(1):1-14. doi:10.1111/j.1365-2427.2008.02086.x.

Lewis, W. M., (2008). 1 - Physical and Chemical Features of Tropical Flowing Waters. In: Dudgeon, D. (ed), *Tropical stream ecology.* Academic Press, London, 1-21. doi:https://doi.org/10.1016/B978-012088449-0.50003-0.

Lewis, W. M., Jr, (1986). Nitrogen and Phosphorus Runoff Losses from a Nutrient-Poor Tropical Moist Forest. *Ecology,* 67(5):1275-1282. doi:10.2307/1938683.

Lewis, W. M., Jr., (1974). Primary Production in the Plankton Community of a Tropical Lake. *Ecological Monographs,* 44(4):377-409. doi:10.2307/1942447.

Lewis, W. M., J. M. Melack, W. H. McDowell, M. McClain & J. E. Richey, (1999). Nitrogen yields from undisturbed watersheds in the Americas. *Biogeochemistry,* 46(1):149-162. doi:10.1007/bf01007577.

Lewis, W. M. J., (1987). Tropical limnology. *Annual Review of Ecology and Systematics,* 18(1):159-184.

Lewis, W. M. J., (1996). Tropical lakes: how latitude makes a difference. In: F. Schiemer & K. T. Boland (eds), *Perspectives in tropical limnology.* SPB Academic Publishing, Amsterdam, 43-64.

Li, H., M. Arias, A. Blauw, H. Los, A. E. Mynett & S. Peters, (2010). Enhancing generic ecological model for short-term prediction of Southern North Sea algal dynamics with remote sensing images. *Ecological Modelling*, 221(20):2435-2446. doi:http://dx.doi.org/10.1016/j.ecolmodel.2010.06.020.

Liggett, J. A. & J. A. Cunge, (1975). Numerical methods of solution of the unsteady flow equations. In: K. Mahmood & V. Yevjevich (eds), *Unsteady Flow in Open Channels*. Water Resources, Fort Collins, Colorado, 89-182.

Loh, M., W. Vital, V. Vu, R. Navarrete, P. Calle, V. Shervette, A. Torres & W. Aguirre, (2014). Isolation of sixteen microsatellite loci for Rhoadsia altipinna (Characiformes: Characidae) from an impacted river basin in western Ecuador. *Conservation Genetics Resources*, 6(1):229-231. doi:10.1007/s12686-013-0062-y.

Lung'Ayia, H. B. O., A. M'Harzi, M. Tackx, J. Gichuki & J. J. Symoens, (2000). Phytoplankton community structure and environment in the Kenyan waters of Lake Victoria. *Freshwater Biology*, 43(4):529-543. doi:10.1046/j.1365-2427.2000.t01-1-00525.x.

Magoulick, D. D. & R. M. Kobza, (2003). The role of refugia for fishes during drought: a review and synthesis. *Freshwater Biology*, 48(7):1186-1198. doi:10.1046/j.1365-2427.2003.01089.x.

Maldonado-Ocampo, J., A. Ortega-Lara, J. Usma Oviedo, G. Galvis Vergara, F. Villa-Navarro, L. Vásquez Gamboa, S. Prada-Pedreros & C. Ardila Rodrígues, (2005). Peces de Los Andes de Colombia (In Spanish). Instituto de Investigación de Recursos Biológicos Alexander von Humboldt, Bogotá, D.C., Colombia, 346.

Maldonado-Ocampo, J. A., J. S. Usma Oviedo, F. A. Villa-Navarro, A. Ortega-Lara, S. Prada-Pedreros, L. F. Jimenez, U. Jaramillo-Villa, A. Arango, T. S. Rivas & G. C. Sanchez Garces, (2012). Peces Dulceacuícolas del Chocó Biogeográfico de Colombia. WWF Colombia, Instituto de Investigación de Recursos Biológicos Alexander von Humboldt (IAvH), Universidad del Tolima, Autoridad Nacional de Acuicultura y Pesca (AUNAP), Pontificia Universidad Javeriana, Bogotá D.C., 400.

Masese, F. O., M. Muchiri & P. O. Raburu, (2009). Macroinvertebrate assemblages as biological indicators of water quality in the Moiben River, Kenya. *African Journal of Aquatic Science*, 34(1):15-26. doi:10.2989/AJAS.2009.34.1.2.727.

Matthews, G. V. T., (1993). The Ramsar Convention on Wetlands: its History and Development. Ramsar Convention Bureau, Gland, Switzerland.

McClain, M. E. & R. J. Naiman, (2008). Andean Influences on the Biogeochemistry and Ecology of the Amazon River. *BioScience*, 58(4):325-338. doi:10.1641/b580408.

McClain, M. E., A. L. Subalusky, E. P. Anderson, S. B. Dessu, A. M. Melesse, P. M. Ndomba, J. O. D. Mtamba, R. A. Tamatamah & C. Mligo, (2014). Comparing flow regime, channel hydraulics, and biological communities to infer flow–ecology relationships in the Mara River of Kenya and Tanzania. *Hydrological Sciences Journal*, 59(3-4):801-819. doi:10.1080/02626667.2013.853121.

McQueen, D. J., M. R. S. Johannes, J. R. Post, T. J. Stewart & D. R. S. Lean, (1989). Bottom-Up and Top-Down Impacts on Freshwater Pelagic Community Structure. *Ecological Monographs*, 59(3):289-309. doi:10.2307/1942603.

MEA, (2005). Ecosystems and Human Well-being : Wetlands and Water Synthesis : a report of the Millennium Ecosystem Assessment. World Resources Institute, Washington, DC., 80.

Meerhoff, M., J. M. Clemente, F. T. De Mello, C. Iglesias, A. R. Pedersen & E. Jeppesen, (2007a). Can warm climate-related structure of littoral predator assemblies weaken the clear water state in shallow lakes? *Global Change Biology*, 13(9):1888-1897. doi:10.1111/j.1365-2486.2007.01408.x.

Meerhoff, M., C. Iglesias, F. T. De Mello, J. M. Clemente, E. Jensen, T. L. Lauridsen & E. Jeppesen, (2007b). Effects of habitat complexity on community structure and predator avoidance behaviour of littoral zooplankton in temperate versus subtropical shallow lakes. *Freshwater Biology*, 52(6):1009-1021. doi:10.1111/j.1365-2427.2007.01748.x.

Menden-Deuer, S. & E. J. Lessard, (2000). Carbon to volume relationships for dinoflagellates, diatoms, and other protist plankton. *Limnology and Oceanography*, 45(3):569-579. doi:https://doi.org/10.4319/lo.2000.45.3.0569

Meschiatti, A., M. Arcifa & N. Fenerich-Verani, (2000). Fish Communities Associated with Macrophytes in Brazilian Floodplain Lakes. *Environmental Biology of Fishes*, 58(2):133-143. doi:10.1023/a:1007637631663.

Meybeck, M., (1982). Carbon, nitrogen, and phosphorus transport by world rivers. *Am. J. Sci*, 282(4):401-450. doi:10.2475/ajs.282.4.401.

Mitsch, W. J. & J. G. Gosselink, (2007). Wetlands, 4th edn. John Wiley & Sons, Inc, Hoboken, New Jersey, USA, 582.

Moraes, M., C. F. Rezende & R. Mazzoni, (2013). Feeding ecology of stream-dwelling Characidae (Osteichthyes: Characiformes) from the upper Tocantins River, Brazil. *Zoologia (Curitiba)*, 30:645-651. doi:10.1590/S1984-46702013005000003

Moss, B., S. Kosten, M. Meerhof, R. Battarbee, E. Jeppesen, N. Mazzeo, K. Havens, G. Lacerot, Z. Liu & L. De Meester, (2011). Allied attack: climate change and eutrophication. *Inland waters*, 1(2):101-105. doi:10.5268/IW-1.2.359.

Mynett, A., L. Hong & Q. Chen, (2007). Spatio-temporal simulation in ecosystem dynamics: hydroinformatics support to advanced ecohydraulics modelling. Paper presented at the International Workshop on Advances in Hydroinformatics. 4 – 7 June 2007, Niagara Falls, Canada.

Mynett, A. E., (2002). Environmental Hydroinformatics: the way ahead (opening keynote). *5th Int. Conf. on Hydroinformatics*, 1:31-36. IWA Publishing, Cardiff, UK.

Mynett, A. E., (2003). Environmental hydroinformatics capabilities in natural and urban water systems; Proc. Int. Conf. AGUA 2003, Cartagena, Colombia, October 2003.

Mynett, A. E., (2004). Hydroinformatics tools for ecohydraulics modelling, opening keynote, Proc 6th Int. Conf. on Hydroinformatics - Liong, Phoon & Babovic (eds), World Scientific Publishing, ISBN 981-238-787-0, (June 2004).

Mynett, A. E. & Q. Chen, (2004). Hydroinformatics Techniques in Eutrophication Modelling, Proc 6th Int. Symposium on Ecohydraulics –Garcia (ed), Madrid (Sep. 2004).

Nash, J. E. & J. V. Sutcliffe, (1970). River flow forecasting through conceptual models part I — A discussion of principles. *Journal of Hydrology*, 10(3):282-290. doi:http://dx.doi.org/10.1016/0022-1694(70)90255-6.

Newman, R. M., (1991). Herbivory and Detritivory on Freshwater Macrophytes by Invertebrates: A Review. *Journal of the North American Benthological Society*, 10(2):89-114. doi:10.2307/1467571.

Novak, P., V. Guinot, A. Jeffrey & D. E. Reeve, (2010). Hydraulic modelling–an Introduction. Spon Prees, New York, USA.

Ortiz-Zayas, J. R., W. M. Lewis, J. F. Saunders, J. H. McCutchan & F. N. Scatena, (2005). Metabolism of a tropical rainforest stream. *Journal of the North American Benthological Society*, 24(4):769-783. doi:10.1899/03-094.1.

Padisák, J., É. Soróczki-Pintér & Z. Rezner, (2003). Sinking properties of some phytoplankton shapes and the relation of form resistance to morphological diversity of plankton — an experimental study. In: Martens, K. (ed), *Aquatic Biodiversity: A Celebratory Volume in Honour of Henri J. Dumont.* Springer Netherlands, Dordrecht, 243-257. doi:10.1007/978-94-007-1084-9_18.

Parra, O., M. González, V. Dellarossa, P. Rivera & M. Orellana, (1982). Manual taxonómico del fitoplancton de aguas continentales con especial referencia al fitoplancton de Chile (In Spanish). Universidad de Concepción, Concepción, Chile, 70.

Pedrós-Alió, C., R. Massana, M. Latasa, J. Garcí-Cantizano & J. M. Gasol, (1995). Predation by ciliates on a metalimnetic Cryptomonas population: feeding rates, impact and effects of vertical migration. *Journal of Plankton Research,* 17(11):2131-2154. doi:10.1093/plankt/17.11.2131.

Penman, H. L. & B. A. Keen, (1948). Natural evaporation from open water, bare soil and grass. *Proceedings of the Royal Society of London. Series A. Mathematical and Physical Sciences,* 193(1032):120-145. doi:10.1098/rspa.1948.0037.

Pennak, R. W., (1989). Fresh-water invertebrates of the United States. Protozoa to Mollusca, 3rd edn. Wiley, New York, 628.

Poff, N. L., M. M. Brinson & J. W. Day, (2002). Aquatic ecosystems and global climate change. Technical Report, Pew Center on Global Climate Change, Arlington, USA, 1-36.

Prado, M., (2009). Aspectos biologicos y pesqueros de los principales peces de aguas continentales de la provincia de Los Ríos durante Julio del 2009. *Boletin Científico y Técnico del Instituto Nacional de Pesca,* 5:11-13. Guayaquil, Ecuador.

Prado, M., W. Revelo, R. Castro, R. Bucheli, G. Calderón & P. Macías, (2012). Caracterización química y biológica de sistemas hídricos en la Provincia de Los Ríos-Ecuador (In Spanish). Instituto Nacional de Pesca, Guayaquil, Ecuador, 100.

Prescott, G. W., (1982). Algae of the Western Great Lakes Area with an illustrated key to the genera of desmids and freshwater diatoms. Otto Koeltz Sci. Publ., Koeningstein, Germany, 977.

Ptacnik, R., L. Lepistö, E. Willén, P. Brettum, T. Andersen, S. Rekolainen, A. Lyche Solheim & L. Carvalho, (2008). Quantitative responses of lake phytoplankton to eutrophication in Northern Europe. *Aquatic Ecology,* 42(2):227-236. doi:10.1007/s10452-008-9181-z.

Quevedo, O., (2008). Ficha Ramsar del Humedal Abras de Mantequilla - Ecuador (In Spanish), Guayaquil, Ecuador.

Ramsar, (1971). Final Text adopted by the International Conference on the Wetlands and Waterfowl. Ramsar, Iran.

Ramsar, (2010a). Ramsar's liquid assets: 40 years of the Convention on Wetlands. Gland, Switzerland, 32.

Ramsar, (2010b). River basin management: Integrating wetland conservation and wise use into river basin management. *Ramsar handbooks for the wise use of wetlands*. Vol. 9, 4th edn. Ramsar Convention Secretariat, Gland, Switzerland, 90.

Ramsar, (2010c). Wise use of wetlands: Concepts and approaches for the wise use of wetlands. *Ramsar Handbooks for the wise use of wetlands*. Vol. 1, 4th edn. Ramsar Convention Secretariat, Gland, Switzerland, 60.

Ramsar, (2013). The Ramsar Convention Manual: a guide to the Convention on Wetlands *(Ramsar, Iran, 1971)*, 6th edn. Ramsar Convention Secretariat Gland, Switzerland.

Ramsar, (2014). The List of Wetlands of International Importance. In: http://www.ramsar.org/pdf/sitelist.pdf.

Ramsar, (2016a). An Introduction to the Convention on Wetlands (previously 'The Ramsar Convention Manual'), 5th edn. Ramsar Convention Secretariat, Gland, Switzerland, 110.

Ramsar, (2016b). Wetlands: a natural safeguard against disasters. Fact Sheet 9. In: https://www.ramsar.org/sites/default/files/fs_9_drr_eng_22fev.pdf.

Redfield, A. C., (1934). On the proportions of organic derivatives in sea water and their relation to the composition of plankton. In: Daniel, R. J. (ed), *James Johnstone Memorial Volume*. University of Liverpool Press, Liverpool 176-192.

Reis, R. E., J. S. Albert, F. Di Dario, M. M. Mincarone, P. Petry & L. A. Rocha, (2016). Fish biodiversity and conservation in South America. *Journal of Fish Biology*, 89(1):12-47. doi:10.1111/jfb.13016.

Reis, R. E., S. O. Kullander & C. J. Ferraris (eds), 2003. Check list of the freshwater fishes of South and Central America. Edipucrs, Porto Alegre, Brasil, 729.

Revelo, W., (2010). Aspectos Biológicos y Pesqueros de los principales peces del Sistema Hídrico de la Provincia de Los Ríos, durante 2009 (In Spanish). *Boletin Cientifico y Técnico del Instituto Nacional de Pesca,* 20(6):53-84. Guayaquil, Ecuador.

Reynolds, C. S., (1995). Successional change in the planktonic vegetation: species, structures, scales. In: Joint, I. (ed), *Molecular ecology of aquatic microbes. NATO ASI Series (Series G: Ecological Sciences).* vol 38. Springer, Berlin, Heidelberg, 115-132. doi:10.1007/978-3-642-79923-5_7.

Reynolds, C. S., (2006). The Ecology of Phytoplankton. Cambridge University Press, Cambridge, UK, 550.

Reynolds, C. S., V. Huszar, C. Kruk, L. Naselli-Flores & S. Melo, (2002). Towards a functional classification of the freshwater phytoplankton. *Journal of Plankton Research,* 24(5):417-428. doi:10.1093/plankt/24.5.417.

Roberts, T. R., (1973). The glandulocaudine characid fishes of the Guayas basin in western Ecuador. Bulletin of The Museum of Comparative Zoology at Harvard College. vol 144, In: http://www.biodiversitylibrary.org/page/4228014#page/519/mode/1up.

Rohwer, F. C., W. P. Johnson & E. R. Loos, (2002). Blue-winged Teal (Spatula discors), version 2.0. In:The Birds of North America (A. F. Poole and F. B. Gill, editors). Cornell Lab of Ornithology, Ithaca, New York, USA. https://doi.org/10.2173/bna.625.

Roldán, G., (1996). Guía para el estudio de los macroinvertebrados acuáticos del departamento de Antioquia (In Spanish). FENColciencias, Bogotá, Colombia, 217.

Roldán, G., (2003). Bioindicación de la calidad del agua en Colombia, uso del BMWP/Colombia (In Spanish). FEN-Universidad Antioquia, Medellín, Colombia, 170.

Salmaso, N., G. Morabito, R. Mosello, L. Garibaldi, M. Simona, F. Buzzi & D. Ruggiu, (2003). A synoptic study of phytoplankton in the deep lakes south of the Alps (lakes Garda, Iseo, Como, Lugano and Maggiore). *Journal of Limnology,* 62(2):207-227. doi:10.4081/jlimnol.2003.207.

Sandgren, C. D., (1988). The ecology of chrysophyte flagellates: their growth and perennation strategies as freshwater phytoplankton. In: Sandgren, C. D. (ed), *Growth and reproductive strategies of freshwater phytoplankton.* Cambridge University Press, Cambridge.

Saunders, D. L. & J. Kalff, (2001). Nitrogen retention in wetlands, lakes and rivers. *Hydrobiologia*, 443(1):205-212. doi:10.1023/a:1017506914063.

Saunders, T. J., M. E. McClain & C. A. Llerena, (2006). The biogeochemistry of dissolved nitrogen, phosphorus, and organic carbon along terrestrial-aquatic flowpaths of a montane headwater catchment in the Peruvian Amazon. *Hydrological Processes*, 20(12):2549-2562. doi:10.1002/hyp.6215

Schamberger, M., A. , a. H. Farmer & J. W. Terrell., (1982). Habitat suitability index models: Introduction. Fish and Wildlife Service, U.S. Department of the Interior. FWS/OBS-82/10, Washington, D.C.

Schneider, M., I. Kopecki, J. Eberstaller, C. Frangez & J. A. Tuhtan, (2012). Application of CASiMiR-GIS for the simulation of brown trout habitat during rapid flow changes. Paper presented at the Proceedings of the 9th International Symposium on Ecohydraulics (ISE 2012), Vienna, Austria.

Seitzinger, S. P., R. V. Styles, E. W. Boyer, R. B. Alexander, G. Billen, R. W. Howarth, B. Mayer & N. Van Breemen, (2002). Nitrogen retention in rivers: model development and application to watersheds in the northeastern U.S.A. In: Boyer, E. W. & R. W. Howarth (eds), *The Nitrogen Cycle at Regional to Global Scales*. Springer Netherlands, Dordrecht, 199-237. doi:10.1007/978-94-017-3405-9_6.

Semina, H. J., (1978). Using the standar microscope.Treatment of an aliquot sample. In: Sournia, A. (ed), *Phytoplankton manual*. UNESCO, Paris, 344.

Shamir, U. & J. T. A. Verhoeven, (2013). Management of wetlands in river basins: The WETwin project. *Environmental Science & Policy*, 34(0):1-2. doi:http://dx.doi.org/10.1016/j.envsci.2013.11.001.

Sicko-Goad, L. M., C. L. Schelske & E. F. Stoermer, (1984). Estimation of intracellular carbon and silica content of diatoms from natural assemblages using morphometric techniques. *Limnol. Oceanogr*, 29(6):1170-1178. doi:10.4319/lo.1984.29.6.1170

Smits, J., (2007). Development of DELFT3D ECO. Calibration for a tropical stratified reservoir in Singapore. WL-Delft Hydraulics, Delft, the Netherlands.

Smits, J. G. C. & J. K. L. van Beek, (2013). ECO: A Generic Eutrophication Model Including Comprehensive Sediment-Water Interaction. *PLOS ONE*, 8(7):e68104. doi:10.1371/journal.pone.0068104.

Snoeijs, P., S. Busse & M. Potapova, (2002). The importance of diatom cell size in community analysis. *Journal of Phycology,* 38(2):265-281. doi:10.1046/j.1529-8817.2002.01105.x

Socolofsky, S. A. & G. H. Jirka, (2005). Special topics in mixing and transport processes in the environment. Coastal and Ocean Engineering Division,Texas A&M University, Texas, USA, 184.

Sommer, U., (1986). The periodicity of phytoplankton in Lake Constance (Bodensee) in comparison to other deep lakes of central Europe. *Hydrobiologia,* 138(1):1-7. doi:10.1007/bf00027228.

Sommer, U. (ed) 1989. Plankton ecology: Succession in plankton communities. Springer-Verlag, Berlin, Heidelberg, 369. doi:10.1007/978-3-642-74890-5.

Spiteri, C., B. v. Maren, T. v. Kessel & J. Dijkstra, (2011). Effect Chain Modelling to Support Ems-Dollard Management. *Journal of Coastal Research*:226-233. doi:10.2112/SI61-001.19.

Standley, C., P. Vounatsou, L. Gosoniu, A. Jorgensen, M. Adriko, N. Lwambo, C. Lange, N. Kabatereine & J. Stothard, (2012). The distribution of Biomphalaria (Gastropoda: Planorbidae) in Lake Victoria with ecological and spatial predictions using Bayesian modelling. *Hydrobiologia,* 683(1):249-264. doi:10.1007/s10750-011-0962-3.

Sterner, R. W., (2009). Nutrient Stoichiometry in Aquatic Ecosystems. In: Likens, G. E. (ed), *Encyclopedia of Inland Waters.* Academic Press, Oxford, 820-831. doi:https://doi.org/10.1016/B978-012370626-3.00113-7.

Szymkiewicz, R., (2010). Numerical modeling in open channel hydraulics, vol 83. Springer, Dordrecht.

Takeoka, H., (1984). Fundamental concepts of exchange and transport time scales in a coastal sea. *Continental Shelf Research,* 3(3):311-326. doi:10.1016/0278-4343(84)90014-1.

Talling, J. F., (1992). Environmental regulation in African shallow lakes and wetlands. *Rev. hydrobiol. trop,* 25(2):87-144.

Tank, J. L., E. J. Rosi-Marshall, N. A. Griffiths, S. A. Entrekin & M. L. Stephen, (2010). A review of allochthonous organic matter dynamics and metabolism in streams. *Journal of the North American Benthological Society,* 29(1):118-146. doi:0.1899/08-170.1.

Teixeira-de Mello, F., M. Meerhoff, Z. Pekcan-Hekim & E. Jeppesen, (2009). Substantial differences in littoral fish community structure and dynamics in subtropical and temperate shallow lakes. *Freshwater Biology*, 54(6):1202-1215. doi:10.1111/j.1365-2427.2009.02167.x.

Teresa, F. B. & L. Casatti, (2013). Development of habitat suitability criteria for Neotropical stream fishes and an assessment of their transferability to streams with different conservation status. *Neotropical Ichthyology*, 11(2):395-402. doi:10.1590/S1679-62252013005000009

The_CornellLab-of-Ornithology, (2018). All About Birds, Bird Guide. In: https://www.allaboutbirds.org/guide/search/.

Van Asselen, S., P. H. Verburg, J. E. Vermaat & J. H. Janse, (2013). Drivers of Wetland Conversion: a Global Meta-Analysis. *PLOS ONE*, 8(11):1-13. doi:10.1371/journal.pone.0081292.

Van Breemen, N., E. W. Boyer, C. L. Goodale, N. A. Jaworski, K. Paustian, S. P. Seitzinger, K. Lajtha, B. Mayer, D. van Dam, R. W. Howarth, K. J. Nadelhoffer, M. Eve & G. Billen, (2002). Where did all the nitrogen go? Fate of nitrogen inputs to large watersheds in the northeastern U.S.A. *Biogeochemistry*, 57(1):267-293. doi:10.1023/a:1015775225913.

Van de Wolfshaar, K. E., A. C. Ruizeveld de Winter, M. W. Straatsma, N. G. M. van den Brink & J. J. de Leeuw, (2010). Estimating spawning habitat availability in flooded areas of the river Waal, the Netherlands. *River Research and Applications*, 26(4):487-498. doi:10.1002/rra.1306.

Van den Bossche, M., (2009). Performance of biological indices based on macroinvertebrate communities in the assessment of the Abras de Mantequilla wetland (Ecuador). MSc, Gent Universiteit, Gent, Belgium, 94.

Van der Valk, A., (2012). The biology of freshwater wetlands, 2nd edn. Oxford University Press Inc., New York.

Vannote, R., G. Minshall, K. Cummins, J. Sedell & C. Cushing, (1980). The River Continuum Concept. *Canadian Journal of Fisheries and Aquatic Sciences*, 37(1):130-137. doi:10.1139/f80-017.

Velez, C., (2006). Integrated Water Quality and Ecosystem Modelling a Case Study for Sonso Lagoon, Colombia. MSc thesis, Water Sciences and Engineering, Hydroinformatics Department, UNESCO-IHE, Delft, 137.

Vilella, F. S., F. G. Becker & S. M. Hartz, (2002). Diet of Astyanax species (Teleostei, Characidae) in an Atlantic Forest River in Southern Brazil. *Brazilian Archives of Biology and Technology*, 45(2):223-232. doi:10.1590/S1516-89132002000200015.

Wetzel, R. G., (2001a). 12 - The nitrogen cycle. In: Wetzel, R. G. (ed), *Limnology (Third Edition)*. Academic Press, San Diego, 205-237. doi:http://dx.doi.org/10.1016/B978-0-08-057439-4.50016-2.

Wetzel, R. G., (2001b). 13 - The phosphorus cycle. In: Wetzel, R. G. (ed), *Limnology (Third Edition)*. Academic Press, San Diego, 239-288. doi:http://dx.doi.org/10.1016/B978-0-08-057439-4.50017-4.

Wetzel, R. G., (2001c). 14 - Iron, Sulfur, and silica cycles. In: Wetzel, R. G. (ed), *Limnology (Third Edition)*. Academic Press, San Diego, 289-330. doi:http://dx.doi.org/10.1016/B978-0-08-057439-4.50018-6.

Wetzel, R. G., (2001d). 15 - Planktonic communities algae and cyanobacteria. In: Wetzel, R. G. (ed), *Limnology (Third Edition)*. Academic Press, San Diego, 331-393. doi:http://dx.doi.org/10.1016/B978-0-08-057439-4.50019-8.

Wetzel, R. G., (2001e). 23 - Detritus: organic carbon cycling and ecosystem metabolism. In: Wetzel, R. G. (ed), *Limnology (Third Edition)*. Academic Press, San Diego, 731-783. doi:http://dx.doi.org/10.1016/B978-0-08-057439-4.50027-7.

Winemiller, K. O., (1996). Dynamic Diversity in Fish Assemblages of Tropical Rivers. In: Martin, L. C. & A. S. Jeffrey (eds), *Long-Term Studies of Vertebrate Communities*. Academic Press, San Diego, 99-134.

Winemiller, K. O., (2004). Floodplain river food webs: generalizations and implications for fisheries management. In: Welcomme, R., L. & T. Petr (eds), *Proceedings of the second international symposium on the management of large rivers for fisheries*. vol 2. FAO and Mekong River Commission, Phnom Penh, Kingdom of Cambodia, 285-309.

Winemiller, K. O. & D. B. Jepsen, (1998). Effects of seasonality and fish movement on tropical river food webs. *Journal of Fish Biology*, 53:267-296. doi:10.1111/j.1095-8649.1998.tb01032.x.

Woolhouse, M. E. J., (1992). Population Biology of the Freshwater Snail Biomphalaria pfeifferi in the Zimbabwe Highveld. *Journal of Applied Ecology*, 29(3):687-694. doi:10.2307/2404477.

Woolhouse, M. E. J. & S. K. Chandiwana, (1990). Population dynamics model for Bulinus globosus, intermediate host for Schistosoma haematobium, in river habitats. *Acta Tropica,* 47(3):151-160. doi:https://doi.org/10.1016/0001-706X(90)90021-Q.

Zsuffa, I. & J. Cools, (2011). DSIR analyses at the study sites of the WETWIN project. Deliverables: D3.2 with contributions to D5.1, D7.2, D4.3. Version 32. Prepared under contract from the European Commission, Seventh Framework Programme. In: http://www.wetwin.eu/downloads/D3-2.pdf.

APPENDIX A

A.1 The 1 D De Saint-Venant equations

De Saint-Venant equations, formulated in the 19th century by two mathematicians: De Saint-Venant and Boussinesq, are derived from Navier-Stokes equations for shallow water flow conditions (De Saint-Venant, 1871). The shallow water equations in unidirectional form are also called Saint-Venant equations, and are derived from depth integrating the Navier-Stokes equations, when the horizontal length scale is greater than the vertical length scale. In this condition, conservation of mass implies that the scale of vertical velocity of the fluid is small compared to the scale of horizontal velocity. The vertical integration allows the vertical velocity to be removed from the equations (Clint Dawson and Mirabito, 2008). It is common to analyze flow in a river using a one-dimensional model based on the De Saint-Venant equations of open channel flow to determine the hydrodynamics (Novak et al., 2010). The 1D unsteady flow in an open channel is given by De Saint-Venant equations:

1. **Continuity equation** (based on conservation of mass): in any control volume consisting of the fluid (water) under consideration, the net change of mass in the control volume due to inflow and outflow is equal to the net rate of change of mass in the control volume.

$$\frac{\partial Q}{\partial x} + \frac{\partial A}{\partial t} = 0$$

2. **Momentum equation** (based on conservation of momentum): the change in momentum of a body of water in a flowing channel is equal to the resultant of all external forces acting on that body. The conservation of momentum law states that the rate of change of momentum in the control volume is equal to the net forces acting on the control volume.

$$\underbrace{\frac{1}{A}\frac{\partial Q}{\partial t}}_{(a)} + \underbrace{\frac{1}{A}\frac{\partial}{\partial x}\left(\frac{Q^2}{A}\right)}_{(b)} + \underbrace{g\frac{\partial h}{\partial x}}_{(c)} - \underbrace{g(S_o}_{(d)} - \underbrace{S_f)}_{(e)} = 0$$

Where:

Q = discharge through the channel

A = area of cross-section of flow

x = longitudinal space co-ordinate in horizontal plane

h = water depth

S_0 = channel bottom slope

S_f = friction slope

t = time

Terms: (a) is the local acceleration, (b) the convective acceleration, (c) is the pressure gradient, (d) the gravity, and (e) friction.

A.1.1 A 1D model application

A 1D river model was set along the Vinces and Nuevo rivers in order to serve as a carrier taking information from the upstream basin (Quevedo-Vinces) to the Nuevo River and subsequently to the wetland connection point (Arias-Hidalgo, 2012). The model was built in HEC-RAS, a tool developed by the US Army Corp of Engineers (Brunner, 2010), carried out and unsteady computation in 1D (x direction) applying the De Saint-Venant equations. Topographic data sources were pre-processed in ArcGIS to produce a Digital Elevation Model (DEM). For the upstream first 70 Km, the DEM included the bathymetry from feasibility studies of the Baba project (Efficacitas, 2006). The following 50 Km were derived from the Shuttle Radar Topographic Mission (SRTM) data (available at http://srtm.csi.cgiar.org/). For the lower part and the wetland area, a raster file processed by the Ecuadorian Army Geographic Institute (IGM) in scale 1:10000 was incorporated to the overall elevation model (spatial resolution 5m). In total, 174 km along the Vinces and 25 km along the Nuevo River (Arias-Hidalgo, 2012).

Cross sections were generated with HEC-GEORAS (Ackerman, 2009) using the DEM. The space step was fixed in 200 meters to reduce computational costs. Roughness (Manning) coefficients were assigned to every cross-section, in both main channel and floodplains, according to the existent land use (e.g. vegetation/crops) and literature (Chow, 1959). Values range from 0.03 to 0.04 for the river reaches and a maximum of 0.06 for floodplains, especially along the Nuevo River (dense vegetation). The total simulation period was 12 months (January 2nd, December 30th, 2006) (Arias-Hidalgo, 2012).

Boundary conditions were set: as upstream BC, a flow hydrograph on the location of the future Baba Multipurpose project (upstream BC for Vinces River, Km 174). As lateral inflows, the flow hydrographs from the Lulu & San Pablo Rivers, main tributaries of the Vinces River (Figure A-1). As downstream BC, normal depths, resulting from given values of friction slopes in the Manning equation, were assigned in both Vinces (Vinces town) and Nuevo River (Hcda. Lolita). The wetlands along the left bank of the Nuevo River (West Abras, Central Abras and Main Abras) were simulated as storage areas, and initial water depths were assigned to each of them (Figure A-2).

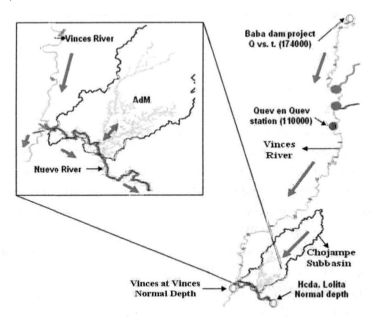

Figure A-1 HECRAS model geometric schematization. Boundary conditions (yellow dots), calibration point (blue dot). Inflows from rivers: Lulu (orange) and San Pablo (blue) (Arias-Hidalgo, 2012).

Figure A-2 Wetland water bodies along the Nuevo River. Example of natural weir (red rectangle) (Arias-Hidalgo, 2012).

Based on existing topography, the stage-volume curves were obtained for each wetland (Table A1) (Arias-Hidalgo, 2012). 'Main Abras' is the water body evaluated in the present research.

Table A-1 Stage-volume curves for the Abras wetlands (Arias-Hidalgo, 2012).

West Abras		Central Abras		Main Abras	
Stage (m)	Volume (m3)	Stage (m)	Volume (m3)	Stage (m)	Volume (m3)
10.500	0	10.000	0	5.970	0
10.790	58	10.320	200358	6.770	36280
11.140	131	10.790	497996	7.740	364870
11.560	227	11.090	689979	8.900	4193935
12.070	367	11.900	1212514	9.624	8156494
12.670	55651	12.420	1560393	10.290	11801611
13.400	336301	12.800	1816831	11.950	25749776
14.270	769527	13.810	2498762	13.960	46415740
14.500	919077	14.000	2603640	16.360	72353576
15.320	1452253	14.500	2930895	19.240	104084552
16.570	2301906	15.000	3258405	22.700	142454688
18.080	3337804	16.000	3912660	26.840	188495280

A.2 The 2D De Saint-Venant equation

The 2D De Saint-Venant equations are obtained from Reynolds Navier-Stokes equations (also referred to as shallow water equations) by depth averaging. They are suitable to apply for floodplains. The equations are used to govern the surface flow, and are obtained from the continuity and momentum equations by depth averaging technique. The basic assumptions of the derivation of the 2D De Saint-Venant are the hydrostatic pressure distribution and small channel slope (Liggett and Cunge, 1975). The governing equations for surface are obtained as follows for:

Continuity equation:

$$\frac{\partial A}{\partial t} + \frac{\partial Q}{\partial x} + \frac{\partial Q}{\partial y} = 0$$

Momentum equation for x momentum

$$\frac{\partial Q}{\partial t} + \frac{\partial (Q^2 / A)}{\partial x} + \frac{\partial Q}{\partial y} + gA\left(\frac{\partial h}{\partial x} - S_{ox} + S_{fx}\right) = 0$$

Momentum equation for y momentum

$$\frac{\partial Q}{\partial t} + \frac{\partial Q}{\partial x} + \frac{\partial (Q^2 / A)}{\partial y} + gA\left(\frac{\partial h}{\partial y} - S_{oy} + S_{fy}\right) = 0$$

Where:

Q = volumetric water discharge

A = area (m²)

h = water surface elevation (depth or height) (m)

x = longitudinal space co-ordinate in horizontal plane

y = longitudinal space co-ordinate in vertical plane

g = acceleration of gravity (m/s2)

S_{0x} and S_{fx} = water surface gradient and friction resistance in x direction

S_{0y} and S_{fy} = water surface gradient and friction resistance in y direction

t = time step

These equations are the based for numerical models for flow simulation such as DELFT3D, as follows.

A.2.1 A 2D application

In this thesis, a 2D model (depth-averaged) was built with the hydrodynamic module DELFT3D-FLOW. A depth-averaged approach is appropriate when the fluid is vertically homogeneous and well mixed. The module runs in two-dimensional mode (one computational layer), which corresponds to solve the depth-average equations (Deltares, 2013a). Delft3D-FLOW solves the Navier Stokes equations for an incompressible fluid, under the shallow water and the Boussinesq assumptions. In the vertical momentum equation the vertical accelerations are neglected, which leads to the hydrostatic pressure equation (Deltares, 2013a). The depth-averaged continuity equation is derived by integration the continuity equation for incompressible fluids over the total depth, taken into account the kinematic boundary conditions at water surface and bed level. The hydrodynamics conditions calculated with DELFT3D-FLOW module can be used as input to the other modules of DELFT3D: WAVE (short wave propagation); WAQ (water quality); PART (particle tracking); ECO (ecological modelling); and SED (sediment transport) (Figure A-3) (Deltares, 2013a). In this thesis, three modules have been applied: FLOW, WAQ and ECO. The opening windows interface for FLOW is depicted in Figure A-4. The grid of the model with boundary conditions and observations points is presented in Figure 2-17 and the inundation patterns of the wetland in Figure A-6.

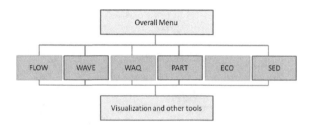

Figure A-3 DELFT3D- System Architecture. The modules in green boxes are the ones applied in the present thesis. Adapted from Deltares (2013a)

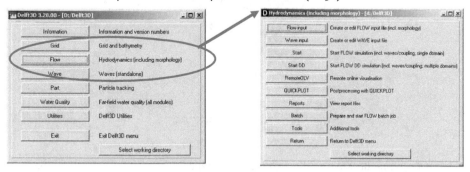

Figure A-4 Main window DELFT3D menu (a), selection window for Hydrodynamics (b). Source (Deltares, 2013a)

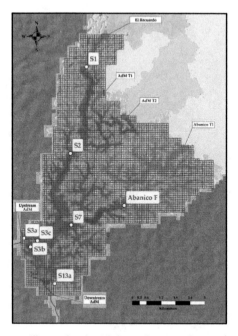

Figure A-5 Abras de Mantequilla wetland grid; boundary conditions (red lines): Upstream AdM (Nuevo river-main inflow to the wetland); Upper Chojampe (El Recuerdo, AdmT1, AdMT2, AbanicoT1); Downstream AdM (wetland outflow). Observation points (white dots). Source: Galecio (2013).

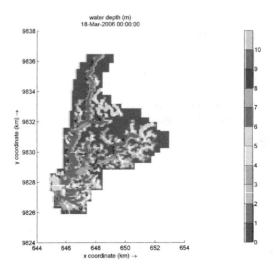

Figure A-6   Abras de Mantequilla maximum inundation area of the verification year (2006) occurring in middle March. Dark red indicates land. Scale displays the water depth in meters (m). Depths in the main channel between 6 and 7 m (in blue)

APPENDIX B

B.1 The 1D advection-dispersion equation

B.1.1 The 1D equation

The advection-dispersion equation is one of the most challenging equations in mathematical physics, as it represents a superposition of two very different processes: advection and dispersion. Advection is the movement of the constituent dissolved in water due to the motion of water itself. The dispersion causes the mass of the dissolved constituent to move from places of higher concentrations towards diminishing concentrations (Szymkiewicz, 2010). The equation is derived following Fick's first law and holds after the initial mixing period or for the far field where the longitudinal shear flow dispersion becomes a dominant mechanism of pollutant mixing in rivers (Deng et al., 2004; French, 1986). The fundamental form of the one dimensional advection-dispersion equation can be expressed as:

$$\frac{\partial C}{\partial t} + U \frac{\partial C}{\partial x} = K \frac{\partial^2 C}{\partial x^2}$$

Where:

C = passive scalar (e.g., temperature or concentration of contaminants)

U = advective fluid velocity or the drift in the x direction; (transport velocity)

K = dispersion coefficient

t = time

B.1.2 A 1D application

The advection-dispersion equation has been extensively used to solve a range of problems in physical, chemical, and biological sciences, involving dispersion or diffusion, such as mixing in inland and coastal waters. Rivers are constantly receiving discharges from industrial and public wastes, thus rivers efficiently transport these pollutants downstream (Socolofsky and Jirka, 2005). When a conservative pollutant is released in a river, advective transport and dispersion determined the movement and change in concentration of the pollutant. Three stages described this transport process: firstly, the pollutant is diluted by the flow because of its initial momentum; secondly, the pollutant is mixed throughout the cross-section by turbulent transport processes; and thirdly, once the cross-section and the vertical mixing is complete, longitudinal dispersion diminish any longitudinal variation in the pollutant concentration (French, 1986). Figure B1 shows the schematic solution of the advective diffusion equation in one dimension.

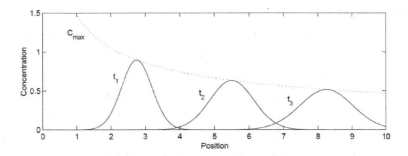

Figure B1 Schematic solution of the advective-diffusion equation in one dimension. The dotted line plots the maximum concentration as the cloud moves downstream (Socolofsky and Jirka, 2005)

B.2 The 2D advection-dispersion equation

B.2.1 The 2D equation

The 2D advection-dispersion equation is the base of 2D water quality models. An example of a 2D model using this equation is DELFT-WAQ. This model solves the advection dispersion-reaction equation on a predefined computational grid for a wide range of substances (Deltares, 2014). There are two different parts: (i) solving the equations for advective and dispersion transport of substances in the water body; and (ii) model the water quality kinetics of chemistry, biology and physics that determines the behaviour of substances and organisms.

$$\frac{\partial C}{\partial t} = -u\frac{\partial C}{\partial x} - v\frac{\partial C}{\partial y} + \frac{\partial}{\partial x}(Dx\frac{\partial C}{\partial x}) + \frac{\partial}{\partial y}(Dy\frac{\partial}{\partial y}) + S + P$$

Source: Adopted from Blauw et al., (2008); Smits and van Beek (2013)

Where:

C = concentration (g m^{-3})

u, v = components of the velocity vector (m s^{-1})

Dx, Dy = components of the dispersion tensor (m^2 s^{-1})

x, y = coordinates in two spatial dimensions (m)

S = sources and sinks of mass due to loads and boundaries (g m^{-3} s^{-1})

P = sources and sinks of mass due to processes (g m^{-3} s^{-1})

B.2.2 A 2D application

The DELFT-WAQ module from the DELFT3D suite allows to model the transport of substances applying the advection-dispersion equation. The substances and processes to be modelled with the water quality module are selected from its Processes Library, which includes a comprehensive set of substances and processes, covering a wide range of water quality parameters. The model does not compute the flow, so it needs to be coupled to a hydrodynamic flow model. Default processes allow to simulate for instance transport of conservative tracers, nitrification, denitrification, growth of algae, sedimentation of nutrients, and other processes related to eutrophication (Deltares, 2013a; Deltares, 2013d; Deltares, 2014). The substances and processes to be modelled with the water quality module are selected from its Processes Library (Deltares, 2014). An example of a transport application is presented as follows. In order to evaluate the dynamics of AdM wetland, as well as the contribution of each inflow, a tracer analysis was implemented in the DELFT-WAQ application. For this purpose, six conservative tracers were used via the 'age and fraction' substance. One conservative tracer was assigned to each boundary condition and one additional for the initial conditions. Each conservative tracer at the boundary conditions was set up to be 1 g/m³ in the assign boundary and zero in the rest of the boundaries. The conservative tracers for the initial conditions and the outflow were set up to zero. Figure B2 shows the transport pattern of the two main inflows of the AdM wetland system during the peak of the wet season 2006 (verification year), showing how far can reach the influence of El Recuerdo inflow in this specific day, while Nuevo River inflow influence is minor. Notice that this pattern corresponds to a specific day, while during other day or even hours the patterns can be inversed, thus indicating the dynamic characteristics of the AdM wetland system.

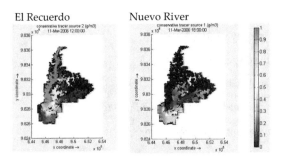

Figure B-2 Transport patterns at the two main inflows of AdM for the verification year 2006, during an specific day (March 11th) of the peak of the wet season. Scale from 0 to 1 for the conservative-tracers

APPENDIX C

Table C1 Physico-chemical variables in water. sediment (sed) per sampling sites. Low inundation conditions (LIC) (February 2011). Wetland sites (a), river sites (b)

FEBRUARY 2011		a) WETLAND SITES (from upper to lower location)							
WATER VARIABLES	UNITS	S5	S6	S1	S2	S7	S3C	MIN	MAX
pH		6.9	6.9	7.3	6.8	7	6.8	6.8	7.3
Temperature	(°C)	27.9	25.9	26.4	28.1	30.7	27.3	25.9	30.7
Conductivity	µS/cm	21	26	33	34	32	30	21	34
Turbidity	NTU	256	23	14	3	5	11	3	256
Hardness	mg/CaCO₃/l	9.3	10.1	12.4	14	14	12.4	9.3	14
Alkalinity	mg/CaCO₃/l	30.5	32.5	48.7	48.7	44.7	38.6	30.5	48.7
DO	mg/l	2.2	2.5	1.2	1.7	5.5	3.6	1.2	5.5
BOD	mg/l	0.5	0.5	0.2	0.1	2.2	0.1	0.1	2.2
COD	mg/l	27.4	68.6	34.3	34.3	17.1	17.1	17.1	68.6
TSS	mg/l	19	16	23	21	19	13	13	23
TS	mg/l	53	61	88	70	75	67	53	88
NO_2_N	mg/l	0.009	0.013	0.008	0.004	0.004	0.003	0.003	0.013
NO_3_N	mg/l	0.35	0.47	0.08	0.05	0.0002	0.22	0.0002	0.47
NH_4_N	mg/l	0.03	0.04	0.04	0.02	0.01	0.01	0.01	0.04
DIN	mg/l	0.38	0.52	0.13	0.07	0.02	0.24	0.02	0.52
N_Organic	mg/l	0.4	0.2	0.2	0.2	0.5	0.4	0.2	0.5
N_Total	mg/l	0.8	0.7	0.3	0.3	0.5	0.6	0.3	0.8
$PO4\_P$	mg/l	0.09	0.1	0.08	0.07	0.05	0.04	0.04	0.10
P_Organic	mg/l	0.0	0.0	0.01	0.04	0.04	0.04	0	0.04
P_Total	mg/l	0.09	0.1	0.09	0.11	0.09	0.08	0.08	0.11
N:P (DIN:PO4)	ratio	9.3	11.5	3.6	2.2	0.9	13.3	0.9	13.3
SiO_4_Si	mg/l	7.2	8.7	10.7	11.5	9.7	11.5	7.2	11.5
Organic carbon	mg/l	4	5	2.2	2.2	3.4	1.8	1.8	5
Chlorides	mg/l	5.9	6.9	2.9	2.9	3.9	3.9	2.9	6.9
Sulphates	mg/l	4	3.7	2.3	0.8	0.9	1.1	0.8	4
Sulphides	mg/l	0.1	0.2	0.2	0.1	0.1	0.2	0.1	0.2
Chlorophyll_a	µg/l	0.9	0.8	1.8	4.7	4.5	8.3	0.8	8.3
Secchi	m	0.7	0.7	0.9	1.4	1.1	1.1	0.65	1.4
Velocity	m/sec	0.05	0.3	0.3	0.2	0.4	0.4	0.05	0.4
SEDIMENT VARIABLES									
Sulphides(sed)	mg/Kg	2.4	1.6	1.6	1.6	1.6	4.4	1.6	4.4
Sand	%	13	29	11	21	19	22	11	29
Silt	%	31	49	32	49	48	49	30.6	49
Clay	%	56	22	56	30	33	29	22	56
NO_2_N (sed)	mg/Kg	0.004	0.003	0.006	0.005	0.002	0.001	0.001	0.006
NO_3_N (sed)	mg/Kg	0.01	0.002	0.02	0.01	0.01	0.01	0.002	0.02
NH_4_N (sed)	mg/Kg	0.48	0.61	0.18	0.75	0.8	0.41	0.18	0.8
DIN (sed)	mg/Kg	0.5	0.6	0.2	0.8	0.8	0.4	0.2	0.8
N_organic (sed)	mg/Kg	0.8	0.9	1	0.3	0.4	1.8	0.3	1.8
N_Total (sed)	mg/Kg	1.3	1.5	1.2	1.1	1.2	2.2	1.1	2.2
$PO4\_P$ (sed)	mg/Kg	0.12	0.15	0.15	0.19	0.07	0.01	0.01	0.19
P_organic (sed)	mg/Kg	0.29	0.31	0.18	0.37	0.32	0.15	0.15	0.37
P_Total (sed)	mg/Kg	0.4	0.5	0.3	0.6	0.4	0.2	0.17	0.56
Organic matter (sed)	%	19	14	14	25	18	17	14	25
Carbonates (sed)	%	0.6	1.2	0.5	0.5	0.6	0.5	0.5	1.2
Organic carbon (sed)	%	7.2	5.6	6.3	7.6	7.4	7	5.6	7.6
COD (sed)	mg/Kg	237	208	195	327	417	273	195	417

Table C1 cont..

FEBRUARY 2011					b) River sites				
WATER Variables	Units	S3A	S3B	S4	S11	S9	S13	Min	Max
pH		7.4	7	7.2	7.2	6.8	6.7	6.7	7.4
Temperature	(°C)	24.7	24.9	25.3	25.3	25.4	25.1	24.7	25.4
Conductivity	µS/cm	29	26	26	27	31	32	26	32
Turbidity	NTU	117	102	158	141	19	43	19	158
Hardness	mg/CaCO$_3$/l	10.9	9.3	10.9	10.9	10.9	12.4	9.3	12.4
Alkalinity	mg/CaCO$_3$/l	29.4	29.4	30.5	30.5	36.5	34.5	29.4	36.5
DO	mg/l	5.7	5.8	5.2	5.3	5.2	4.3	4.3	5.8
BOD	mg/l	0.2	0.2	0.2	0.4	0.4	0.5	0.2	0.5
COD	mg/l	20.6	34.3	17.1	34.3	51.4	17.1	17.1	51.4
TSS	mg/l	78	80	29	71	75	24	24	80
TS	mg/l	102	99	106	90	90	60	60	106
NO$_2$_N	mg/l	0.006	0.005	0.005	0.006	0.008	0.006	0.005	0.008
NO$_3$_N	mg/l	0.29	0.42	0.25	0.46	0.3	0.39	0.25	0.46
NH$_4$_N	mg/l	0.02	0.02	0.01	0.02	0.03	0.03	0.01	0.03
DIN	mg/l	0.31	0.45	0.27	0.48	0.34	0.42	0.27	0.48
N_Organic	mg/l	0.5	0.4	0.4	0.4	0.4	0.4	0.4	0.5
N_Total	mg/l	0.8	0.9	0.7	0.9	0.7	0.9	0.7	0.9
PO4_P	mg/l	0.02	0.02	0.02	0.02	0.08	0.03	0.02	0.08
P_Organic	mg/l	0.06	0.05	0.05	0.04	0.01	0.03	0.01	0.1
P_Total	mg/l	0.07	0.07	0.07	0.06	0.09	0.06	0.06	0.09
N:P (DIN:PO4)	ratio	34.3	49.8	29.9	53.1	9.4	31.0	9.4	53.1
SiO$_4$_Si	mg/l	8.7	12.6	10.8	10.4	12.9	10.8	8.7	12.9
Organic carbon	mg/l	7	7.8	8.1	7.8	8.6	3.3	3.3	8.6
Chlorides	mg/l	3.9	3.9	3.9	3.9	3.9	4.9	3.9	4.9
Sulphates	mg/l	8.5	10.5	8.3	8.1	1.5	1.1	1.1	10.5
Sulphides	mg/l	0.2	0.2	0.2	0.2	0.1	0.1	0.1	0.2
Chlorophyll_a	µg/l	1.2	0.6	0.5	0.3	1.0	1.9	0.3	1.9
Secchi	m	0.2	0.2	0.2	0.2	1	0.2	0.2	1
Velocity	m/sec	0.8	0.6	0.8	0.54	0.4	0.5	0.4	0.8
SEDIMENT VARIABLES									
Sulphides(sed)	mg/Kg		2.2	4	1.6	1.6		1.6	4
Sand	%		39	100	99	26		26	100
Silt	%		45	0.4	0.4	50		0.4	50
Clay	%		16	0.1	0.1	24		0.1	24.4
NO$_2$_N (sed)	mg/Kg		0.002	0.001	0.003	0.003		0.001	0.003
NO$_3$_N (sed)	mg/Kg		0.01	0.01	0.01	0.01		0.01	0.01
NH$_4$_N (sed)	mg/Kg		0.41	0.28	0.27	0.27		0.27	0.41
DIN (sed)	mg/Kg		0.4	0.3	0.3	0.3		0.3	0.4
N_organic (sed)	mg/Kg		1.9	1.8	1.6	2.1		1.6	2.1
N_Total (sed)	mg/Kg		2.3	2.1	1.9	2.4		1.9	2.4
PO4_P (sed)	mg/Kg		0.04	0.02	0.04	0.08		0.02	0.08
P_organic (sed)	mg/Kg		0.17	0.05	0.05	0.13		0.05	0.17
P_Total (sed)	mg/Kg		0.2	0.1	0.1	0.2		0.07	0.21
Organic matter (sed)	%		13	2	2	14		1.5	14
Carbonates (sed)	%		0.6	0.5	0.5	0.6		0.5	0.6
Organic carbon (sed)	%		5.8	0.8	0.7	5.8		0.7	5.8
COD (sed)	mg/Kg		367	198	250	253		198	367

Table C2 Physico-chemical variables in water. sediment (sed) per sampling sites (January 2012). Wetland sites (a), river sites (b)

JANUARY 2012		a) WETLAND SITES (from upper to lower location)							
WATER VARIABLES	UNITS	S1 (s)	S1 (m)	S1 (b)	S1 (shore)	S7	S13a	Min	Max
pH		7.1	7.1	7.1	7.1	7.3	7.7	7.1	7.7
Temperature	(oC)	28.5	26.6	26.5	27.2	27.3	25.7	25.7	28.5
Conductivity	µS/cm	90.3	90.2	90.3	90.5	85.2	75.3	75.3	90.5
Turbidity	NTU	15	12	14	21	14	55	12.0	55.0
Hardness	mg/CaCO3/l	14.3	14.3	14.3	14.3	12.7	12.7	12.7	14.3
Alkalinity	mg/CaCO3/l	42.4	42.4	42.4	42.4	42.4	34.3	34.3	42.4
DO	mg/l	4.7	1.0	1.3	1.4	3.7	3.9	1.0	4.7
BOD	mg/l	2.6	0.19	2	1.37	3.54	0.2	0.2	3.5
COD	mg/l	103.3	118.3	81.6	31.7	66.6	26.7	27	118
TSS	mg/l	17	12	12	14	14	24	12	24
TS	mg/l	159	187	152	147	152	140	140	187
NO2_N	mg/l	0.015	0.028	0.020	0.021	0.003	0.007	0.003	0.028
NO3_N	mg/l	0.09	0.20	0.20	0.17	0.11	0.32	0.09	0.32
NH4_N	mg/l	0.012	0.029	0.031	0.018	0.016	0.054	0.012	0.054
DIN	mg/l	0.12	0.26	0.25	0.21	0.13	0.38	0.12	0.38
N_Organic	mg/l	1.1	1.6	1.9	1.4	1.7	1.9	1.1	1.9
N_Total	mg/l	1.2	1.9	2.2	1.6	1.8	2.3	1.2	2.3
PO4_P	mg/l	0.12	0.15	0.16	0.14	0.05	0.03	0.03	0.16
P_Organic	mg/l	0.10	0.07	0.07	0.08	0.15	0.11	0.07	0.15
P_Total	mg/l	0.21	0.22	0.23	0.23	0.19	0.14	0.14	0.23
SiO4_Si	mg/l	12.0	8.9	8.0	9.4	10.6	10.4	8.0	12.0
Sulphates	mg/l	2.3	3.0	2.8	2.9	2.3	1.9	1.9	3.0
Sulphides	mg/l	0.2	0.3	0.4	0.4	0.1	0.2	0.1	0.4
Chlorophyll_a	µg/l	13.3	3.7	3.2	12.3	42.0	13.0	3.2	42.0
Secchi	m	0.8						0.8	0.8
Water depth	m	0	3	6	2.5	4.5	2.5	0.0	6.0
Velocity	m/sec						0.4	0.4	0.4
SEDIMENT VARIABLES									
Sulphides(sed)	mg/Kg	0.2				0.5	0.5	0.2	0.5
Sand	%	2				4	32	2	32
Silt	%	48				79	55	48	79
Clay	%	49				16	13	13	49
NO2_N (sed)	mg/Kg	0.006				0.003	0.002	0.002	0.006
NO3_N (sed)	mg/Kg	0.017				0.020	0.020	0.017	0.020
NH4_N (sed)	mg/Kg	0.10				0.05	0.08	0.05	0.10
DIN (sed)	mg/Kg	0.13				0.08	0.10	0.08	0.13
N_organic (sed)	mg/Kg	4.8				4.9	4.8	4.8	4.9
N_Total (sed)	mg/Kg	5.0				5.0	4.9	4.9	5.0
PO4_P (sed)	mg/Kg	0.16				0.11	0.03	0.03	0.16
P_organic (sed)	mg/Kg	0.43				0.30	0.22	0.22	0.43
P_Total (sed)	mg/Kg	0.6				0.4	0.2	0.2	0.6
Organic matter (sed)	%	19				19	18	18	19
Carbonates (sed)	%	0.7				0.6	2.5	0.6	2.5
Organic carbon (sed)	%	9.6				7.7	8.4	7.7	9.6
COD (sed)	mg/Kg	460				470	478	460	478

Table C2 cont..

JANUARY 2012				b) RIVER SITES			
WATER VARIABLES	UNITS	S3a	S3b	S11	S13b	Min	Max
pH		7.1	7.4	7.7	7.0	7.0	7.7
Temperature	(oC)	25.0	25.2	24.4	25.6	24.4	25.6
Conductivity	µS/cm	71.2	72.2	31.9	95.1	31.9	95.1
Turbidity	NTU	60	55	62.1	23	23.0	62.1
Hardness	mg/CaCO3/l	12.7	12.7	14.3	12.7	12.7	14.3
Alkalinity	mg/CaCO3/l	30.3	32.3	30.3	34.3	30.3	34.3
DO	mg/l	5.1	5.1	5.7	3.5	3.5	5.7
BOD	mg/l	0.6	1.13	2.69	1.28	0.6	2.7
COD	mg/l	42	70	85	167	42	167
TSS	mg/l	64	39	81	28	28	81
TS	mg/l	220	173	194	135	135	220
NO2_N	mg/l	0.005	0.006	0.004	0.006	0.004	0.006
NO3_N	mg/l	0.27	0.33	0.32	0.26	0.26	0.33
NH4_N	mg/l	0.016	0.009	0.008	0.020	0.01	0.02
DIN	mg/l	0.30	0.35	0.34	0.29	0.29	0.35
N_Organic	mg/l	1.1	1.1	1.4	1.1	1.1	1.4
N_Total	mg/l	1.4	1.5	1.7	1.4	1.4	1.7
PO4_P	mg/l	0.02	0.03	0.03	0.04	0.02	0.04
P_Organic	mg/l	0.12	0.10	0.11	0.11	0.10	0.12
P_Total	mg/l	0.14	0.13	0.14	0.15	0.13	0.15
SiO4_Si	mg/l	9.7	9.6	12.3	10.0	9.6	12.3
Sulphates	mg/l	1.9	1.2	1.1	2.0	1.1	2.0
Sulphides	mg/l	0.2	0.2	0.3	0.2	0.2	0.3
Chlorophyll_a	µg/l	6.7	7.6	9.0	12.8	6.7	12.8
Secchi	m			0.2	0.6	0.2	0.6
Water depth	m	2.8	2.5	5	5	2.5	5.0
Velocity	m/sec	0.4				0.4	0.4
SEDIMENT VARIABLES							
Sulphides(sed)	mg/Kg	0.1	0.4		0.3	0.1	0.4
Sand	%	87	38		90	38	90
Silt	%	9	51		7	7	51
Clay	%	5	10		1	1	10
NO2_N (sed)	mg/Kg	0.002	0.002		0.000	0.000	0.002
NO3_N (sed)	mg/Kg	0.018	0.023		0.016	0.016	0.023
NH4_N (sed)	mg/Kg	0.05	0.03		0.04	0.03	0.05
DIN (sed)	mg/Kg	0.07	0.06		0.05	0.05	0.07
N_organic (sed)	mg/Kg	3.0	3.8		4.5	3.0	4.5
N_Total (sed)	mg/Kg	3.0	3.9		4.6	3.0	4.6
PO4_P (sed)	mg/Kg	0.01	0.01		0.01	0.01	0.01
P_organic (sed)	mg/Kg	0.18	0.20		0.42	0.18	0.42
P_Total (sed)	mg/Kg	0.2	0.2		0.4	0.2	0.4
Organic matter (sed)	%	5	13		3	3.2	13.2
Carbonates (sed)	%	2.5	0.6		0.6	0.6	2.5
Organic carbon (sed)	%	1.5	5.7		1.8	1.5	5.7
COD (sed)	mg/Kg	347	410		220	220	410

Table C3 Physico-chemical variables in water. sediment (sed) per sampling sites (March 2012). Wetland sites (a), river sites (b)

March 2012 WATER VARIABLES	UNITS	a) WETLAND SITES (FROM UPPER TO LOWER LOCATION)									
		S5 (s)	S6 (s)	S1 (s)	S1 (m)	S1 (b)	S1 veg	S1 night	S2(s)	S2(b)	S2 night
pH		6.1	6.6	7.2	7.1	6.99	6.9	6.9	7.1	6.93	6.8
Temp	(oC)	28.7	28.9	26.9	26.8	26.7	29.6	29.1	30.1	30	29.1
Conductivity	µS/cm	61.3	73.2	68.9	68.9	67.4	63.5	62.9	65.9	62.2	57.2
Turbidity	NTU	19	15	14	17	17	12	11	8.7	109	10
Hardness	mg/CaCO3/l	14.3	13.5	11.1	12.7	14.3	11.1	12.7	14.3	12.7	14.3
Alkalinity	mg/CaCO3/l	36.4	44.4	36.4	34.3	34.3	38.4	38.4	38.4	38.4	34.3
DO	mg/l	2.6	3.0	3.8	2.6	2.5	1.6	2.8	4.0	0.7	2.7
BOD	mg/l	0.6	0.2	1.5	0.5	0.3	1.2	0.2	3.7	0.4	0.4
COD	mg/l	11.7	11.7	25.0	60.0	110.0	31.7	68.3	48.3	78.3	13.3
TSS	mg/l	18	10	18	19.1	14.4	26	15	22	30.8	16
TS	mg/l	116	164	70	68	50	90	60	52	58	40
NO2_N	mg/l	0.005	0.005	0.007	0.006	0.006	0.008	0.007	0.009	0.004	0.008
NO3_N	mg/l	0.30	0.24	0.17	0.22	0.17	0.04	0.21	0.13	0.05	0.07
NH4_N	mg/l	0.005	0.005	0.005	0.002	0.005	0.005	0.005	0.005	0.006	0.007
DIN	mg/l	0.32	0.25	0.19	0.22	0.18	0.05	0.22	0.14	0.06	0.08
N_Organic	mg/l	1.7	1.1	1.9	2.0	1.7	1.4	2.2	1.4	1.9	2.5
N_Total	mg/l	2.0	1.4	2.1	2.2	1.9	1.4	2.4	1.5	2.0	2.6
PO4_P	mg/l	0.08	0.08	0.06	0.06	0.06	0.05	0.06	0.06	0.03	0.06
P_Organic	mg/l	0.02	0.00	0.02	0.01	0.02	0.03	0.02	0.02	0.11	0.01
P_Total	mg/l	0.10	0.08	0.08	0.08	0.08	0.08	0.08	0.08	0.14	0.07
N:P (DIN:PO4)	ratio	8.9	6.9	7.0	8.1	6.6	2.2	8.1	5.2	4.4	3.0
SiO4_Si	mg/l	10.2	12.8	9.0	11.4	9.9	9.9	9.8	7.2	9.5	9.8
Sulphates	mg/l	2.6	2.3	2.0	2.0	1.7	1.7	2.1	2.2	2.2	2.2
Sulphides	mg/l	0.4	0.8	0.9	0.6	0.9	1.0	0.6	0.8	0.6	0.4
Chlorophyll_a	µg/l	6	3	4	2	2	9	3	12	24	3
Secchi	m	0.5	1	1					1		
Total Water depth	m	3	6	7			2.7	4	7		7
Velocity	m/sec	0.2	0.6								
SEDIMENT VARIABLES											
Sulphides(sed)	mg/Kg	0.2	0.4	0.3					0.4		
Sand	%	13	29	23					19		
Silt	%	67	52	62					65		
Clay	%	20	2	15					15		
NO2_N (sed)	mg/Kg	0.006	0.010	0.007					0.003		
NO3_N (sed)	mg/Kg	0.000	0.000	0.013					0.002		
NH4_N (sed)	mg/Kg	0.35	0.31	0.23					0.26		
DIN (sed)	mg/Kg	0.4	0.3	0.2					0.3		
N_organic (sed)	mg/Kg	4.8	4.8	4.9					5.0		
N_Total (sed)	mg/Kg	5.1	5.1	5.2					5.3		
PO4_P (sed)	mg/Kg	0.10	0.10	0.14					0.14		
P_organic (sed)	mg/Kg	0.2	0.3	2.1					0.4		
P_Total (sed)	mg/Kg	0.3	0.4	2.2					0.5		
Organic matter (sed)	%	26	17	22					29		
Carbonates (sed)	%	0.9	1.0	0.8					0.9		
Organic carbon (sed)	%	9.5	6.3	9.1					8.6		
COD (sed)	mg/Kg	338	393	267					368		

Table C3 cont..

MARCH 2012		a) WETLAND SITES					b) RIVER SITES				
WATER VARIABLES	UNITS	S7 (s)	S7 (b)	S13A (s)	Min	Max	S4	S11	S13B	Min	Max
pH		7.1	6.51	6.8	6.1	7.2	7.4	7.2	7.2	7.2	7.4
Temp	(oC)	30.9	29.1	28.7			27.1	30.2	30.3		
Conductivity	μS/cm	61.6	69.2	67.9	57.2	73.2	60.9	57.3	74.6	57.3	74.6
Turbidity	NTU	9	27	28	8.7	109	71	106	10	10.0	106
Hardness	mg/CaCO3/l	11.1	11.1	9.6	9.6	14.3	9.6	9.6	9.6	9.6	9.6
Alkalinity	mg/CaCO3/l	36.4	36.4	34.3	34.3	44.4	30.3	30.3	34.3	30.3	34.3
DO	mg/l	6.1	2.3	3.5	0.7	6.1	6.4	6.0	2.9	2.9	6.4
BOD	mg/l	2.4	0.5	1.7	0.2	3.7	1.2	1.0	0.3	0.3	1.2
COD	mg/l	45.0	31.7	20.0	11.7	110	26.7	68.3	40.0	6.7	68.3
TSS	mg/l	55	36.4	28	10	55	87	97	27	27	97
TS	mg/l	78	106	88	40	164	166	136	74	74	166
NO2_N	mg/l	0.005	0.009	0.006	0.004	0.009	0.004	0.004	0.006	0.004	0.006
NO3_N	mg/l	0.11	0.11	0.31	0.04	0.31	0.29	0.32	0.20	0.20	0.35
NH4_N	mg/l	0.003	0.005	0.005	0.002	0.007	0.003	0.003	0.005	0.002	0.005
DIN	mg/l	0.12	0.12	0.32	0.05	0.32	0.30	0.33	0.21	0.21	0.36
N_Organic	mg/l	2.0	2.2	2.0	1.1	2.5	0.8	1.4	1.1	0.8	1.4
N_Total	mg/l	2.1	2.4	2.3	1.4	2.6	1.1	1.7	1.3	1.1	1.8
PO4_P	mg/l	0.07	0.04	0.03	0.03	0.08	0.03	0.02	0.04	0.02	0.04
P_Organic	mg/l	0.00	0.04	0.03	0.00	0.11	0.05	0.06	0.03	0.03	0.06
P_Total	mg/l	0.07	0.08	0.07	0.07	0.14	0.08	0.08	0.07	0.06	0.08
N:P (DIN:PO4)	ratio	3.8	6.6	23.6	2.2	23.6	22.1	36.5	11.6	11.6	39.9
SiO4_Si	mg/l	11.0	10.1	9.5	7.2	12.8	14.2	13.0	8.1	8.1	14.2
Sulphates	mg/l	2.2	1.7	2.1	1.7	2.6	1.2	1.4	2.5	1.2	2.5
Sulphides	mg/l	0.6	0.8	0.6	0.4	1.0	0.6	0.6	0.4	0.4	0.8
Chlorophyll_a	μg/l	24	7	14	2	24	5	4	10	4	10
Secchi	m	1		0.5	0.5	1.0	0.2			0.2	0.2
Total Water depth	m	7.2		3.9	2.7	7.2	5	6.2	5.8	5.0	6.2
Velocity	m/sec			0.8	0.2	0.8	0.8	0.9	0.5	0.4	0.9
SEDIMENT VARIABLES											
Sulphides(sed)	mg/Kg	0.6		0.4	0.2	0.6	0.3	0.4	0.3	0.1	0.4
Sand	%	8		12	8	29	46.9	100.0	38.7	32.4	100.0
Silt	%	89		80	52	89	45.5	0.0	48.9	0.01	64.3
Clay	%	2		7	2	20	7.7	0.0	12.4	0.03	12.4
NO2_N (sed)	mg/Kg	0.004		0.001	0.001	0.010	0.003	0.014	0.001	0.001	0.014
NO3_N (sed)	mg/Kg	0.003		0.005	0.000	0.013	0.000	0.008	0.000	0.000	0.008
NH4_N (sed)	mg/Kg	0.49		0.51	0.23	0.51	0.14	0.27	0.28	0.13	0.28
DIN (sed)	mg/Kg	0.5		0.5	0.2	0.5	0.1	0.3	0.3	0.1	0.3
N_organic (sed)	mg/Kg	4.9		4.5	4.5	5.0	3.6	3.6	5.0	3.4	5.0
N_Total (sed)	mg/Kg	5.4		5.0	5.0	5.4	3.7	3.9	5.3	3.6	5.3
PO4_P (sed)	mg/Kg	0.14		0.09	0.09	0.14	0.05	0.14	0.07	0.04	0.14
P_organic (sed)	mg/Kg	0.3		0.3	0.2	2.1	0.1	0.0	0.1	0.0	0.1
P_Total (sed)	mg/Kg	0.4		0.4	0.3	2.2	0.1	0.2	0.2	0.1	0.2
Organic matter (sed)	%	18		28	17	29	7	11	4	4	14
Carbonates (sed)	%	1.0		1.8	0.8	1.8	0.9	1.1	0.9	0.9	1.1
Organic carbon (sed)	%	8.3		5.3	5.3	9.5	4.6	5.1	1.4	1.4	5.4
COD (sed)	mg/Kg	410		490	267	490	280	228	287	228	298

Notes: all sites were sampled in the pelagic zone:

(s): sampled at 0 meters, surface level

(m): sampled at the middle of the water column (3m)

(b): sampled at the bottom of the water column (6m)

Night: sampled during the night.

Veg: sampled at vegetation located along the bank (littoral zone)

APPENDIX D

Table D1 Major taxa of each biotic assemblage to the within ZONES similarities, from SIMPER results based on abundance data (square root transformed). Only major taxa (with contribution ≥5% is displayed). For complete taxa list contributing up to 90% of the total similarity (see Appendix D-Tables D5-D9). LIC: low inundation conditions, HIC: high inundation conditions. H: Horizontal tows, V: vertical hauls. SIM: total similarity within each group (intragroup), C %: percentage contribution of each taxa to the intragroup similarity. SIM (Ave) is the average of the similarities of all areas. Range AvSIM: describes the minimum and maximum similarity for all areas.

Group	Period	Upper Taxa	SIM	C%	Middle Taxa	SIM	C%	Lower Taxa	SIM	C%	River Taxa	SIM	C%	All areas SIM (Ave)	Range Av SIM	U	M	L	R
Phyto (bot)	LIC	Cryptomonas sp; Trachelomonas sp; Synedra sp	13.1	33; 33; 33				Oscillatoria sp; Synedra sp; Cryptomonas sp	22.2	16; 16; 16	Synedra sp	20.0	68	18.4	13-22	3		8	3
Phyto (bot)	HIC	Cryptomonas sp; Nitzschia recta; Cyclotella comta	22.1	48; 11; 9	Cryptomonas sp; Ph.curvicauda	14.7	37; 22	Cryptomonas sp	47.7	46	M.granulata	27.7	34	28.1	15-48	8	4	4	4
Phyto (net)	(H) LIC	F.longissima	41.0	23				Microcystis sp; M.granulata	55.6	41; 29	F.longissima	38.5	57	45.1	39-56	11		3	6
Phyto (net)	(H) HIC	F.longissima	24.1	73	M.granulata	38.6	53	M.granulata	40.5	55	F.longissima	29.3	42	33.1	24-40	4	3	5	3
Phyto (net)	(V)HIC	F.longissima	18.9	49	M.granulata; F.longissima	4.8	50; 50							11.8	5-19	8	2		
Zooplankton	(H) LIC	Lecane sp	41.2	33				P. cf cuadricornis	28.2	57	Difflugia sp (Rhizopoda)	25.7	60	31.7	26-41	8		5	3
Zooplankton	(H)HIC	Ch.sphaericus; M.venezolanus	27.6	29; 12	Ch.sphaericus; L.occidentalis	51.1	25; 16	Ch.sphaericus; M.venezolanus	49.8	27; 22	Mesocyclops; M.venezolanus	34.2	20; 14	40.7	28-51	12	11	9	8
Zooplankton	(V)HIC	Ch.sphaericus; M.venezolanus	28.1	15; 13	Mesocyclops	47.9	25							38.0	28-48	14	7		
Macro	LIC	Planorbidae; Belostomatidae; Dytiscidae	39.7	43; 13; 9	Planorbidae; Chironomidae; Culicidae	44.5	14; 12; 12	Hydrophilidae; Noteridae; Libellulidae	33.0	16; 16; 13	Baetidae	28.7	54	36.5	29-45	8	13	8	7
Macro	HIC	Baetidae; Planorbidae; Chironomidae; Gerridae	62.1	21; 10; 9; 5	Chironomidae; Baetidae; Planorbidae; Noteridae	43.5	13; 11; 11; 10	Lynceidae; Chironomidae; Scirtidae; Baetidae; Planorbidae	68.6	9; 8; 8; 7; 6	Baetidae; Chironomidae	47.5	22; 15	55.4	43-69	18	14	2	11
Fish	HIC	Astyanax festae	51.1	35	Astyanax festae	56.1	39	Hemibrycon polyodon	54.4	28	Pseudocurimata boulengeri; H.polyodon	51.1	30; 23	53.2	51-56	6	4	5	5

*: No of taxa responsible for within area similarity (C=90% contribution to SIM) SIM

Table D2 ANOSIM (1-way) results based on similarity matrices derived from square root abundance data of planktonic communities. LIC: low inundation conditions-Feb 2011; HIC: high inundation conditions-March 2012. H: horizontal tows; V: vertical hauls; HV (pooled H and V). Factors: A) Location: River/Wetland; B) Zones (U: upper, M: middle, L: low, R: river); C) Diel variation: Day/Night; D) littoral/pelagic-limnetic; E) Sampling effort: Horizontal/Vertical. Significant results in bold (P: significance level).

	Sampling effort H/V	Condition	A) River/Wetland Global R	P	B) Zones Global R	P	Pairwise	R	P	C) Day/Night Global R	P	D) Littoral/Pelagic Global R	P	E) Horizontal/Vertical Global R	P	F) CONDITION LIC/HIC[c] Global R	P
Phyto		LIC	0.17	0.09	0.14	0.17				-0.01	0.51	0.36	0.23	---	---		
		HIC	0.17	0.13	0.12	0.16				-0.04	0.56	0.04	0.44	---	---		
		LIC+HIC[b]	**0.27**	**0.005**	0.13	0.05	U,R	0.28	0.058							0.48	0.001
Phyto (net)	H	LIC	-0.03	0.48	**0.7**	**0.002**	L,R	0.96	0.06	---		---		---			
	H	HIC	-0.44	1	**0.49**	**0.028**	U,L	0.86	0.06	-0.21	0.87	---		---			
							U,M	0.79	0.06								
	V	HIC	---		**0.36**	**0.05**	U,L	0.83	0.1	**0.29**	**0.05**	0.07	0.24	---			
	HV[a]	HIC	-0.26	0.96	**0.24**	**0.01**	U,L	**0.51**	**0.004**	0.10	0.17	**0.35**	**0.002**	**0.37**	**0.001**		
							U,M	0.34	0.008								
	H	LIC+HIC[b]	-0.25	0.99	**0.32**	**0.003**	U,L	**0.67**	**0.003**	-0.14	0.76	---		---			
							U,M	0.49	0.025								
							L,R	0.47	0.01								
							M,R	0.53	0.024								
	HV[a]	LIC+HIC[b]	-0.26	0.99	**0.21**	**0.006**	**U,M**	**0.25**	**0.017**	0.17	**0.07**	**0.34**	**0.003**	**0.37**	**0.001**	0.03	0.35
							U,L	**0.58**	**0.002**								
							L,R	**0.50**	**0.002**								
Zoo	H	LIC	0.12	0.21	-0.034	0.53				---				---			
	H	HIC	**0.43**	0.08	-0.006	0.45	---			-0.03	0.47	0.15	0.11	---			
	V	HIC	---		0.149	0.19				-0.15	0.88			---			
	HV[a]	HIC	**0.34**	**0.05**	0.091	0.19	M,R	**0.64**	**0.048**	-0.04	0.59	**0.24**	**0.02**	**0.35**	**0.001**		
	H	LIC+HIC[b]	**0.25**	**0.03**	-0.004	0.44				-0.05	0.59	---		---		0.80	0.001
	HV[a]	LIC+HIC[b]	**0.38**	**0.004**	0.12	0.07	U,R	**0.31**	**0.012**	-0.05	0.64	0.13	0.11	**0.34**	**0.001**	0.65	0.001
							M,R	**0.57**	**0.002**								

NOTES : (a) HV=pooled horizontal and vertical samples; (b) pooled samples of both conditions (LIC+HIC); (c) FACTOR F "CONDITION" applies only for the combined analysis: LIC+HIC

Table D3 ANOSIM (1-way) results based on similarity matrices derived from square root abundance data of littoral communities. LIC: low inundation conditions-Feb 2011; HIC: high inundation conditions-March 2012. Factors: A) Location: River/Wetland; B) Zones (U: upper, M: middle, L: low, R: river); C) Diel variation: Day/Night. Significant results in bold (P: significance level).

Biotic group	Condition during sampling	A) River/Wetland Global R	P	B) Zones Global R	P	Pairwise	R	P	C) Day/Night Global R	P	F) Condition LIC/HIC[c] Global R	P
Macroinvertebrates	LIC	**0.38**	**0.02**	0.215	0.12							
	HIC	**0.75**	**0.009**	**0.49**	**0.005**	U.L	0.54	0.048	-0.01	0.44		
						U.R	0.69	0.018				
	LIC+HIC[b]	**0.52**	**0.003**	**0.38**	**0.003**	U,R	0.56	0.004	0.04	0.43	**0.37**	**0.004**
						L,R	0.73	0.012				
Fish	HIC	0.11	0.26	**0.37**	**0.003**	U.M	0.42	0.008	0.008	0.46		
						M.L	0.45	0.054				
						M.R	0.81	0.018				
	LIC+HIC[b]	**0.25**	**0.05**	**0.27**	**0.006**	L,M	0.32	0.043	-0.2	0.83	**0.42**	**0.012**
						U,M	0.3	0.026				
						M,R	0.59	0.005				

NOTES: (b) pooled samples of both conditions (LIC+HIC). (c) FACTOR F "CONDITION" applies only for the combined analysis: LIC+HIC

Table D4 BVSTEP AND RELATE ROUTINES. a) BVSTEP routine results: Environmental driving variables for each biotic assemblage during high inundation conditions (HIC). Resemblance measure based on Euclidean distances. Driving variables appeared in order of importance (the first variable has the higher weight). b) RELATE routine tested matching between Biotic matrix (based on Bray Curtis similarity) and Environmental matrix (built with variables selected by BVSTEP routine and based on Euclidean distances). Both routines (a & b) based on Spearman Rank correlation method (ρ); Significance level (P %) and 999 permutations.

Biotic Group	a)BVSTEP			b)RELATE	
	Driving variables	(ρ)	P (%)	(ρ)	P (%)
Phytoplankton (bot) (all variables) and Forc nutrients	N_Inorg, PO4_P, SiO4-Si, Temp, Turb, COD, Abund_macro, Rich_fish, Sand, pH_YSI, Sulphides, CODsed	0.4	0.11*	0.54	0.001
Phytoplankton (bot) (onlywater variables)	Turb, Temp, TSS, COD, Sulphides, SiO4-Si	0.53	0.04	0.53	0.001
Phytoplankton (V) (all variables)	Abund_zoo, Silt, Abund_macro, Depth	0.658	0.08*	0.65	0.002
Zooplankton (H) (all variables)	TS, Rich_macro, Cond, N_org, Div_macro, Sand, Abund_macro	0.90	0.001	0.90	0.001
Zooplankton (H) (onlywater variables)	TS, N_total, pH, N_org, P_org	0.82	0.015	0.83	0.001
Zooplankton (V)	Temp, SiO4-Si, Depth, Rich_phytonet, PO4_P, N_org, Rich_fish	0.62	0.05	0.62	0.001
Macroinvertebrates (allsites-all variables)	Depth, Div_zoo, N_org_sed, N_org	0.80	0.02	0.80	0.001
Macroinvertebrates (wetland sites) -allvariables/ excluding S4)	P_total, BOD, N_org, N_total_sed, Div_zoo, Sulphate	0.78	0.002 Bioenv (0.002)	0.73	0.001
Fish (all variables WSI)	Div_macro, Org_mat_sed, Rich_macro, BOD, Temp_sed, Rich_zoo	0.56	0.07*	0.57	0.001
Fish (only with SED variables)	Temp_sed, Org_mat_sed	0.46	0.03		

*: Correlation coefficient (ρ) no significant when P >0.05. Thus, no significant rank correlation between the selected driving variables and the response biotic matrix.

WSI=water, sediment and biotic indices variables

bot: collected with Niskin bottle (phytobot),H:horizontal tows, Div: diversity, Rich: richness

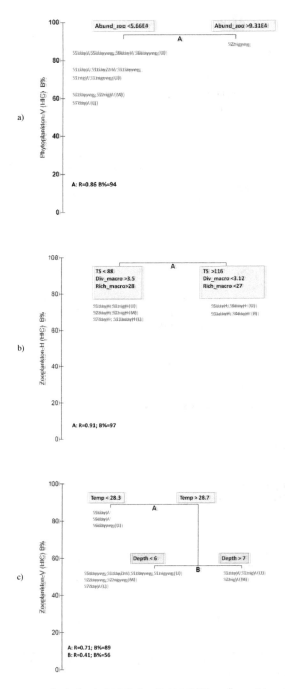

Figure D1 Linkage tree analysis (LINKTREE) for PLANKTON communities during high inundation conditions (HIC). a) Phytoplankton V (vertical hauls); b) Zooplankton H (horizontal tows); b) Zooplankton V (vertical hauls). The divisive clustering of sampling sites are driven by the explained thresholds of the environmental variables (with SIMPROF test P< 0.05). R: ANOSIM R statistic provides a measure of the degree of separation between 2 groups. B%: is the absolute subgroup separation, in relation to the maximum separation of the first split.

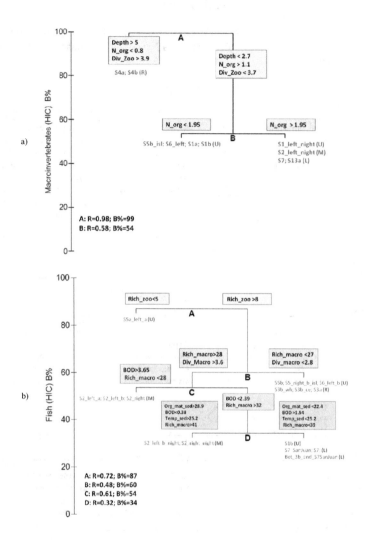

Figure D2 Linkage tree analysis (LINKTREE) for macroinvertebrates and fish communities during high inundation conditions (HIC). The divisive clusterings of sampling sites are driven by the explained thresholds of the environmental variables (without SIMPROF test). R: ANOSIM R statistic provides a measure of the degree of separation between 2 groups. B%: is the absolute subgroup separation, in relation to the maximum separation of the first split.

SIMILARITIES

Table D5 SIMPER results of within "ZONES" similarities for Phytoplankton (collected with Niskin bottle). LIC: low inundation conditions; HIC: high inundation conditions. Based on abundance data (square root transformed).

Phytoplankton (Niskin bottle) - LIC

Group U — Average similarity: 13.1

Species	Av. Abund	Av. Sim	Sim/SD	Contrib %	Cum %
Cryptomonas sp.	80.5	4.4	0.6	33.3	33.3
Trachelomonas sp.	100.0	4.4	0.6	33.3	66.7
Synedra sp.	66.7	4.4	0.6	33.3	100

Group M — Less than 2 samples in group

Group R — Average similarity: 19.9

Species	Av. Abund	Av. Sim	Sim/SD	Contrib %	Cum %
Synedra sp.	102.9	13.7	0.6	68.4	68.4
Pseudanabaena sp.	40.0	2.6	0.3	12.9	81.3
Nitzchia acicularis	40.0	2.2	0.3	10.8	92.1

Group L — Average similarity: 22.2

Species	Av. Abund	Av. Sim	Sim/SD	Contrib %	Cum %
Oscillatoria sp.	66.7	3.7	0.6	16.6	16.6
Synedra sp.	66.7	3.7	0.6	16.6	33.2
Cryptomonas sp.	317.4	3.6	0.6	16.3	49.5
Ank_acicularis	157.7	2.1	0.6	9.4	58.9
Scene_quadricauda	94.3	2.1	0.6	9.4	68.3
Trachelomonas sp.	104.9	2.1	0.6	9.4	77.7
Cymbella sp.	66.7	2.0	0.6	9.1	86.7
Melo_granulata	80.5	1.5	0.6	6.6	93.4

Phytoplankton (Niskin bottle) - HIC

Group U — Average similarity: 22.1

Species	Av. Abund	Av. Sim	Sim/SD	Contrib %	Cum %
Cryptomonas sp.	165.1	10.5	0.9	47.5	47.5
Nitzchia recta	57.1	2.5	0.6	11.3	58.8
Cyclotella comta	57.1	1.9	0.4	8.6	67.4
Frag_longissima	42.9	1.2	0.4	5.3	72.7
Scen_bijugus	42.9	1.1	0.4	5.2	77.8
Phacus suecicus	42.9	1.0	0.4	4.5	82.3
Hantzchia amphioxys	42.9	0.9	0.4	4.2	86.5
Nitzschia acicularis	42.9	0.9	0.4	4.2	90.7

Group M — Average similarity: 14.7

Species	Av. Abund	Av. Sim	Sim/SD	Contrib %	Cum %
Cryptomonas sp.	173.2	5.5	0.6	37.4	37.4
Phacus curvicauda	66.7	3.2	0.6	21.6	59.0
Rhopalodia gibba	91.1	3.0	0.6	20.5	79.5
Melosira granulata	91.1	3.0	0.6	20.5	100

Group L — Average similarity: 47.7

Species	Av. Abund	Av. Sim	Sim/SD	Contrib %	Cum %
Cryptomonas sp.	329.4	22.1	10.8	46.2	46.2
Melosira granulata	244.6	17.2	4.4	36.0	82.2
Actinastrun hantzschii	66.7	3.0	0.6	6.3	88.5
Nitzschia longissima	80.5	2.8	0.6	5.8	94.2

Group R — Average similarity: 27.7

Species	Av. Abund	Av. Sim	Sim/SD	Contrib %	Cum %
Melosira granulata	141.4	9.5	0.9	34.4	34.4
Nitzschia sp.	75.0	6.5	0.9	23.5	57.9
Stauroneis phoenicenteron	75.0	6.5	0.9	23.5	81.4
Fragi_longissima	50.0	2.6	0.4	9.4	90.8

Table D6 SIMPER results of within "ZONES" similarities for Phytoplankton (net). H: horizontal tows; V: vertical hauls. LIC: low inundation conditions; HIC: high inundation conditions. Based on abundance data (square root transformed).

Phytoplankton H - LIC

Group U — Average similarity: 41

Species	Av. Abund	Av. Sim	Sim/SD	Contrib %	Cum %
Fragilaria longissima	359.3	9.2	3.8	22.5	22.5
Nitzschia palea	146.2	4.7	2.2	11.4	34.0
Navicula sp.	161.2	4.5	1.3	11.0	44.9
Gomphonema sp.	80.9	3.4	4.0	8.2	53.2
Fragilaria sp.	70.9	3.3	6.1	8.1	61.3
Gomhonema gracile	153.5	3.3	0.6	8.1	69.4
Nitzschia recta	55.9	3.2	4.1	7.7	77.1
Scenedesmus bijugus	45.0	2.2	4.1	5.4	82.5
Phacus sp.	37.2	1.3	0.6	3.2	85.7
Anabaena constricta	82.1	1.3	0.6	3.1	88.8
Mougeotia jogensis	71.9	1.1	0.6	2.8	91.6

Group L — Average similarity: 55.6

Species	Av. Abund	Av. Sim	Contrib %	Cum %
Microcystis sp.	55.9	23.0	41.4	41.4
Melosira granulata	68.1	16.3	29.3	70.7
Fragilaria sp.	39.5	16.3	29.3	100.0

Group R — Average similarity: 38.5

Species	Av. Abund	Av. Sim	Sim/SD	Contrib %	Cum %
Fragilaria longissima	217.0	21.8	3.3	56.7	56.7
Gomphonema sp.	63.4	3.7	0.9	9.6	66.3
Cymbella tumida	40.9	2.9	0.9	7.6	73.9
Nitzschia palea	46.7	2.4	0.9	6.3	80.1
Oscillatoria sp.	44.1	2.4	0.9	6.3	86.4
Fragilaria sp.	52.3	1.6	0.4	4.3	90.7

Phytoplankton H - HIC

Group U — Average similarity: 24.1

Species	Av. Abund	Av. Sim	Sim/SD	Contrib %	Cum %
Frag_long	116.7	17.7	3.2	73.3	73.3
Phorm_sp	19.1	1.7	0.4	7.2	80.6
Navic_rad	19.1	1.7	0.4	7.2	87.8
Ulna_ulna	19.1	1.7	0.4	7.2	95.0

Group L — Average similarity: 40.5

Species	Av. Abund	Av. Sim	Contrib %	Cum %
Melo_granu	328.2	22.1	54.5	54.5
Desmo_rab	52.2	4.6	11.4	65.9
Frag_long	46.2	4.6	11.4	77.3
Nitz_pa	52.2	4.6	11.4	88.6
Nitz_sp	38.2	4.6	11.4	100

Group M — Average similarity: 38.6

Species	Av. Abund	Av. Sim	Contrib %	Cum.%
Melo_granu	85.5	20.4	52.8	52.8
Anab_sp	52.2	9.1	23.6	76.4
Eun_lun	38.2	9.1	23.6	100

Group R — Average similarity: 29.3

Species	Av. Abund	Av. Sim	Contrib %	Cum.%
Frag_long	101.6	12.4	42.3	42.3
Melo_granu	104.7	10.7	36.6	78.9
Frag_sp	52.2	6.2	21.1	100

Phytoplankton V - HIC

Group U — Average similarity: 18.9

Species	Av. Abund	Av. Sim	Sim/SD	Contrib %	Cum %
Frag_long	433.2	9.2	1.1	48.6	48.6
Nitz_rec	205.1	2.4	0.6	12.8	61.4
Cymb_vent	174.7	1.1	0.4	5.9	67.3
Gomp_abbre	109.7	0.8	0.3	4.3	71.6
Hantz_amphi	135.3	0.7	0.4	3.9	75.5
Frag_sp	105.6	0.7	0.3	3.5	79.0
Pseuda_moni	113.5	0.6	0.3	3.3	82.3
Neidium_aff	102.5	0.6	0.3	2.9	85.2
Gomp_acu	109.8	0.5	0.3	2.6	87.9
Stauro_phoeni	95.3	0.5	0.3	2.5	90.3

Group M — Average similarity: 4.8

Species	Av. Abund	Av. Sim	Sim/SD	Contrib %	Cum %
Frag_long	231.1	2.4	0.6	50	50
Melo_granu	231.1	2.4	0.6	50	100

Table D7 SIMPER results of within "ZONES" similarities for Zooplankton. H: horizontal tows; V: vertical hauls. LIC: low inundation conditions; HIC: high inundation conditions. Based on abundance data (square root transformed).

Zooplankton H - LIC

Group U	Average similarity: 41.2					Group L	Average similarity: 28.2			
Species	Av.Abund	Av.Sim	Sim/SD	Contrib%	Cum.%	Species	Av.Abund	Av.Sim	Contrib%	Cum.%
						Platyias				
Lecane sp.	36.3	13.6	6.9	32.9	32.9	cf cuadricornis	58.8	16.0	56.9	56.9
Arcella sp.	33.1	5.3	0.6	12.8	45.7	Chydorus sphaericus	12.8	5.1	18.0	74.9
						Mesocyclops				
Trichotria sp.	24.2	5.3	0.6	12.8	58.5	venezolanus	5.1	2.0	7.0	81.9
Difflugia sp.	24.2	4.2	0.6	10.2	68.7	Mesocyclops sp_cop	5.1	2.0	7.0	88.9
Platyias										
cf cuadricornis	24.2	4.1	0.6	9.9	78.6	Leydigia cf leydigii	3.1	1.2	4.1	93.0
Chydorus										
sphaericus	10.4	2.7	8.8	6.5	85.1	Group R	Average similarity: 25.7			
Simocephalus										
cf serrulatus	4.1	1.3	6.9	3.3	88.4	Species	Av.Abund	Av.Sim	Sim/SD	Contrib% Cum.%
Oxyurella sp.	3.5	1.0	6.9	2.4	90.8	Difflugia sp.	27.3	15.3	0.9	59.5 59.5
						Arcella sp.	18.2	4.2	0.4	16.5 76.0
						Lecane sp.	21.9	4.2	0.4	16.5 92.5

Zooplankton H - HIC

Group U	Average similarity: 27.8					Group L	Average similarity: 49.8			
Species	Av.Abund	Av.Sim	Sim/SD	Contrib%	Cum.%	Species	Av.Abund	Av.Sim	Contrib%	Cum.%
Chy_sph	9.9	8.1	2.0	29.4	29.4	Chy_sph	17.7	13.2	26.6	26.6
Mesoc_ven	6.9	3.4	0.9	12.4	41.8	Mesoc_ven	16.2	11.1	22.2	48.8
Cali_wi	1.6	2.8	1.5	10.1	51.9	Simo_acu	8.4	5.1	10.3	59.1
Mesoc_sp_cop	2.8	1.8	0.8	6.5	58.4	Leydig_ley	4.9	3.0	5.9	65.0
Simo_sp_juv	2.9	1.7	0.8	6.0	64.4	Lynceus_sp	3.7	3.0	5.9	71.0
Simo_acu	4.1	1.6	0.6	5.8	70.1	Macrot_sp	2.7	3.0	5.9	76.9
Tanyp_larv	0.7	1.3	0.6	4.7	74.9	Skisto_sp	2.7	3.0	5.9	82.9
Macrot_sp	3.1	1.3	0.7	4.6	79.4	fish_egg	1.9	2.1	4.2	87.1
Kurz_sp	3.0	0.9	0.9	3.3	82.7	fish_larv	1.9	2.1	4.2	91.3
Lynceus_sp	1.6	0.9	0.8	3.2	85.9					
Laton_occ_juv	5.4	0.8	0.4	3.0	88.9					
Physoc_sp	1.5	0.8	0.8	2.8	91.6					

Group M	Average similarity: 51.1				Group R	Average similarity: 34.2			
Species	Av.Abund	Av.Sim	Contrib%	Cum.%	Species	Av.Abund	Av.Sim	Contrib%	Cum.%
Chy_sph	14.3	13.0	25.4	25.4	Mesoc_sp_cop	4.4	6.9	20.3	20.3
Laton_occ_juv	9.9	8.0	15.7	41.1	Mesoc_ven	2.7	4.9	14.3	34.6
Mesoc_ven	15.4	6.9	13.6	54.7	Baet	1.1	3.5	10.1	44.8
Leydig_ley	3.0	2.8	5.5	60.2	Chyronom	1.1	3.3	9.7	54.4
Macrot_sp	2.9	2.8	5.5	65.8	Herpet_sp	1.1	3.3	9.7	64.1
Simo_acu	3.3	2.8	5.5	71.3	Platy_cua	1.5	3.3	9.7	73.8
Skisto_sp	3.0	2.8	5.5	76.9	Zoea_Cari	1.1	3.3	9.7	83.4
Laton_occ	3.5	2.5	4.8	81.7	Dero	0.9	2.8	8.3	91.7
Herpet_sp	2.3	2.0	3.9	85.6					
Physoc_sp	2.3	2.0	3.9	89.5					

Zooplankton V - HIC

Group U	Average similarity: 28.1					Group M	Average similarity: 47.9			
Species	Av.Abund	Av.Sim	Sim/SD	Contrib%	Cum.%	Species	Av.Abund	Av.Sim	Sim/SD	Contrib% Cum%
Chy sph	21.8	4.4	1.1	15.5	15.5	Mesoc sp cop	115.1	12.1	1.8	25.2 25.2
Mesoc ven	22.0	3.8	1.3	13.4	28.9	Mesoc ven	109.5	11.6	1.6	24.3 49.5
Simo sp juv	16.7	2.6	0.9	9.3	38.2	Moi mi	110.8	8.2	1.1	17.1 66.6
Platy cua	11.8	2.3	1.0	8.3	46.5	Laton occ juv	54.5	7.5	1.6	15.6 82.2
Mesoc sp cop	22.6	2.2	0.9	8.0	54.5	Cerio ri	27.9	1.7	0.6	3.5 85.8
Dero	6.5	2.1	0.8	7.6	62.0	Chy sph	23.5	1.5	0.6	3.1 88.9
Laton occ juv	26.4	2.0	0.7	7.0	69.0	Macrot sp	20.4	1.1	0.6	2.2 91.1
Ilyocry spi	14.5	1.5	0.7	5.4	74.4					
Laton occ	17.0	1.2	0.8	4.4	78.7					
Macrot sp	10.0	1.1	0.7	3.9	82.7					
Moi mi	5.8	0.7	0.4	2.4	85.1					
Physoc sp	4.9	0.6	0.5	2.3	87.4					
Simo vetu	5.5	0.5	0.4	1.9	89.2					
Moi sp	3.4	0.4	0.4	1.5	90.8					

Table D8 SIMPER results of within "ZONES" similarities for Macroinvertebrates. LIC: low inundation conditions; HIC: high inundation conditions. Based on abundance data (square root transformed).

Macroinvertebrates - LIC

Group U	Average similarity: 39.7					Group M	Average similarity: 44.5			
Species	Av.Abund	Av.Sim	Sim/SD	Contrib%	Cum.%	Species	Av.Abund	Av.Sim	Contrib%	Cum.%
Planorbiidae	9.8	17.1	3.3	43.1	43.1	Planorbiidae	6.5	6.4	14.3	14.3
Belostomatidae	2.7	5.3	9.2	13.2	56.3	Chironomidae	5.9	5.4	12.0	26.3
Dytiscidae	2.5	3.8	5.9	9.5	65.8	Culicidae	5.7	5.2	11.6	37.9
Baetidae	3.2	3.5	1.5	8.7	74.5	Baetidae	5.4	3.1	7.0	44.9
Hydrophilidae	2.2	3.0	4.4	7.7	82.1	Hydrophilidae	3.8	3.1	7.0	51.8
Hyalellidae	6.7	1.7	0.6	4.2	86.4	Coenagrionidae	2.0	2.4	5.4	57.2
Chironomidae	2.7	1.2	0.6	3.0	89.4	Dytiscidae	2.9	2.4	5.4	62.6
Lymnaeidae	1.1	0.8	0.6	2.1	91.5	Hydracarina	2.7	2.4	5.4	68.0
						Libellulidae	2.3	2.4	5.4	73.4
						Tabanidae	1.7	2.4	5.4	78.8
						Noteridae	3.7	2.0	4.4	83.2
						Scirtidae	1.8	2.0	4.4	87.6
						Aeshnidae	1.0	1.4	3.1	90.7

Group L	Average similarity: 33.0				Group R	Average similarity: 28.7				
Species	Av.Abund	Av.Sim	Contrib%	Cum.%	Species	Av.Abund	Av.Sim	Sim/SD	Contrib%	Cum.%
Hydrophilidae	2.3	5.1	15.5	15.5	Baetidae	2.5	15.4	1.1	53.7	53.7
Noteridae	4.1	5.1	15.5	31.1	Veliidae	0.8	4.6	0.6	16.1	69.8
Libellulidae	1.9	4.4	13.5	44.5	Culicidae	1.0	2.2	0.9	7.7	77.6
Tabanidae	1.9	4.4	13.5	58.0	Hydrophilidae	1.6	1.1	0.4	3.8	81.4
Baetidae	3.6	3.6	11.0	68.9	Chironomidae	2.1	1.0	0.4	3.5	84.9
Belostomatidae	1.2	2.6	7.8	76.7	Hebridae	0.5	0.9	0.4	3.0	87.9
Coenagrionidae	1.7	2.6	7.8	84.5	Noteridae	1.1	0.8	0.4	2.7	90.6
Naucoridae	1.2	2.6	7.8	92.2						

Macroinvertebrates - HIC

Group U	Average similarity: 62.1					Group M	Average similarity: 43.5			
Species	Av.Abund	Av.Sim	Sim/SD	Contrib%	Cum.%	Species	Av.Abund	Av.Sim	Contrib%	Cum.%
Baetidae	13.7	13.3	3.6	21.5	21.5	Chironomidae	9.3	5.5	12.6	12.6
Planorbidae	6.9	6.0	4.0	9.7	31.2	Baetidae	9.0	4.9	11.3	23.9
Chironomidae	7.4	5.9	4.7	9.5	40.7	Planorbidae	5.6	4.7	10.8	34.6
Gerridae	5.1	3.3	1.8	5.3	46.0	Noteridae	4.8	4.1	9.5	44.1
Hydrophilidae larv	3.0	3.1	5.9	5.1	51.0	Culicidae	5.1	3.9	8.9	53.0
Pyralidae	3.0	2.9	4.3	4.7	55.7	pupas	3.3	2.5	5.6	58.7
Culicidae	3.0	2.7	4.6	4.3	60.0	Pyralidae	2.8	2.5	5.6	64.3
Scirtidae larv	2.4	2.5	5.8	4.1	64.1	Libellulidae	3.3	2.2	5.1	69.4
Hyalellidae	3.4	2.5	2.7	4.0	68.1	Tabanidae	2.4	2.0	4.6	74.0
pupas	2.1	2.3	3.7	3.6	71.7	Aracnida	1.9	1.7	4.0	78.0
Libellulidae	2.8	2.3	5.3	3.6	75.4	Stratiomyidae	1.9	1.7	4.0	82.0
Pleidae	2.2	2.0	4.3	3.2	78.6	Gerridae	3.4	1.4	3.3	85.3
Noteridae	3.0	2.0	1.1	3.2	81.7	Hyalellidae	2.6	1.4	3.3	88.5
Caenidae	2.1	1.5	1.0	2.4	84.1	Coenagrionidae	1.8	1.0	2.3	90.8
Mesoveliidae	1.5	1.3	5.1	2.2	86.2					
Coenagrionidae	1.9	1.1	1.0	1.8	88.1					
Hydrophilidae adult	1.5	1.1	1.1	1.8	89.9					
Belostomatidae	1.6	1.1	1.1	1.7	91.6					

Macroinvertebrates HIC cont.....

Group L	Average similarity: 68.6				Group R	Average similarity: 47.5				
Species	Av.Abund	Av.Sim	Contrib%	Cum.%	Species	Av.Abund	Av.Sim	Sim/SD	Contrib%	Cum.%
Lynceidae	8.0	6.1	8.9	8.9	Baetidae	6.4	10.4	3.6	21.9	21.9
Chironomidae	9.0	5.8	8.5	17.3	Chironomidae	7.4	7.0	3.6	14.8	36.7
Scirtidae larv	6.6	5.6	8.1	25.4	Hydrophilidae larva	2.4	5.6	3.1	11.9	48.5
Baetidae	9.7	5.1	7.4	32.9	Culicidae	3.0	3.1	3.1	6.5	55.1
Planorbidae	4.9	4.0	5.9	38.7	Aracnida	1.3	2.9	1.8	6.2	61.3
Cladocera	4.7	3.9	5.7	44.4	Dytiscidae larv	1.3	2.9	1.8	6.2	67.4
Culicidae	4.8	3.7	5.5	49.9	Belostomatidae	1.4	2.7	3.7	5.8	73.2
Pyralidae	4.5	3.7	5.5	55.4	Pyralidae	1.5	2.7	3.7	5.8	78.9
Orthoptera	3.8	3.1	4.6	59.9	Scirtidae larv	1.0	2.5	2.9	5.2	84.1
Oligochaeta	4.3	3.0	4.4	64.3	Notonectidae	0.9	1.6	0.6	3.4	87.5
pupas	3.8	2.5	3.7	68.0	pupas	1.3	1.6	0.6	3.4	90.9
Libellulidae	3.5	2.4	3.5	71.4						
Aracnida	2.8	2.2	3.2	74.7						
Hydrophilidae larv	3.2	2.1	3.0	77.7						
Gerridae	3.6	1.9	2.7	80.4						
Caenidae	2.1	1.2	1.7	82.1						
Curculionidae	1.6	1.2	1.7	83.8						
Hyalellidae	1.7	1.2	1.7	85.6						
Naucoridae	1.6	1.2	1.7	87.3						
Stratiomyidae	1.6	1.2	1.7	89.0						

Table D9 SIMPER results of within "ZONES" similarities for littoral FISH. HIC: high inundation conditions. Based on abundance data (square root transformed).

Fish - HIC

Group U	Average similarity: 51.1			
Species	Av.Abund	Av.Sim	Sim/SD	Contrib%	Cum.%
Astyanax festae	3.8	17.7	4.2	34.7	34.7
Hemibrycon polyodon	2.3	8.6	1.1	16.8	51.4
Hoplias microlepis	1.4	7.8	2.1	15.2	66.6
Pseudocurimata boulengeri	1.9	6.1	1.1	11.8	78.5
Landonia latidens	2.0	4.5	1.1	8.9	87.4
Hyphessobrycon ecuadoriensis	2.0	4.4	0.9	8.7	96.0

Group M	Average similarity: 56.1			
Species	Av.Abund	Av.Sim	Sim/SD	Contrib%	Cum.%
Astyanax festae	7.8	21.7	3.9	38.7	38.7
Landonia latidens	6.3	19.6	3.6	35.0	73.7
Hoplias microlepis	1.8	6.3	3.9	11.2	84.8
Pseudocurimata boulengeri	1.4	3.4	1.1	6.0	90.9

Group L	Average similarity: 54.4			
Species	Av.Abund	Av.Sim	Sim/SD	Contrib%	Cum.%
Hemibrycon polyodon	5.7	15.1	8.6	27.8	27.8
Astyanax festae	6.4	14.0	1.6	25.7	53.5
Pseudocurimata boulengeri	4.9	13.0	7.7	23.9	77.4
Landonia latidens	3.7	6.1	0.6	11.3	88.6
Hyphessobrycon ecuadoriensis	1.8	2.6	0.6	4.7	93.4

Group R	Average similarity: 51.1			
Species	Av.Abund	Av.Sim	Sim/SD	Contrib%	Cum.%
Pseudocurimata boulengeri	2.3	15.3	21.5	29.9	29.9
Hemibrycon polyodon	2.1	11.9	3.9	23.3	53.3
Astyanax festae	2.9	8.0	0.6	15.7	69.0
Hoplias microlepis	1.2	7.4	23.7	14.4	83.4
Hyphessobrycon ecuadoriensis	1.4	4.8	0.6	9.5	92.9

Dissimilarities

Factor **"Zones"** (Tables D10 - D11)

Table D10 SIMPER results of between "ZONES/areas" dissimilarities for Macroinvertebrates. HIC: high inundation conditions. Based on abundance data (square root transformed). Factor "ZONES" (U: upper; M: middle; L: low; R: river).

Macroinvertebrates
HIC

Groups U & M
Average dissimilarity = 44.5

Species	Group U Av. Abund	Group M Av. Abund	Av. Diss	Diss/ SD	Contrib %	Cum. %
Baetidae	13.7	9.0	3.6	1.3	8.0	8.0
Chironomidae	7.4	9.3	2.2	1.4	4.9	13.0
Lynceidae	2.0	4.2	2.0	1.1	4.6	17.5
Gerridae	5.1	3.4	1.7	1.3	3.9	21.4
Hydrophilidae_larvae	3.0	0.7	1.4	1.9	3.1	24.4
Cladocera	0.0	3.0	1.4	0.9	3.0	27.5
Scirtidae larva	2.4	2.4	1.3	2.9	3.0	30.5
Planorbidae	6.9	5.6	1.3	1.2	3.0	33.5
Culicidae	3.0	5.1	1.3	2.3	2.9	36.4
Oligochaeta	2.6	1.0	1.3	1.0	2.8	39.2
Noteridae	3.0	4.8	1.2	1.3	2.7	41.8
Caenidae	2.1	0.0	1.2	1.4	2.6	44.4
Tabanidae	0.7	2.4	1.1	1.3	2.4	46.8
Hyalellidae	3.4	2.6	1.0	1.2	2.3	49.1

Groups U & L
Average dissimilarity = 42.2

Species	Group U Av. Abund	Group L Av. Abund	Av. Diss	Diss/ SD	Contrib %	Cum. %
Lynceidae	2.0	8.0	3.1	1.8	7.3	7.3
Baetidae	13.7	9.7	2.6	1.5	6.1	13.4
Cladocera	0.0	4.7	2.3	7.1	5.5	18.9
Scirtidae_larv	2.4	6.6	2.1	4.7	5.0	23.9
Noteridae	3.0	3.7	1.8	1.7	4.3	28.2
Chironomidae	7.4	9.0	1.7	1.4	4.1	32.2
Oligochaeta	2.6	4.3	1.5	1.8	3.5	35.7
Noteridae	1.3	2.7	1.4	1.0	3.3	39.0
Gerridae	5.1	3.6	1.3	1.5	3.2	42.2
Orthoptera	1.3	3.8	1.3	2.1	3.0	45.2
Ephemeroptera_NI	0.6	2.6	1.2	1.1	2.9	48.1
Planorbidae	6.9	4.9	1.2	1.2	2.8	50.9

Groups M & L
Average dissimilarity = 39.7

Species	Group M Av. Abund	Group L Av. Abund	Av. Diss	Diss/ SD	Contrib %	Cum. %
Lynceidae	4.2	8.0	2.3	1.0	5.9	5.9
Baetidae	9.0	9.7	2.0	1.0	5.1	11.0
Scirtidae_larv	2.4	6.6	1.8	2.1	4.6	15.6
Chironomidae	9.3	9.0	1.8	1.6	4.4	20.0
Noteridae	4.8	3.7	1.7	2.6	4.3	24.3
Oligochaeta	1.0	4.3	1.6	1.7	4.1	28.5
Cladocera	3.0	4.7	1.6	1.3	3.9	32.4
Noteridae	0.0	2.7	1.2	0.8	3.1	35.5
Hydrophilidae_larvae	0.7	3.2	1.2	1.7	3.0	38.5
Ephemeroptera NI	1.5	2.6	1.2	1.0	3.0	41.5
Caenidae	0.0	2.1	1.0	2.1	2.5	44.0
Gerridae	3.4	3.6	0.9	1.2	2.3	46.3
Orthoptera	2.1	3.8	0.9	1.2	2.3	48.6
Hydrophilidae_adult	1.7	0.7	0.9	1.1	2.2	50.8

Groups U & R
Average dissimilarity = 56

Species	Group U Av. Abund	Group R Av. Abund	Av. Diss	Diss/ SD	Contrib %	Cum. %
Baetidae	13.7	6.4	6.1	1.6	10.9	10.9
Planorbidae	6.9	1.1	4.7	2.0	8.3	19.2
Chironomidae	7.4	7.4	4.2	1.6	7.5	26.7
Gerridae	5.1	0.8	3.2	1.9	5.7	32.4
Hyalellidae	3.4	0.5	2.3	1.5	4.1	36.5
Oligochaeta	2.6	2.5	2.2	1.4	3.9	40.4
Noteridae	3.0	0.7	2.0	1.5	3.6	44.0
Pleidae	2.2	0.0	1.7	2.5	3.1	47.1
Libellulidae	2.8	0.9	1.7	1.8	3.0	50.1

Cont...

Groups M & R Average dissimilarity =61.3							Groups L & R Average dissimilarity =59.7						
	Group M	Group R						Group L	Group R				
Species	Av. Abund	Av. Abund	Av. Diss	Diss/ SD	Contrib %	Cum. %	Species	Av. Abund	Av. Abund	Av. Diss	Diss/ SD	Contrib %	Cum. %
Chironomidae	9.3	7.4	4.5	1.7	7.3	7.3	Lynceidae	8.0	0.5	4.8	3.8	8.0	8.0
Planorbidae	5.6	1.1	3.3	2.6	5.5	12.7	Chironomidae	9.0	7.4	3.9	2.9	6.5	14.5
Noteridae	4.8	0.7	3.1	2.4	5.1	17.9	Scirtidae_ larvae	6.6	1.0	3.5	7.3	5.9	20.4
Baetidae	9.0	6.4	2.9	1.3	4.7	22.5	Cladocera	4.7	0.0	2.9	7.3	4.9	25.3
Lynceidae	4.2	0.5	2.4	1.0	3.9	26.4	Baetidae	9.7	6.4	2.9	1.3	4.9	30.1
Culicidae	5.1	3.0	2.1	1.8	3.5	29.9	Planorbidae	4.9	1.1	2.4	2.7	4.0	34.2
Scirtidae_larv	2.4	1.0	2.1	1.2	3.4	33.3	Noteridae	3.7	0.7	2.4	1.0	4.0	38.2
Tabanidae	2.4	0.0	1.9	2.0	3.2	36.4	Orthoptera	3.8	0.7	2.0	3.9	3.3	41.5
Libellulidae	3.3	0.9	1.9	1.8	3.1	39.5	Oligochaeta	4.3	2.5	1.9	1.8	3.3	44.7
Gerridae	3.4	0.8	1.8	2.1	3.0	42.5	Pyralidae	4.5	1.5	1.9	4.9	3.2	47.9
Suborder_Cladocera	3.0	0.0	1.7	0.9	2.7	45.2	Libellulidae	3.5	0.9	1.8	1.5	3.0	50.9
Oligochaeta	1.0	2.5	1.6	1.0	2.6	47.8							
Hyalellidae	2.6	0.5	1.5	1.7	2.4	50.2							

Table D11 SIMPER results of between "ZONES/areas" dissimilarities for littoral fish. LIC: low inundation conditions; HIC: high inundation conditions. Based on abundance data (square root transformed). Factor "ZONES" (U: upper; M: middle; L: low; R: river).

FISH - LIC

Groups U & M Average dissimilarity =23							Groups U & L Average dissimilarity = 26.5					
	Group U	Group M						Group U	Group L			
Species	Av. Abund	Av. Abund	Av. Diss	Contrib %	Cum. %		Species	Av. Abund	Av. Abund	Av. Diss	Contrib %	Cum %
Astyanax festae	18.6	14.5	3.2	14.1	14.1		Rhoadsia altipinna	4.7	11.4	4.4	16.7	16.7
Ichthyoelephas humeralis	4.7	1.0	2.9	12.7	26.8		Brycon dentex	6.6	0.0	4.4	16.5	33.2
Landonia latidens	6.6	3.5	2.5	10.9	37.7		Hyphessobrycon ecuadoriensis	7.9	4.0	2.6	9.9	43.1
Iotabrycon praecox	5.1	2.0	2.4	10.7	48.4		Pseudopoecilia fria	2.0	5.7	2.4	9.2	52.3
Pseudocurimata boulengeri	2.5	0.0	1.9	8.4	56.8							

Groups M & L Average dissimilarity = 36.7							Groups U & R Average dissimilarity = 56.8					
	Group M	Group L						Group U	R			
Species	Av. Abund	Av. Abund	Av. Dlss	Contrib %	Cum		Species	Av. Abund	Av. Abund	Av. Diss	Contrib %	Cum
Astyanax festae	14.5	20.5	4.7	12.7	12.7		Rhoadsia altipinna	4.7	21.7	13.9	24.5	24.5
Rhoadsia altipinna	6.2	11.4	4.0	11.0	23.7		Astyanax festae	18.6	7.1	9.4	16.6	41.1
Ichthyoelephas humeralis	1.0	6.1	3.9	10.7	34.4		Landonia latidens	6.6	1.7	4.0	7.1	48.2
Hyphessobrycon ecuadoriensis	8.6	4.0	3.6	9.7	44.1		Ichthyoelephas humeralis	4.7	0.0	3.8	6.8	55.0
Brycon dentex	4.6	0.0	3.5	9.7	53.7							

Groups M & R Average dissimilarity = 46.9							Groups L & R Average dissimilarity = 54					
	Group M	Group R						Group L	R			
Species	Av. Abund	Av. Abund	Av. Diss	Contrib %	Cum		Species	Av. Abund	Av. Abund	Av. Diss	Contrib %	Cum
Rhoadsia altipinna	6.2	21.7	15.4	32.8	32.8		Astyanax festae	20.5	7.1	10.8	20.0	20.0
Astyanax festae	14.5	7.1	7.3	15.7	48.4		Rhoadsia altipinna	11.4	21.7	8.3	15.4	35.4
Pseudopoecilia fria	3.5	0.0	3.4	7.3	55.7		Ichthyoelephas humeralis	6.1	0.0	4.9	9.0	44.4
							Pseudopoecilia fria	5.7	0.0	4.5	8.4	52.8

Cont..

FISH - HIC

Groups U & M	Average dissimilarity = 55.9					Groups U & L	Average dissimilarity = 53.7				
	Group U	Group M					Group U	Group L			
Species	Av. Abund	Av. Abund	Av. Diss	Diss/ SD	Contrib %	Cum %	Species	Av. Abund	Av. Abund	Av. Diss	Diss/ SD	Contrib %	Cum %
Landonia latidens	2.0	6.3	11.5	1.7	20.6	20.6	Astyanax festae	3.8	6.4	7.9	1.4	14.7	14.7
Astyanax festae	3.8	7.8	9.9	1.2	17.7	38.3	Landonia latidens	2.0	3.7	7.5	1.3	13.9	28.6
Hyphessobrycon ecuadoriensis	2.0	2.0	5.7	1.2	10.2	48.5	Hemibrycon polyodon	2.3	5.7	7.3	1.8	13.5	42.1
Hemibrycon polyodon	2.3	0.0	5.7	1.6	10.1	58.6	Pseudocurimata boulengeri	1.9	4.9	6.8	1.5	12.7	54.8

Groups M & L	Average dissimilarity = 54.2					Groups U & R	Average dissimilarity = 44.4				
	Group M	Group L					Group U	Group R			
Species	Av. Abund	Av. Abund	Av. Diss	Diss/ SD	Contrib %	Cum %	Species	Av. Abund	Av. Abund	Av. Diss	Diss/ SD	Contrib %	Cum %
Hemibrycon polyodon	0.0	5.7	10.1	5.8	18.7	18.7	Astyanax festae	3.8	2.9	7.8	1.4	17.6	17.6
Astyanax festae	7.8	6.4	6.8	1.5	12.6	31.2	Landonia latidens	2.0	1.0	6.3	1.4	14.1	31.7
Pseudocurimata boulengeri	1.4	4.9	6.4	2.3	11.7	42.9	Hyphessobrycon ecuadoriensis	2.0	1.4	5.8	1.6	13.1	44.8
Landonia latidens	6.3	3.7	5.5	1.4	10.2	53.1	Hemibrycon polyodon	2.3	2.1	4.4	1.6	9.8	54.6

Factor **"Sampling effort"** (Horizontal tows- Vertical hauls)

Table D12 SIMPER results of between "SAMPLING EFFORT" dissimilarities for Phytoplankton and Zooplankton (net). LIC: low inundation conditions; HIC: high inundation conditions. Based on abundance data (square root transformed). Factor "SAMPLING EFFORT" (H: horizontal tows; V: vertical hauls).

Phytoplankton - HIC							Zooplankton - HIC					
Groups H & V							Groups H & V					
Average dissimilarity = 90.7							Average dissimilarity = 79.8					
	Group H	Group V						Group H	Group V			
Species	Av. Abund	Av. Abund	Av. Diss	Diss/ SD	Contrib %	Cum %	Species	Av. Abund	Av. Abund	Av. Diss	Diss/ SD	Contrib %	Cum %
Frag_long	80.1	353.3	10.2	1.2	11.2	11.2	Moi_mi	3.3	40.2	8.1	0.8	10.2	10.2
Melo_granu	103.7	130.0	6.9	0.8	7.6	18.8	Mesoc_sp_cop	2.7	45.6	8.0	1.2	10.1	20.3
Nitz_rec	9.2	196.5	4.7	1.0	5.2	24.0	Mesoc_ven	9.6	43.0	7.8	1.2	9.8	30.0
Melo_var	7.7	41.6	3.7	0.3	4.1	28.1	Laton_occ_juv	5.2	36.3	7.4	1.3	9.3	39.3
Gomp_abbre	9.2	89.3	3.2	0.7	3.5	31.6	Chi_sph	10.8	20.5	4.7	1.6	5.9	45.2
Cymb_vent	0.0	120.9	3.0	0.6	3.3	34.9	Laton_occ	2.8	17.1	3.5	1.1	4.4	49.7
Gomp_acu	16.2	89.4	2.8	0.7	3.1	38.0							
Frag_sp	15.9	73.1	2.4	0.7	2.7	40.7							
Nitz_frus	9.2	78.7	2.3	0.6	2.6	43.3							
Cyclo_bod	9.2	88.2	2.2	0.7	2.5	45.7							
Nitz_amp	0.0	122.3	2.1	0.4	2.3	48.0							
Neidium_aff	0.0	71.0	2.0	0.5	2.2	50.2							

Factor "LOCATION" (River or Wetland) (Tables D13 - D16)

Table D13 SIMPER results of between "RIVER/WETLAND" dissimilarities for Phytoplankton (collected with Niskin bottle). LIC: low inundation conditions; HIC: high inundation conditions. LIC+HIC: pooled samples of both conditions. Based on abundance data (square root transformed). Factor "LOCATION" (R: river; W: wetland).

Phytoplankton (Niskin bottle)

Groups W & R (LIC)
Average dissimilarity = 87.7

Species	Group W Av. Abund	Group R Av. Abund	Av. Diss	Diss/SD	Contrib %	Cum. %
Cryptomonas sp.	170.5	0	8	1.0	9.4	9.4
Synedra sp.	42.9	122.9	7.7	1.2	8.8	18.2
Trachelomonas sp.	87.8	0	5.6	1.0	6.4	24.6
Oscillatoria sp.	48.8	20	4.1	0.8	4.6	29.2
Cyclotella sp.	42.9	0	3.9	0.8	4.5	33.7
Ankistrodesmus acicularis	67.6	20	3.9	0.7	4.5	38.2
Pseudanabaena sp.	39.0	40	3.9	0.9	4.5	42.6
Nitzschia acicularis	40.4	40	3.9	0.9	4.5	47.1
Nitzschia sp.	39.0	20	3.7	0.7	4.2	51.3
Melosira granulata	48.8	20	3.2	0.8	3.7	55.0

Groups W & R (HIC)
Average dissimilarity = 79.9

Species	Group W Av. Abund	Group R Av. Abund	Av. Diss	Diss/SD	Contrib %	Cum. %
Cryptomonas sp.	204.9	70.7	10.3	1.3	12.9	12.9
Melosira granulata	85.2	141.4	7.6	1.3	9.5	22.4
Stauroneis phoenicenteron	18.6	75	4.1	1.4	5.2	27.6
Nitzschia sp.	7.7	75	4.1	1.4	5.2	32.8
Cyclotella comta	53.9	25	3.5	0.9	4.4	37.2
Fragilaria longissima	34.0	50	3.3	1.0	4.1	41.3
Nitzschia longissima	41.7	35.4	3.1	0.9	3.9	45.2
Trachelomonas sp.	26.3	25	2.5	0.7	3.2	48.4
Scenedesmus bijugus	38.5	0	2.4	0.7	3.0	51.4

Groups W & R (LIC+HIC)
Average dissimilarity = 89.2

Species	Group W Av. Abund	Group R Av. Abund	Av. Diss	Diss/SD	Contrib %	Cum. %
Cryptomonas sp.	203.0	28.3	11.6	1.3	13.0	13.0
Melosira granulata	71.0	76.6	6.0	1.0	6.8	19.7
Synedra sp.	15.8	61.5	4.9	0.8	5.5	25.2
Trachelomonas sp.	50.3	10	3.3	0.8	3.7	28.9
Nitzschia sp.	19.6	40	3.0	0.8	3.4	32.3
Cyclotella comta	36.8	10	2.8	0.7	3.1	35.4
Fragilaria longissima	23.2	20	2.3	0.6	2.6	38.0
Nitzschia longissima	28.5	14	2.2	0.6	2.5	40.5
Pseudanabaena sp.	19.6	20	2.1	0.6	2.4	42.8
Stauroneis phoenicenteron	12.7	30	2.1	0.7	2.3	45.2
Oscillatoria sp.	18.0	20	2.1	0.6	2.3	47.5
Nitzschia acicularis	14.9	20	1.9	0.6	2.2	49.6
Cyclotella sp.	26.3	10	1.9	0.6	2.1	51.7

Table D14 SIMPER results of between "RIVER/WETLAND" dissimilarities for Zooplankton collected with horizontal tows (H). LIC: low inundation conditions; HIC: high inundation conditions. LIC+HIC: pooled samples of both conditions. Based on abundance data (square root transformed). Factor "LOCATION" (R: river; W: wetland).

Zooplankton H

Groups W & R (LIC)
Average dissimilarity = 73.2

Species	Group W Av. Abund	Group R Av. Abund	Av. Diss	Diss/ SD	Contrib %	Cum. %
Platyjas cf cuadricornis	31.7	9.1	9.0	1.0	12.2	12.2
Arcella sp.	22.6	18.2	7.7	1.0	10.5	22.7
Platyjas cf patulus.	20.7	0	7.6	0.9	10.4	33.1
Lecane sp.	24.2	21.9	7.2	1.1	9.9	43.0
Difflugia sp.	18.2	27.3	6.4	0.9	8.7	51.7

Groups W & R (HIC)
Average dissimilarity = 74.2

Species	Group W Av. Abund	Group R Av. Abund	Av. Diss	Diss/ SD	Contrib %	Cum. %
Chy_sph	13.0	1.9	9.7	2.1	13.1	13.1
Mesoc_ven	11.4	2.7	7.9	1.5	10.7	23.7
Laton_occ_juv	6.3	0.5	5.0	1.4	6.7	30.4
Simo_acu	5.0	0	4.0	1.4	5.4	35.8
Mesoc_sp_cop	2.2	4.4	3.5	1.3	4.8	40.6
Moi_mi	4.0	0.7	3.2	0.7	4.4	45.0
Laton_occ	3.3	1.0	2.6	1.0	3.6	48.5
Macrot_sp	2.9	0	2.4	1.7	3.2	51.8

Groups W & R (LIC+HIC)
Average dissimilarity = 80.9

Species	Group W Av. Abund	Group R Av. Abund	Av. Diss	Diss/ SD	Contrib %	Cum. %
Difflugia sp.	7.8	21.8	9.5	1.0	11.7	11.7
Lecane sp.	10.4	17.6	7.8	0.9	9.6	21.3
Arcella sp.	9.7	14.5	7.2	0.9	8.9	30.2
Platyjas cuadricornis	14.1	7.9	7.2	0.7	8.8	39.1
Chydorus sphaericus	12.4	2.9	5.2	1.3	6.4	45.5
Mesocyclops venezolanus	7.9	1.8	3.9	0.9	4.8	50.3

Table D15 SIMPER results of between "RIVER/WETLAND" dissimilarities for Zooplankton. HV: horizontal and vertical samples pooled and included in the analysis. LIC: low inundation conditions; HIC: high inundation conditions. LIC+HIC: pooled samples of both conditions. Based on abundance data (square root transformed). Factor "LOCATION" (R: river; W: wetland).

Zooplankton HV

Groups W & R (HIC)
Average dissimilarity = 82.5

Species	Group W Av. Abund	Group R Av. Abund	Av. Diss	Diss/ SD	Contrib %	Cum. %
Mesoc_ven	31.0	2.7	8.5	1.4	10.3	10.3
Chy_sph	17.6	1.9	7.0	1.5	8.5	18.8
Laton_occ_juv	24.9	0.5	7.0	1.2	8.5	27.3
Moi_mi	26.4	0.7	6.7	0.7	8.1	35.4
Mesoc_sp_cop	29.1	4.4	6.7	1.1	8.1	43.5
Laton_occ	11.8	1.0	3.4	1.1	4.1	47.7
Simo_sp_juv	9.1	1.0	3.3	1.0	4.1	51.7

Groups W & R (LIC+HIC)
Average dissimilarity = 86.1

Species	Group W Av. Abund	Group R Av. Abund	Av. Diss	Diss/ SD	Contrib %	Cum. %
Difflugia sp.	4.0	21.8	8.2	0.9	9.5	9.5
Lecane sp.	5.4	17.6	6.4	0.8	7.5	17.0
Platyjas cuadricornis	12.6	7.9	5.8	0.7	6.8	23.8
Mesocyclops venezolanus	24.8	1.8	5.8	1.0	6.8	30.5
Arcella sp.	5.0	14.5	5.7	0.8	6.6	37.2
Chydorus sphaericus	16.3	2.9	4.8	1.3	5.5	42.7
Mesocyclops sp. (copepodite)	23.4	2.8	4.7	0.8	5.4	48.1
Latonopsis cf occidentalis (juvenil)	19.4	0.2	4.6	0.9	5.4	53.5

Table D16 SIMPER results of between "RIVER/WETLAND" dissimilarities for Macroinvertebrates and littoral Fish. LIC: low inundation conditions; HIC: high inundation conditions. LIC+HIC: pooled samples of both conditions. Based on abundance data (square root transformed). Factor "LOCATION" (R: river; W: wetland).

Macroinvertebrates

Groups W & R (LIC) — Average dissimilarity = 73.1

Species	Group W Av. Abund	Group R Av. Abund	Av. Diss	Diss/SD	Contrib %	Cum. %
Planorbiidae	7.8	0.6	9.9	1.8	13.5	13.5
Hyalellidae	3.6	0.4	4.4	0.6	6.0	19.5
Chironomidae	3.3	3.1	4.2	1.3	5.8	25.3
Hyalellidae	2.3	0	3.8	0.4	5.2	30.5
Noteridae	2.7	1.3	3.7	0.9	5.1	35.5
Culicidae	2.1	1.7	2.8	1.3	3.8	39.3
Baetidae	3.6	3.2	2.7	1.5	3.7	43.0
Dytiscidae	2.5	0.7	2.5	1.7	3.5	46.5
Hydrophilidae	2.7	1.8	2.5	1.4	3.4	49.9
Belostomatidae	2.0	0.5	2.3	1.3	3.2	53.0

Groups W & R (HIC) — Average dissimilarity = 57.9

Species	Group W Av. Abund	Group R Av. Abund	Av. Diss	Diss/SD	Contrib %	Cum. %
Baetidae	11.8	6.4	4.7	1.3	8.0	8.0
Chironomidae	8.2	7.4	4.2	1.8	7.2	15.3
Planorbidae	6.2	1.1	3.9	1.9	6.7	21.9
Gerridae	4.4	0.8	2.6	1.7	4.4	26.4
Noteridae	3.6	0.7	2.4	1.5	4.1	30.4
Lynceidae	3.8	0.5	2.3	1.1	4.0	34.4
Oligochaeta	2.7	2.5	2.0	1.4	3.4	37.9
Scirtidae (larvae)	3.3	1.0	1.8	1.4	3.2	41.0
Hyalellidae	2.9	0.5	1.8	1.3	3.1	44.1
Libellulidae	3.1	0.9	1.8	1.8	3.0	47.1
Culicidae	3.9	3.0	1.7	1.7	2.9	50.0
Pleidae	1.7	0	1.3	1.9	2.3	52.3

Groups W & R (LIC+HIC) — Average dissimilarity = 58.9

Species	Group W Av. Abund	Group R Av. Abund	Av. Diss	Diss/SD	Contrib %	Cum. %
Planorbidae	6.5	1.0	4.8	1.9	8.1	8.1
Baetidae	9.1	5.1	4.5	1.4	7.7	15.8
Chironomidae	6.7	6.3	3.8	1.3	6.4	22.2
Hyalellidae	3.4	0.4	2.7	0.6	4.6	26.8
Noteridae	4.3	1.6	2.6	1.4	4.4	31.2
Gerridae	3.5	0.5	2.4	1.2	4.1	35.3
Scirtidae (larvae)	2.7	0.5	2.0	1.3	3.4	38.7
Culicidae	3.4	2.8	2.0	1.5	3.3	42.0
Oligochaeta	1.8	1.4	1.7	1.1	2.8	44.8
Lynceidae	2.5	0.2	1.6	0.8	2.7	47.5
Pyralidae	2.6	1.0	1.5	1.7	2.6	50.1
Libellulidae	2.5	1.0	1.5	1.5	2.5	52.7

FISH

a) Groups W & R (LIC) — Average dissimilarity = 52.6

Species	Group W Av. Abund	Group R Av. Abund	Av. Diss	Diss/SD	Contrib %	Cum. %
Rhoadsia altipinna	7.4	21.7	12.5	3.4	23.8	23.8
Astyanax festae	17.9	7.1	9.2	5.3	17.5	41.3
Ichthyoelephas humeralis	3.9	0.0	3.2	1.6	6.2	47.5
Pseudopoecilia fria	3.7	0.0	3.2	2.2	6.1	53.6
Landonia latidens	5.3	1.7	3.0	2.6	5.6	59.2

b) Groups W & R (HIC) — Average dissimilarity = 54.2

Species	Group W Av. Abund	Group R Av. Abund	Av. Diss	Diss/SD	Contrib %	Cum. %
Astyanax festae	5.9	2.9	10.0	1.4	18.4	18.4
Landonia latidens	4.1	1.0	9.5	1.5	17.5	35.9
Hemibrycon polyodon	2.2	2.1	5.6	2.1	10.4	46.3
Hyphessobrycon ecuadoriensis	1.9	1.4	5.1	1.4	9.4	55.7

c) Groups W & R (LIC+HIC) — Ave dissimilarity = 58.5

Species	Group W Av. Abund	Group R Av. Abund	Av. Diss	Diss/SD	Contrib %	Cum. %
Astyanax festae	6.3	3.0	9.2	1.4	15.8	15.8
Landonia latidens	3.7	0.9	7.9	1.3	13.5	29.2
Rhoadsia altipinna	1.5	2.4	7.0	0.9	11.9	41.1
Hemibrycon polyodon	1.8	1.6	5.0	1.6	8.6	49.7
Hyphessobrycon ecuadoriensis	2.2	1.8	4.5	1.5	7.7	57.4

Table D17 SIMPER results of between "CONDITIONS" dissimilarities for Phytoplankton (collected with Niskin bottle). Based on abundance data (square root transformed). Factor "CONDITION" (LIC: low inundation conditions; HIC: high inundation conditions). Species that contribute up to 50% of the total dissimilarity are presented.

Phytoplankton (Niskin bottle)

Groups HIC & LIC

Average dissimilarity = 91.8

Species	Group HIC Av.Abund	Group LIC Av.Abund	Av.Diss	Diss/SD	Contrib%	Cum.%
Cryptomonas sp.	173.3	99.5	10.8	1.3	11.7	11.7
Melosira granulata	98.4	36.8	6.1	0.9	6.6	18.3
Synedra sp.	0.0	76.2	5.7	0.9	6.2	24.5
Trachelomonas sp.	26.0	51.2	3.3	0.8	3.6	28.1
Cyclotella comta	47.1	0.0	3.0	0.7	3.3	31.4
Nitzschia sp.	23.5	31.1	2.8	0.7	3.0	34.4
Fragilaria longissima	37.7	0.0	2.5	0.7	2.7	37.2
Oscillatoria sp.	5.9	36.8	2.3	0.7	2.5	39.7
Pseudanabaena sp.	5.9	39.4	2.3	0.7	2.5	42.2
Nitzschia longissima	40.2	0.0	2.3	0.7	2.5	44.7
Nitzschia acicularis	0.0	40.2	2.2	0.6	2.4	47.1
Ankistrodesmus acicularis	0.0	47.8	2.0	0.5	2.2	49.2
Cyclotella sp.	17.7	25.0	1.9	0.7	2.1	51.3

Table D18 SIMPER results of between "CONDITIONS" dissimilarities for Zooplankton collected with horizontal tows (H). Based on abundance data (square root transformed). Factor "CONDITION" (LIC: low inundation conditions; HIC: high inundation conditions). Species that contribute up to 70% of the total dissimilarity are presented.

Zooplankton H

Groups HIC & LIC

Average dissimilarity = 88.6

Species	Group HIC Av.Abund	Group LIC Av.Abund	Av.Diss	Diss/SD	Contrib%	Cum.%
Difflugia sp.	0.0	24.2	10.9	1.1	12.4	12.4
Lecane sp.	0.0	25.9	9.7	1.1	11.0	23.3
Platyias cuadricornis	1.1	25.2	9.7	0.9	10.9	34.2
Arcella sp.	0.0	23.2	9.3	1.0	10.5	44.7
Platyias patulus patulus	0.7	13.8	5.5	0.7	6.2	50.9
Mesocyclops sp. (nauplius I)	1.6	13.8	4.0	0.6	4.6	55.5
Centropyxis sp	0.0	8.1	3.4	0.5	3.8	59.2
Mesocyclops venezolanus	9.6	2.6	3.1	1.1	3.5	62.7
Chidorus sphaericus	10.8	8.9	3.1	1.2	3.5	66.2
Trichotria sp.	0.0	8.1	2.8	0.5	3.1	69.3
Latonopsis cf occidentalis (juvenile)	5.2	0.0	1.9	0.9	2.2	71.5

Table D19 SIMPER results of between "CONDITIONS" dissimilarities for Macroinvertebrates. Based on abundance data (square root transformed). Factor "CONDITION" (LIC: low inundation conditions; HIC: high inundation conditions). Species that contribute up to 50% of the total dissimilarity are presented.

Macroinvertebrates

Groups HIC & LIC

Average dissimilarity = 57

Species	Group HIC Av.Abund	Group LIC Av.Abund	Av.Diss	Diss/SD	Contrib%	Cum.%
Baetidae	10.5	4.0	5.1	1.6	9.0	9.0
Chironomidae	8.0	4.4	3.6	1.3	6.3	15.3
Planorbidae	4.9	4.8	3.0	1.2	5.3	20.6
Hyalellidae	2.3	2.8	2.7	0.7	4.7	25.3
Gerridae	3.5	1.4	2.5	1.4	4.4	29.7
Noteridae	3.9	2.8	2.1	1.2	3.7	33.4
Culicidae	3.6	2.5	2.0	1.6	3.5	36.9
pupas	2.4	0.0	1.9	2.3	3.2	40.1
Oligochaeta	2.6	0.3	1.8	1.2	3.1	43.2
Lynceidae/Lynceus sp	3.0	0.0	1.8	0.9	3.1	46.3
Scirtidae larva	2.7	1.1	1.8	1.3	3.1	49.4
Pyralidae	2.8	1.1	1.6	1.9	2.9	52.3
Libellulidae	2.5	1.3	1.3	1.4	2.2	54.5

Table D20 SIMPER results of between "CONDITIONS" dissimilarities for littoral Fish. Based on abundance data (square root transformed). Factor "CONDITION" (LIC: low inundation conditions; HIC: high inundation conditions). Species that contribute up to 90% of the total dissimilarity are presented.

Fish

Groups HIC & LIC

Average dissimilarity = 61.2

Species	Group HIC Av.Abund	Group LIC Av.Abund	Av.Diss	Diss/SD	Contrib%	Cum.%
Rhoadsia altipinna	0.9	4.9	9.1	1.2	14.9	14.9
Astyanax festae	5.4	6.8	6.9	1.4	11.3	26.2
Landonia latidens	3.5	2.0	5.2	1.5	8.5	34.6
Hemibrycon polyodon	2.2	0.0	4.2	1.1	6.8	41.5
Hyphessobrycon	1.9	3.0	3.9	1.5	6.4	47.8
Pseudocurimata	2.4	0.7	3.6	1.3	5.9	53.8
Iotabrycon praecox	0.8	1.5	3.5	1.5	5.7	59.5
Brycon dentex	0.0	1.6	3.2	1.5	5.2	64.7
Ichthyoelephas humeralis	0.1	1.5	2.7	1.4	4.3	69.0
Pseudopoecilia fria	0.1	1.3	2.4	1.4	3.9	72.9
Aequidens rivulatus	0.0	0.9	1.7	1.6	2.7	75.6
Pseudocurimata troschelii	0.1	0.9	1.6	1.5	2.6	78.2
Hoplias microlepis	1.4	1.8	1.4	1.4	2.2	80.4
Brachyhypopomus	0.3	0.5	1.2	0.9	2.0	82.4
Brycon atrocaudatus	0.4	0.5	1.2	1.1	1.9	84.3
Phenacobrycon henni	0.6	0.0	1.1	0.4	1.8	86.1
Rhamdia cinerascens	0.0	0.6	1.1	1.0	1.8	87.9
Cichlasoma festae	0.0	0.5	1.1	1.0	1.7	89.6
Lebiasina bimaculata	0.5	0.0	1.0	0.6	1.6	91.2

APPENDIX E

E.1 Total Nitrogen temporal and spatial variability

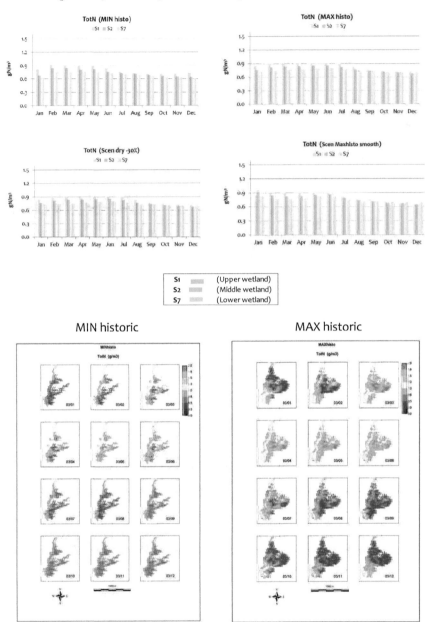

Figure E1 Temporal and spatial simulated concentrations of total nitrogen (TotN) in g/Nm³. Output maps extracted the same day every month. From left to right: January-December. For: MINIMUM and MAXIMUM historical (period 1962-2010)

E.2 Phosphorus - temporal and spatial variability

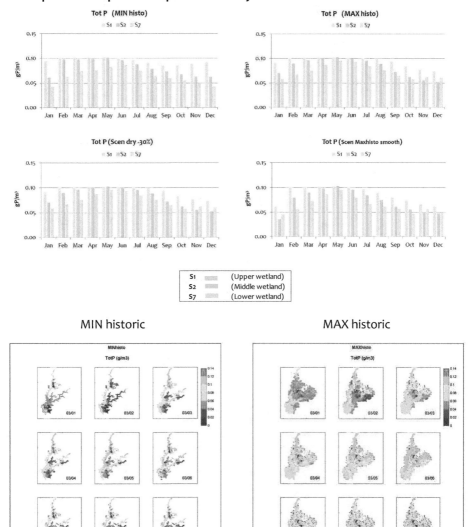

Figure E2 Temporal and spatial simulated concentrations of total phosphorus (TotP) in gP/m³. Output maps extracted the same day every month. From left to right: January-December. For: MINIMUM and MAXIMUM historical (period 1962-2010)

E.3 Total organic carbon - temporal and spatial variability

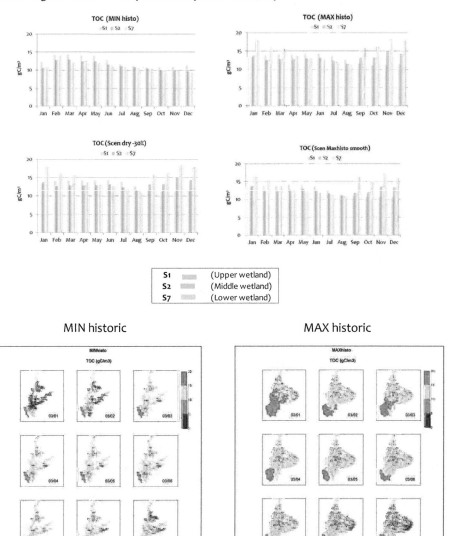

Figure E3 Temporal and spatial simulated concentrations of total carbon (TOC) in gC/m³. Output maps extracted the same day every month. From left to right: January-December. For: MINIMUM and MAXIMUM historical (period 1962-2010)

E.4 Chlorophyll-a - temporal and spatial variability

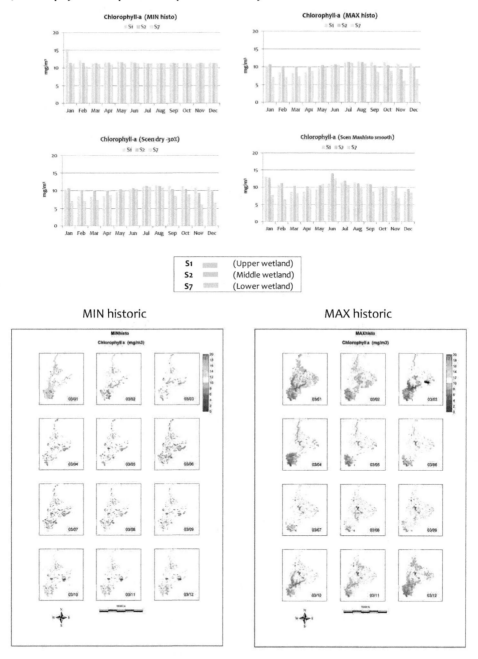

Figure E4 Temporal and spatial simulated concentrations of Chlorophyll-concentrations (mg/m³). Output maps extracted the same day every month. From left to right: January-December. For: MINIMUM and MAXIMUM historical (period 1962-2010)

E.5 Phytoplankton biomass- temporal and spatial variability

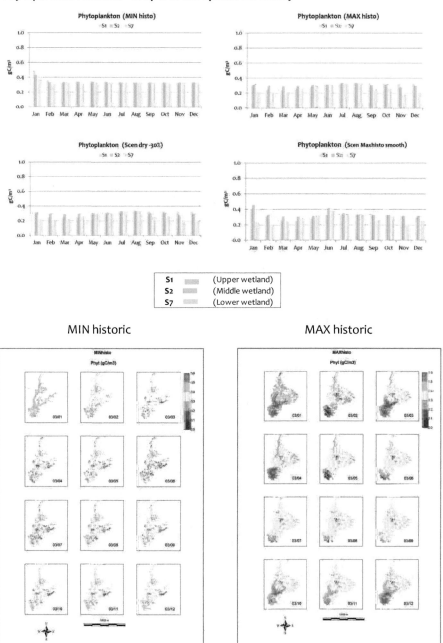

Figure E5 Temporal and spatial simulated concentrations of phytoplankton biomass (Phyt) (gC/m³). Output maps extracted the same day every month. From left to right: January-December. For: MINIMUM and MAXIMUM historical (period 1962-2010)

E.6 Primary consumers - temporal and spatial variability

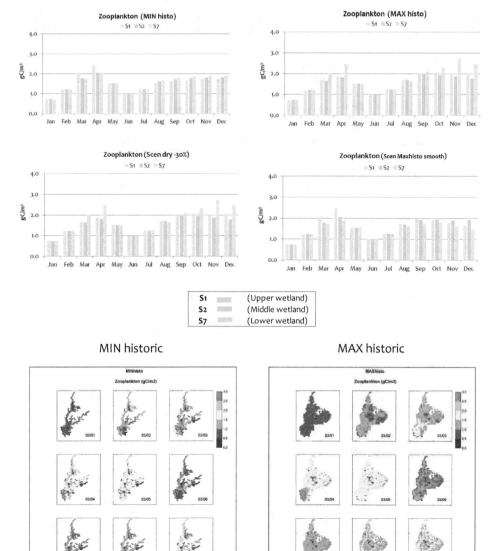

Figure E6 Temporal and spatial simulated concentrations of zooplankton biomass (Zoo) (gC/m³). Output maps extracted the same day every month. From left to right: January-December. For: MINIMUM and MAXIMUM historical (period 1962-2010)

E.7 Nitrogen partitioning

Table E1 Nitrogen compounds partitioning (expressed in yearly average percentage) at upper, middle and low wetland observation points. Results from one-year simulation for different conditions.

| AREAS/CONDITIONS | TOTAL NITROGEN TotN | | TOTAL ORGANIC NITROGEN TON | | PARTICULATE ORGANIC NITROGEN PON | |
	% of total organic nitrogen TON	% of dissolved Inorganic nitrogen DIN	%of dissolved organic nitrogen DON	% of particulate organic nitrogen PON	%of particulate organic nitrogen no algae PONnoa	%of nitrogen in algae AlgN
S1 (Upper wetland)						
Scen dry-30%smooth	84	26	78	22	48	52
1998 (El Niño)	80	20	75	25	57	43
2012	85	15	76	24	49	51
S2 (Middle wetland)						
Scen dry-30%smooth	90	10	80	20	41	59
1998 (El Niño)	86	14	78	22	49	51
2012	89	11	80	20	43	57
S7 (Low wetland)						
Scen dry-30%smooth	96	4	81	19	39	61
1998 (El Niño)	91	8	80	20	46	53
2012	93	7	81	19	42	58

E.8 Phosphorus partitioning

Table E2 Phosphorus compounds partitioning (expressed in yearly average percentage) at upper, middle and low wetland observation points. Results from one-year simulation for different conditions.

| AREAS/CONDITIONS | TOTAL PHOSPHORUS TotP | | TOTAL ORGANIC PHOSPHORUS TOP | | PARTICULATE ORGANIC PHOSPHORUS POP | |
	% of total organic phosphorus TOP	% of phosphates PO4-P	%of dissolved organic phosphorus DOP	% of particulate organic phosphorus POP	%of particulate organic phosphorus no algae POPnoa	%of phosphorus in algae AlgP
S1 (Upper wetland)						
Scen dry-30%smooth	15.9	84.1	38.6	61.4	25.8	74.2
1998 (El Niño)	15.2	84.8	39.3	60.7	29.4	70.6
2012	16.7	83.3	38.3	61.7	26.7	73.3
S2 (Middle wetland)						
Scen dry-30%smooth	17.6	82.4	37.9	62.1	26.2	73.8
1998 (El Niño)	16.4	83.6	38.0	62.0	30.3	69.7
2012	20.1	79.9	38.3	61.7	27.9	72.1
S7 (Low wetland)						
Scen dry-30%smooth	23.9	76.1	37.5	62.5	28.0	72.0
1998 (El Niño)	23.9	76.1	37.5	62.5	28.0	72.0
2012	28.2	71.8	37.9	62.1	32.8	67.2

Table E3 Phosphorus compounds partitioning (expressed in average percentage) at upper, middle and low wetland observation points. Results from 4 months simulation (only wet season)

	TOTAL PHOSPHORUS		TOTAL ORGANIC PHOSPHORUS		PARTICULATE ORGANIC PHOSPHORUS	
	TotP		TOP		POP	
	% of total organic phosphorus	% of phosphates	%of dissolved organic phosphorus	% of particulate organic phosphorus	%of particulate organic phosphorus no algae	%of phosphorus in algae
AREAS/CONDITIONS	TOP	PO4-P	DOP	POP	POPnoa	AlgP
S1 (Upper wetland)						
Scen dry-30%smooth	16.6	83.4	36.6	63.4	30.9	69.1
1998 (El Niño)	15.3	84.7	39.1	60.9	37.9	62.1
2012	16.7	83.3	36.8	63.2	34.9	65.1
S2 (Middle wetland)						
Scen dry-30%smooth	17.7	82.3	36.6	63.4	32.2	67.8
1998 (El Niño)	17.3	82.7	37.2	62.8	37.4	62.6
2012	20.4	79.6	36.9	63.1	37.1	62.9
S7 (Low wetland)						
Scen dry-30%smooth	25.5	74.5	36.9	63.1	35.4	64.6
1998 (El Niño)	25.5	74.5	36.9	63.1	35.4	64.6
2012	36.4	63.6	34.3	65.7	48.7	51.3

E.9 Total organic carbon partitioning

Table E4 Carbon compounds partitioning (expressed in yearly average percentage) at upper, middle and low wetland observation points. Results from one-year simulation for different conditions.

	TOTAL ORGANIC CARBON		PARTICULATE ORGANIC CARBON	
	TOC		POC	
	% of dissolved organic carbon	% of particulate Carbon	%of particulate carbon no algae	% of carbon in algae
AREAS/CONDITIONS	DOC	POC	POCnoa	Phyt
S1 (Upper wetland)				
Scen dry-30%smooth	87	13	64	36
1998 (El Niño)	83	17	75	25
2012	86	14	65	35
S2 (Middle wetland)				
Scen dry-30%smooth	91	9	55	45
1998 (El Niño)	89	11	65	35
2012	91	9	58	42
Abanico (Middle wetland)				
Scen dry-30%smooth	94	6	47	53
1998 (El Niño)	92	8	59	41
2012	93	7	53	47
S7 (Low wetland)				
Scen dry-30%smooth	93	7	52	48
1998 (El Niño)	91	9	61	39
2012	93	7	55	45

Table E5 Carbon compounds partitioning (expressed in average percentage) at upper, middle and low wetland observation points. Results from 4 months simulation (only wet season)

	TOTAL ORGANIC CARBON TOC		PARTICULATE ORGANIC CARBON POC	
	% of dissolved organic carbon	% of particulate Carbon	%of particulate carbon no algae	% of carbon in algae
AREAS/CONDITIONS	---	---	---	---
S1 (Upper wetland)				
Scen dry-30%smooth	78	22	85	15
1998 (El Niño)	78	22	89	11
2012	78	22	86	14
S2 (Middle wetland)				
Scen dry-30%smooth	87	13	76	24
1998 (El Niño)	86	14	81	19
2012	86	14	78	22
Abanico (Middle wetland)				
Scen dry-30%smooth	94	6	54	46
1998 (El Niño)	90	10	72	28
2012	89	11	71	29
S7 (Low wetland)				
Scen dry-30%smooth	91	9	66	34
1998 (El Niño)	88	12	79	21
2012	89	11	74	26

ABOUT THE AUTHOR

Maria Gabriela Alvarez Mieles was born in Guayaquil, Ecuador. She has a Bachelor degree in Biology from the Natural Sciences Faculty at Universidad de Guayaquil (1997), with honours 'Accesit al Contenta' equivalent to magna cum laude. During her early career, she has worked at the Aquatic Pollution department of a National Research Institute in Ecuador (Instituto Nacional de Pesca - INP), participating in monitoring campaigns in some of the main rivers, sea, and estuaries in protected areas.

Later she joined *Efficacitas*, an environmental consultancy office where she participated in several environmental impact assessments, audits and monitoring campaigns for different productive sectors of Ecuador (water, electric, hydrocarbons, mining and industry). The experience at *Efficacitas* highly motivated her to pursue studies abroad with the aim of increasing her knowledge on water-related issues and its interaction with the environment. Thus from 2005-2007 she pursued the MSc in Environmental Sciences at UNESCO-IHE, specialization Planning and Management, with a fellowship from the World Bank. Her MSc thesis was about "*Ecological Indicators-A tool for assessment the present state of a river. A pre-impoundment study in Quevedo-River, Ecuador*".

Following her master studies, she received an internship at DELTARES (formerly Delft Hydraulics), in the Marine and Coastal Department for one of the R&D projects on 'Building with Nature'. During the period 2008-2009, she was engaged at UNESCO-IHE, first contributing to the EU funded research project based on Lake Maryut (Egypt), which involved data analysis of water quality for the construction of a point ecosystem model, and secondly as assistant for the online course in Water and Environmental Law and Policy course (WELP 2008).

In 2009, she had the opportunity to join the WETwin project (EU funded), as a researcher for the Ecuadorian case study (on a wetland in the Guayas River Basin). The project focus aimed to enhance the role of wetlands in integrated water resource management. Her involvement in this project motivated her to pursue a PhD in order to enhance the research for the Abras de Mantequilla wetland.

She started her PhD research in 2010, at UNESCO-IHE, Department of Water Sciences & Engineering, with a fellowship from the Dutch Ministry of Foreign affairs (Netherland fellowship programme NFP-NUFFIC), and financial support from the WETwin project for her fieldwork campaigns. In the initial phase she was hosted at DELTARES gaining additional experience in the use of models.

Unfortunately, two health issues delayed the progress of her PhD research. In 2012, after a preeclampsia, her son was born prematurely and had to remain in hospital for a long time and afterwards required close supervision at home. Then in 2016, just when she was about to submit her draft thesis, she suffered a serious health problem with the central-endocrine system that took more than 7 months to be diagnosed by specialists before it could be treated, and took time to recover afterwards.

Continuing again in 2017 she completed her thesis on environmental monitoring and modelling of the 'Abras de Mantequilla' wetland (RAMSAR site 2000) in Ecuador, applying analytical metrics and ecological modelling to enhance the understanding of tropical wetland systems.

Publications

Book chapter

Alvarez-Mieles, G., G. Corzo & A. E. Mynett, (2019). 6 - Spatial and Temporal Variations' of Habitat Suitability for Fish: A Case Study in Abras de Mantequilla Wetland, Ecuador. In: Corzo, G. & E. A. Varouchakis (eds), *Spatiotemporal Analysis of Extreme Hydrological Events*. Elsevier, 113-141. doi:https://doi.org/10.1016/B978-0-12-811689-0.00006-9.

Journal papers

Alvarez-Mieles, G., K. Irvine, A. V. Griensven, M. Arias-Hidalgo, A. Torres & A. E. Mynett, (2013). Relationships between aquatic biotic communities and water quality in a tropical river–wetland system (Ecuador). *Environmental Science & Policy*, 34(0):115-127. doi:http://dx.doi.org/10.1016/j.envsci.2013.01.011.

Alvarez-Mieles, G., E. Galecio & A. E. Mynett, (2019b). Natural hydrodynamics variability of a tropical river-wetland system. A Ramsar site in Ecuador (submitted). *Wetlands - Journal of the Society of Wetland Scientists*.

Alvarez-Mieles, G., C. Spiteri, & A. E. Mynett, (2019c). Water quality and primary production dynamics in a tropical river-wetland system (submitted). *Journal of Ecohydraulics*.

Alvarez-Mieles, G., K. Irvine & A. E. Mynett, (2019d). Community structure of biotic assemblages in a tropical wetland under different inundation conditions (in preparation). *Limnologica - Ecology and Management of Inland Waters*.

Conference presentations

Alvarez-Mieles, G., Corzo, G., Mynett, A.E. Spatial and temporal variations' analysis of Characidae Habitat, case study in Abras de Mantequilla Wetland in Ecuador. 36th IAHR World Congress, The Hague, the Netherlands, 28th June-3th July, 2015.

Alvarez, G., van Griensven, A., Mynett, A.E. The use of bioindicators to assess the ecosystem health of Abras de Mantequilla wetland. Joint meeting of Society of Wetland Scientists (SWS), WETPOL and Wetland Biogeochemistry Symposium. Prague, 3-8 July 2011 (Oral presentation).

Batdelger O., Dastgheib A., **Alvarez G.,** Arias M., van Griensven A., Mynett A.E. Quantification of the nutrient regulation functions in a tropical riverine wetland system: the case study of the Abras de Mantequilla wetland, Ecuador. SWS Conference (Society of Wetland Scientists), Prague, 3-8 July 2011 (Oral presentation).

Alvarez, G., van Griensven, A., Arias-Hidalgo, M., Mynett, A.E. Aquatic macroinvertebrates communities in an agricultural area of a Tropical River Basin. A case study in the middle catchment of the Guayas River Basin (Ecuador). 3rd International Multidisciplinary Conference on Hydrology and Ecology (HYDROECO-2011). Vienna, Austria, 2-5 May 2011 (Poster presentation)

Alvarez, G., van Griensven, A., Arias-Hidalgo, M., Goethals, P., Mynett, A.E. Aquatic macroinvertebrates as bioindicators for the assessment of the quality of the Guayas River Basin (Ecuador). 7th International Conference on Ecological Informatics (ISEI7), Ghent, Belgium, 13-16 December, 2010 (Conference paper and Oral presentation).

Desmit, X., and **Alvarez, G.** Long-Term Trends in phytoplankton and related parameters in the North Sea. Marine Ecology and Policy Analysis, WL | Delft Hydraulics, Delft, The Netherlands, 2008 (Poster presentation by Desmit, X.).

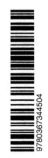

9780367344504